Vehicle Handling Dynamics

Vehicle Handling Dynamics
Theory and Application

Second Edition

Masato Abe
Kanagawa Institute of Technology

AMSTERDAM • BOSTON • HEIDELBERG • LONDON
NEW YORK • OXFORD • PARIS • SAN DIEGO
SAN FRANCISCO • SINGAPORE • SYDNEY • TOKYO
Butterworth-Heinemann is an imprint of Elsevier

Butterworth-Heinemann is an imprint of Elsevier
The Boulevard, Langford Lane, Kidlington, Oxford OX5 1GB, UK
225 Wyman Street, Waltham, MA 02451, USA

Notices
Knowledge and best practice in this field are constantly changing. As new research and experience
broaden our understanding, changes in research methods, professional practices, or medical
treatment may become necessary.

Practitioners and researchers must always rely on their own experience and knowledge in
evaluating and using any information, methods, compounds, or experiments described herein. In
using such information or methods they should be mindful of their own safety and the safety of
others, including parties for whom they have a professional responsibility.

To the fullest extent of the law, neither the Publisher nor the authors, contributors, or editors,
assume any injury and/or damage to persons or property as a matter of products liability, negligence
or otherwise, or from any use or operation of any methods, products, instructions, or ideas contained
in the material herein.

ISBN: 978-0-08-100390-9

British Library Cataloguing in Publication Data
A catalogue record for this book is available from the British Library

Library of Congress Cataloging-in-Publication Data
A catalog record for this book is available from the Library of Congress

For Information on all Butterworth-Heinemann publications
visit our website at http://store.elsevier.com/

Working together
to grow libraries in
developing countries

ELSEVIER   Book Aid
            International

www.elsevier.com • www.bookaid.org

Contents

Preface

This book intends to give readers the fundamental theory and some applications of automotive vehicle dynamics. The book is suitable as a text book of vehicle dynamics for undergraduate and graduate courses in automotive engineering. It is also acceptable as a reference book for researchers and engineers in the field of R&D of vehicle dynamics and control, chassis design and development.

The vehicle motion dealt with in this book is generated by the tire forces, which are produced by the vehicle motion itself. The motion on the ground is possible in any direction by the driver's intention. This is a similar feature to flight dynamics and ship dynamics.

In Chapter 1, the vehicle motion studied in this book is defined. Chapter 2 examines the tire mechanics. The vehicle motion depends on the forces exerted upon tires and this chapter is the base of the book. However, if the reader experiences difficulties in the detailed description of the tire mechanics, they can skip to the next chapter, while still understanding the fundamentals of vehicle dynamics. In Chapter 3, the fundamental theory of vehicle dynamics is dealt with by using a two degree of freedom model. The vehicle motions to external disturbance forces are described using the two degree of freedom model from Chapter 4. This motion is inevitable for a vehicle that can move freely on the ground. In Chapter 5, the effect of the steering system on vehicle motion is studied. The vehicle-body roll effect on the vehicle dynamics is described in the Chapter 6. Chapter 7 looks at the effect of the longitudinal motion on the lateral motion of the vehicle and the fundamental vehicle dynamics with active motion controls is described in the Chapter 8. The vehicle motion is usually controlled by a human driver. The vehicle motion controlled by the human driver is dealt with in the Chapter 9 (Chapter 10 in the second edition) and relations between the driver's evaluation of handling quality and vehicle dynamic characteristics are described in the Chapter 10 (Chapter 11 in the second edition).

For readers who need only to understand the fundamental aspects of the vehicle dynamics and the human driver, it is possible to skip to Chapter 9 after reading from Chapter 1 to Chapter 4. The readers who like to understand and are interested in more in detail of vehicle dynamics should continue to read through the book from the Chapters 5 to 10, depending on their interests.

The original book is written by the author in Japanese and published in Japan. The book was once translated into English by Y. W. Chai when he was a masters-course student of the author. The author has added new parts such as examples in each chapter and problems at the end of the chapters. W. Manning has revised the whole text for the English version.

The publication process started according to a suggestion by the author's old friend, D. A. Crolla. He has consistently continued to give us useful advises from the beginning to the final stage of the publication.

The author has to confess that without any support of the above mentioned three, the publication is not accomplished. The author would like to express his deep gratitude to their contributions to publishing the book. The author is indebted as well to J. Ishio, a former master-course student of the author for his assistance in arranging the examples for each chapter. Also special thanks should go to Yokohama Rubber Co., Ltd. for the preparation of some tire data in the Chapter 2. Finally, author thanks the editorial and production staff of Elsevier Science & Technology Books for their efforts for the publication.

Masato Abe

March 2009

Preface to Second Edition

Five years have passed since the first edition was published. During this period, more and more requirements of understanding the fundamental knowledge of vehicle handling dynamics arise especially for the application to research and development of vehicle active motion controls aiming at vehicle agility and active safety. In view of the situation, the publication of the second edition was pursued in order to make the first edition a still more solid one.

The Chapters 1–8 in the first edition are revised for the second edition by putting the additional parts with correcting existing errors and careless-misses. As a fundamental knowledge of the active vehicle motion control, a description on active front wheel steer controls and an additional note on DYC (Direct Yaw-moment Control) are added in the Chapter 8 and also the new Chapter 9 is provided for the second edition. The Chapter 9 deals with all wheel independent control for full drive-by-wire electric vehicles which is a very updated issue of vehicle dynamics and control for the vehicles of new era.

The previous Chapters 9 and 10 in the first edition are also revised for the Chapters 10 and 11 respectively in the second edition, in which driver-vehicle system behaviors and driver's evaluation of handling qualities are dealt with. The new Chapter 12 is for dealing with a very classical issue which has not been solved yet generally and theoretically in the field of the vehicle handling dynamics. The point is handling quality evaluation and its contribution to the vehicle design for fun-to-drive. The Chapter 12 is a challenge to a fundamental and theoretical approach to this area.

The author thanks the editorial and production staffs of Elsevier Science & Technology Books for their efforts for the publication of the second edition.

Finally the author's old friend, Professor Dave Crolla, who consistently gave us useful suggestions and advices from the beginning to the final stage of the publication of the first edition, regrettably died on 4th September, 2011. The author would like to dedicate this book to the memory of David Anthony Crolla.

Masato Abe
November 2014

Symbols

The following symbols are commonly used throughout from Chapter 3 to Chapter 12 consistently in this book, because they are fundamental symbols for representing the vehicle dynamics and it is rather convenient for the readers to be able to use them consistently. So these symbols are sometimes used without any notice on the symbols. When it is impossible to avoid using these symbols for other meanings than the following, some notice will be given at each part of the chapters where they are used.

m vehicle mass
I vehicle yaw moment inertia
l wheel base
l_f longitudinal position of front wheel(s) from vehicle center of gravity
l_r longitudinal position of rear wheel(s) from vehicle center of gravity
K_f cornering stiffness of front tire
K_r cornering stiffness of rear tire
V vehicle speed
δ front wheel steering angle
β side slip angle
r yaw rate
θ yaw angle
x vehicle longitudinal direction
y vehicle lateral direction and lateral displacement
t time
s Laplace transform variable

The symbols other than the above adopted in each chapter are defined at the first places where they are used.

It should be notified that though, in general, $\ddot{x}$ and $\ddot{y}$ mean the second order time derivative of the variables x and y, they are expediently used in this book for the symbols to represent the vehicle longitudinal and lateral accelerations respectively. In addition, $\delta(s)$, for example, generally means δ as a function of variable, s, however, it represents in this book the Laplace transformation of variable, δ, and this way of representation is applied to all the variables used throughout this book.

VEHICLE DYNAMICS AND CONTROL

1

1.1 DEFINITION OF THE VEHICLE

Ground vehicles can be divided into two main categories: vehicles that are restricted by a track set on the ground (e.g., railway vehicles) and vehicles that are unrestricted by tracks, free to move in any direction on the ground by steering the wheels (e.g., road vehicles).

Aircraft are free to fly in the air, while ships can move freely on the water's surface. In the same way, the road vehicle is free to move by steering its wheels, and it shares similarities with aircraft and ships in the sense that its movements are unrestricted.

From the viewpoint of dynamic motion, the similarity lies in the fact that these three moving bodies receive forces generated by their own movement that are used to accomplish the desired movement. Aircraft depend on the lift force caused by the relative motion of its wings and the air; ships rely on the lift force brought by the relative motion of its body and the water; and ground vehicles rely on the lateral force of the wheels created by the relative motion of the wheels and the road.

In the above described manner, the dynamics and control of the three moving bodies is closely related to their natural function, whereby for an airplane, it is established as flight dynamics, for a ship as ship dynamics, and for a vehicle, similarly, as vehicle dynamics.

The vehicle studied in this book is a vehicle similar to the airplane and ship that is capable of independent motion on the ground using the forces generated by its own motion.

1.2 VIRTUAL FOUR-WHEEL VEHICLE MODEL

For the study of vehicle dynamics and control, a typical vehicle mathematical model is assumed. This vehicle model has wheels that are steerable: two at the front and two at the rear, which are fitted to a rigid body. Passenger cars, trucks, buses, and agricultural vehicles all fall into this category. At first sight, it may seem there are no common dynamics among these vehicles, but by applying a simple four-wheeled vehicle model, as in Figure 1.1, it is possible to obtain fundamental knowledge of the dynamics of all these vehicles.

In the vehicle mathematical model represented in Figure 1.1, the wheels are regarded as weightless, and the rigid body represents the total vehicle weight. The coordinate system is fixed to the vehicle, the x-axis in the longitudinal direction, the y-axis in the lateral direction, and the z-axis in the vertical direction, with the origin at the vehicle's center of gravity.

With this coordinate system, the vehicle motion has six independent degrees of freedom:

1. Vertical motion in the z-direction
2. Left and right motion in the y-direction

FIGURE 1.1

Vehicle dynamics model.

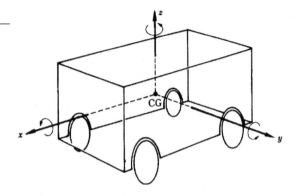

3. Longitudinal motion in the x-direction
4. Rolling motion around the x-axis
5. Pitching motion around the y-axis
6. Yawing motion around the z-axis

These motions can be divided into two main groups. One group consists of motions 1, 3, and 5, which are the motions generated without direct relation to the steering. Motion 1 is the vertical motion caused by an uneven ground/road surface and is related to the vehicle ride. Motion 3 is the longitudinal, straight-line motion of the vehicle due to traction and braking during acceleration or braking. Motion 5 is the motion caused by either road unevenness, acceleration, or braking and is also related to the vehicle ride.

Motions 2 and 6, the yaw and lateral movements, are generated initially by steering the vehicle. Motion 4 is generated mainly by motions 2 and 6 but could occur due to road unevenness as well.

As described earlier, the vehicle studied in this text can move freely in any direction on the ground by steering the vehicle. The main behavior studied here is regarding motions 2, 4, and 6, which are caused by the steering of the vehicle. Motion 2 is the lateral motion, motion 6 is the yawing motion, and motion 4 is the rolling motion.

1.3 **CONTROL OF MOTION**

For normal vehicles, motions are controlled by the driver. The lateral, yaw, and roll motion of the vehicle are generated by the driver's steering and depend on its dynamic characteristics. This does not mean the driver is steering the vehicle meaninglessly. The driver is continuously looking at the path in front of the vehicle, either following his target path or setting a new target path to follow. The driver is observing many things, such as the current position of the vehicle in reference to the target path and the current vehicle motion. The driver is also predicting the imminent vehicle behavior. Based on this information, the driver decides on and makes the suitable steer action. In this manner, the vehicle generates its motion in accordance to a target path that is given or a path set by the driver. Figure 1.2 shows the relation of vehicle motion and control in a block diagram.

The vehicle that is capable of free motion within a plane, without direct restrictions from preset tracks on the ground, only produces a meaningful motion when it is acted on by suitable steering control from the driver.

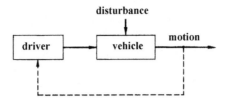

FIGURE 1.2

Vehicle and driver's control.

The primary interest now lies in the inherent dynamic characteristics of the vehicle itself. This becomes clear from the motion of the vehicle to a certain steering input. Next is to study this vehicle's characteristics when it is controlled by a human driver. Finally, the aim is to explore the vehicle dynamic characteristics that make it easier for the driver to control the vehicle.

TIRE MECHANICS

2

2.1 PREFACE

Chapter 1 discussed how this book deals with the independent motion of the vehicle, in the horizontal plane, without restrictions from a preset track on the ground. The force that makes this motion possible is generated by the relative motion of the vehicle to the ground.

The contact between the vehicle and the ground is at the wheels. If the wheel possesses a velocity component perpendicular to its rotation plane, it will receive a force perpendicular to its traveling direction. In other words, the wheel force that makes the vehicle motion possible is produced by the relative motion of the vehicle to the ground, and is generated at the ground. This is similar to the lift force acting vertically on the wing of a body in flight and the lift force acting perpendicularly to the direction of movement of a ship in turning (for the ship, this becomes a force in the lateral direction).

The wheels fitted to the object vehicle not only support the vehicle weight while rotating and produce traction/braking forces, but they also play a major role in making the motion independent from the tracks or guide ways. This is the essential function of our vehicle.

In dealing with the dynamics and control of a vehicle, it is essential to have knowledge of the forces that act on a wheel. Consequently, this chapter deals mainly with the mechanism for generating the force produced by the relative motion of the wheel to the ground and an explanation of its characteristics.

2.2 TIRES PRODUCING LATERAL FORCE
2.2.1 TIRE AND SIDE-SLIP ANGLE

Generally, when a vehicle is traveling in a straight line, the heading direction of the wheel coincides with the traveling direction. In other words, the wheel traveling direction is in line with the wheel rotational plane. However, when the vehicle has lateral motion and/or yaw motion, the traveling direction can be out of line with the rotational plane.

Figure 2.1 is the wheel viewed from the top, where (a) shows the traveling direction in line with the rotation plane, and (b) shows it not in line. The wheel in (b) is said to have side slip. The angle between the wheel traveling direction and the rotational plane, or its heading direction, is called the side-slip angle.

The wheel is also acted on by a traction force if the wheel is moving the vehicle in the traveling direction, or braking force if braking is applied. Also, a rolling resistance force is always

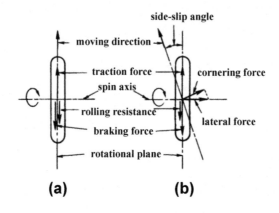

FIGURE 2.1

Vehicle tire in motion, (a) without side slip and (b) with side slip.

at work. If the wheel has side slip, as in (b), a force that is perpendicular to its rotation plane is generated. This force could be regarded as a reaction force that prevents side slip when the wheel produces a side-slip angle. This is an important force that the vehicle depends on for its independent motion. Normally, this force is called the lateral force, whereas the component that is perpendicular to the wheel rotation plane is called the cornering force. When the side-slip angle is small, these two are treated as the same. This force corresponds to the lift force, explained in fluid dynamics, which acts on a body that travels in a fluid at an attack angle, as shown in Figure 2.2.

There are many kinds of wheels, but all produce a force perpendicular to the rotation plane when rotated with side slip. Figure 2.3 shows the schematic comparison of the lateral forces at small side-slip angles for a pneumatic tire wheel, a solid-rubber tire wheel, and an iron wheel.

From here, it is clear that the magnitude of the force produced depends on the type of wheel and is very different. In particular, the maximum possible force produced by an iron wheel is less than one-third of that produced by a rubber tire wheel. Compared to a solid-rubber tire wheel, a pneumatic tire wheel produces a larger force.

For independent motion of the vehicle, the force that acts on a wheel with side slip is desired to be as large as possible. For this reason, the traveling vehicle that is free to move in the plane without external restrictions is usually fitted with pneumatic tires. These are fitted for both the purpose of vehicle ride and for achieving a lateral force that is available for vehicle handling.

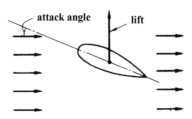

FIGURE 2.2

Lifting force.

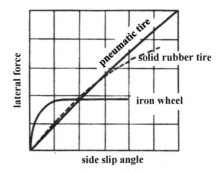

FIGURE 2.3

Lateral forces for several wheels.

In the following text, the pneumatic tire is called a tire, and the mechanism for generating a lateral force that acts on a tire with side slip is explained.

2.2.2 DEFORMATION OF TIRE WITH SIDE SLIP AND LATERAL FORCE

Generally, forces act through the contact surface between the tire and the road. A tire with side slip, as shown by Figure 2.4, is expected to deform in the tire contact surface and its outer circumference: (a) shows the front and side views of the tire deformation; (b) shows the tire contact surface and outer circumference deformation viewed from the top.

At the front of the surface, the deformation direction is almost parallel to the tire's traveling direction. In this part, there is no relative slip to the ground. When the tire slip angle is small, the

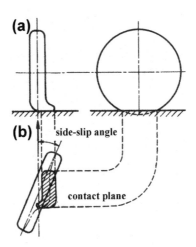

FIGURE 2.4

Tire deflection with side slip, (a) front and side view and (b) plane view.

whole contact surface is similar to this and the rear end of the contact surface has the largest lateral deformation.

When the tire slip angle gets bigger, the front of the surface remains almost parallel to the tire traveling direction. The deformation rate reduces near the center of the contact patch, and the lateral deformation becomes largest at a certain point between the front and rear of the surface. After this maximum point, the tire contact surface slips away from the tire centerline, and the lateral deformation does not increase.

As tire slip angle gets even larger, the point where lateral deformation becomes maximum moves rapidly toward the front. When the slip angle is around 10 to 12°, the contact surface that is parallel to the tire travel direction disappears. The contact surface deformation is nearly symmetric around the wheel's center and consists of nearly all the slip regions.

The lateral deformation of the tire causes a lateral force to act through the contact surface, which is distributed according to the deformation. This lateral force is sometimes called the cornering force when the side-slip angle is small. Looking at the tire lateral deformation, the resultant lateral force may not act on the center of the contact surface. Thus, the lateral force creates a moment around the tire contact surface center. This moment is called the self-aligning torque and acts in the direction that reduces the tire slip angle.

2.2.3 TIRE CAMBER AND LATERAL FORCE

As shown in Figure 2.5, the angle between the tire rotation plane and the vertical axis is called the camber angle. If a tire with a camber angle of ϕ is rotated freely on a horizontal plane, as shown in Figure 2.5, the tire makes a circle with the radius of $R/\sin \phi$ and has its origin at O. If the circular motion is prohibited for a tire with camber angle, and the tire is forced to travel in a straight line only, a force will act on the tire as shown in the figure. This force, due to the camber between the tire and the ground, is called camber thrust.

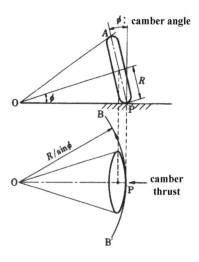

FIGURE 2.5

Tire with camber angle and camber thrust.

2.3 TIRE CORNERING CHARACTERISTICS

The characteristics of the tire that produces lateral force and moment, as elaborated in Section 2.2, are defined as the cornering characteristics. In this section, the tire cornering characteristics will be examined in more detail.

2.3.1 FIALA'S THEORY

The mathematical model proposed by E. Fiala [1] is widely accepted for the aforementioned analysis of the lateral force due to side slip of the tire. It is commonly called Fiala's Theory and is related to the tire cornering characteristics. It is one of the fundamental theories used by many people for explaining tire cornering characteristics [2].

Here, based on Fiala's theory, the tire cornering characteristics will be studied theoretically. The tire's structure is modeled as in Figure 2.6. A is a stiff body equivalent to the rim. B is the pneumatic tube and sidewall that can deform elastically in both vertical and lateral directions. C is the equivalent thin tread base joined to the sidewall at both sides. D is equivalent to the tread rubber. The tread rubber is not a continuous circular body, but it consists of a large number of independent spring bodies around the tire's circumference.

When a force acts in the lateral direction at the ground contact surface, the tire will deform in the lateral direction. The rim is stiff, and it will not be deformed, but the tread base will have a bending deformation in the lateral direction. Moreover, the tread rubber will be deformed by the shear force between the tread base and ground surface. Figure 2.7 shows this kind of deformation in the lateral direction.

Assuming that the tread base deforms equally at the front and rear ends of the ground contact surface, the line that connects these points is the centerline for the tread base and is defined as the x-axis. The y-axis is perpendicular to the x-axis and is positioned at the front endpoint. The x-axis is parallel to the tire rim centerline and also the tread base centerline before deformation. In these axes, the distance along the x-axis from the contact surface front endpoint is x, and the lateral displacement from the x-axis is y. y_1 is the lateral displacement from the x-axis for $0 \leq x \leq l_1$, and

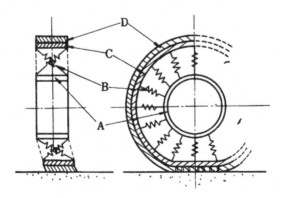

FIGURE 2.6

Tire structural model.

FIGURE 2.7

Tire deflection model.

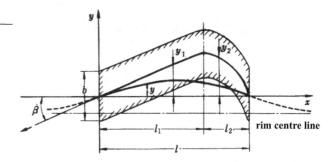

y_2 is the lateral displacement from x-axis for $l_1 < x \le l$. In the region $0 \le x \le l_1$, as described in Section 2.2.2, there is no relative slip between the tire and the ground. The region $l_1 < x \le l$ is where relative slip is produced. β is the side-slip angle of the tire, l is the contact surface length, and b is the contact surface width.

First, consider the lateral deformation, y, of the tread base. If the tread base is extended along the tire circumference, it will look like Figure 2.8. This is a beam with infinite length on top of a spring support that is built up by numerous springs, as B in Figure 2.6.

The deformation of this beam is considered by taking the lateral force acting on the tire as F, the rim centerline as the x-axis, and the line passing through the tire center perpendicular to the x-axis as the y-axis. If the force acts solely on the y-axis (i.e., $x = 0$), the following equation is obtained:

$$EI\frac{d^4y}{dx^4} + ky = w(x) \tag{2.1}$$

Whereby if $x \ne 0$, then $w(x) = 0$, and if $x = 0$, then $w(x) = F$. E is the Young's modulus of the tread material, I is the geometrical moment of inertia of area of the tread base, and k is the spring constant per unit length of the spring support. In solving the previous equation, the lateral displacement, y, is given by the following equation as a general solution:

$$y = \frac{\alpha F}{2k}e^{-ax}[\cos ax + \sin ax] \tag{2.2}$$

FIGURE 2.8

Tire rim deflection model.

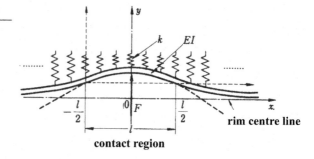

$$\alpha = \frac{1}{\sqrt{2}} \left(\frac{k}{EI}\right)^{\frac{1}{4}} \tag{2.3}$$

The tread base displacement within the ground contact region is assumed to be y at $|ax| \ll 1$. Assuming $\cos ax \approx 1$ and $\sin ax \approx ax$, then y can be approximated to a second-order equation of x.

$$y = \frac{\alpha F}{2k}(1 - \alpha^2 x^2) \tag{2.4}$$

Furthermore, expressing y with a transferred coordinate system such that $y = 0$ at $x = 0$ and $x = l$:

$$y = \frac{\alpha^3 l^2 F}{2k} \frac{x}{l}\left(1 - \frac{x}{l}\right) \tag{2.5}$$

This equation expresses the lateral displacement, y, of the tread base in Figure 2.7.

Next, the lateral displacements, y_1 and y_2, from the ground contact surface centerline are looked at. For the region $0 \le x \le l_1$, there is no relative slip between the tire and the ground. The contact surface deforms relatively in the opposite direction to the tire's lateral traveling direction. The lateral displacement, y_1, for each point on the contact surface along the longitudinal direction can be written as follows:

$$y_1 = \tan \beta x \tag{2.6}$$

The tread base displacement is given by Eqn (2.5) and the tread rubber displacement by Eqn (2.6). As shown in Figure 2.9, a shear strain of $(y_1-y)/d$ occurs between the tread rubber and the tread base. A force per unit length in the lateral direction acts upon each point on the contact surface along the longitudinal direction.

$$f_1 = K_0(y_1 - y) = K_0\left[\tan \beta x - \frac{\alpha^3 l^2 F}{2k} \frac{x}{l}\left(1 - \frac{x}{l}\right)\right] \tag{2.7}$$

$$K_0 = G\frac{b}{d} = \frac{E}{2(1 + v)}\frac{b}{d} \tag{2.8}$$

G is the shear modulus of the tread, and v is the Poisson ratio.

As seen in Figure 2.7, y_1-y becomes larger toward the rear end of the contact surface. If f_1 exceeds the friction force between the tread rubber and the ground, a relative slip will be

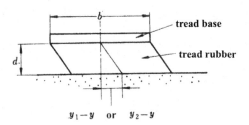

y_1-y or y_2-y

FIGURE 2.9

Shear deformation of tread rubber.

produced between them. The slip region is denoted by $l_1 < x \le l$, and the shear strain of the tread rubber is $(y_2-y)/d$. The force that produces this strain, f_2, is the friction force between the tread rubber and the ground. For simplicity, the tire load is taken as W, and the contact surface pressure distribution, p, along the x-direction is approximated by a second-order equation with the peak pressure at the tire center, as in Figure 2.10.

$$p = 4p_m \frac{x}{l} \left(1 - \frac{x}{l}\right) \tag{2.9}$$

$$p_m = \frac{3W}{2bl} \tag{2.10}$$

Then, f_2 becomes the following:

$$f_2 = K_0(y_2 - y) = \mu pb = 4\mu p_m b \frac{x}{l} \left(1 - \frac{x}{l}\right) \tag{2.11}$$

where μ is the friction coefficient between the tread rubber and the ground, and l_1 is the value of x that satisfies $f_1 = f_2$.

$$K_0 \left[\tan \beta x - \frac{\alpha^3 l^2 F}{2k} \frac{x}{l} \left(1 - \frac{x}{l}\right)\right] = 4\mu p_m b \frac{x}{l} \left(1 - \frac{x}{l}\right) \tag{2.12}$$

thus, l_1 is given by solving this equation by x:

$$l_1 = l - \frac{K_0 l^2 \tan \beta}{\frac{K_0 \alpha^3 l^2}{2k} F + 4\mu p_m b} \tag{2.13}$$

The lateral force that acts on each point on the contact surface along the longitudinal direction, for a small increment of dx, is $f_1 dx$ for $0 \le x \le l_1$ and $f_2 dx$ for $l_1 < x \le l$. The total force acting on the whole contact surface, which is the lateral force, F, is given by the following equation:

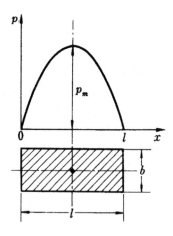

FIGURE 2.10

Contact pressure distribution.

$$F = \int_0^{l_1} f_1 dx + \int_{l_1}^{l} f_2 dx$$

$$= K_0 \int_0^{l_1} \left[\tan \beta x - \frac{\alpha^3 l^2 F}{2k} \frac{x}{l} \left(1 - \frac{x}{l} \right) \right] dx + \int_{l_1}^{l} 4\mu p_m b \frac{x}{l} \left(1 - \frac{x}{l} \right) dx \qquad (2.14)$$

By substituting Eqn (2.13) into Eqn (2.14) and integrating it results in F on both sides. This makes the equation complicated to solve. Fiala, thus, had approximated F in the following way:

$$F = \frac{K_1 l^2}{2} \tan \beta - \frac{1}{8} \frac{K_1^2 l^3}{\mu p_m b} \tan^2 \beta + \frac{1}{96} \frac{K_1^3 l^4}{\mu^2 p_m^2 b^2} \tan^3 \beta \qquad (2.15)$$

$$K_1 = \frac{K_0}{1 + \frac{\alpha^3 l^3}{12k} K_0} \qquad (2.16)$$

This is the fundamental method for expressing the relationship between the tire side-slip angle and the lateral force.

Figure 2.7 shows that the lateral force acting on the contact surface is not symmetric across the center of the contact surface. This causes the lateral force to generate a moment around the vertical axis that passes through the contact surface center. This moment is the self-aligning torque. For a small increment of dx at each point on the contact surface, $(x–l/2) f_1 dx$ for $0 \leq x \leq l_1$ and $(x–l/2) f_2 dx$ for $l_1 < x \leq l$, it is as follows:

$$M = \int_0^{l_1} \left(x - \frac{l}{2} \right) f_1 dx + \int_{l_1}^{l} \left(x - \frac{l}{2} \right) f_2 dx$$

$$= K_0 \int_0^{l_1} \left(x - \frac{l}{2} \right) \left[\tan \beta x - \frac{\alpha^3 l^2 F}{2k} \frac{x}{l} \left(1 - \frac{x}{l} \right) \right] dx + \int_{l_1}^{l} 4\mu p_m b \left(x - \frac{l}{2} \right) \frac{x}{l} \left(1 - \frac{x}{l} \right) dx \qquad (2.17)$$

Substituting Eqn (2.13) for l_1 into Eqn (2.17) gives an equation that is too complicated. Using the approximated equation of F as in Eqn (2.15), Fiala approximated M as follows:

$$M = \frac{K_1 l^3}{12} \tan \beta - \frac{1}{16} \frac{K_1^2 l^4}{\mu p_m b} \tan^2 \beta + \frac{1}{64} \frac{K_1^3 l^5}{\mu^2 p_m^2 b^2} \tan^3 \beta - \frac{1}{768} \frac{K_1^4 l^6}{\mu^3 p_m^3 b^3} \tan^4 \beta \qquad (2.18)$$

This is the fundamental equation that expresses the relation between the tire side-slip angle and the self-aligning torque.

The lateral force per unit side-slip angle, when β is small, is called the cornering stiffness and is given by the following:

$$K = \left(\frac{dF}{d\beta} \right)_{\beta=0} = \frac{K_1 l^2}{2} \qquad (2.19)$$

The tire maximum friction force is found from Eqn (2.10):

$$\mu W = \frac{2}{3}\mu p_m bl \tag{2.20}$$

Defining ψ as follows:

$$\psi = \frac{K}{\mu W}\tan\beta = \frac{3K_1 l}{4\mu p_m b}\tan\beta \tag{2.21}$$

Eqns (2.15) and (2.18) can be written as follows:

$$\frac{F}{\mu W} = \psi - \frac{1}{3}\psi^2 + \frac{1}{27}\psi^3 \tag{2.22}$$

$$\frac{M}{\mu W} = \frac{1}{6}\psi - \frac{1}{6}\psi^2 + \frac{1}{18}\psi^3 - \frac{1}{162}\psi^4 \tag{2.23}$$

Differentiating Eqns (2.22) and (2.23) with respect to ψ, and putting them equal to 0, gives maximum $F/(\mu W)$ at $\psi = 3$, where $F/(\mu W) = 1$. $M/(\mu W)$ becomes largest at $\psi = 3/4$, where it is 27/512.

In other words, the lateral force, F, is largest when the side-slip angle is as follows:

$$\tan\beta = \frac{3\mu W}{K} \tag{2.24}$$

and its maximum value, F_{max}, is the following:

$$F_{max} = \mu W \tag{2.25}$$

while the self-aligning torque is largest at a side-slip angle of the following:

$$\tan\beta = \frac{3\mu W}{4K} \tag{2.26}$$

and its maximum value, M_{max}, is as follows:

$$M_{max} = \frac{27}{512}l\mu W \tag{2.27}$$

Using Eqns (2.22) and (2.23), a dimensionless term of $F/(\mu W)$ for the lateral force and $M/(\mu W)$ for the self-aligning torque can be derived in terms of the side-slip angle, $\psi = K\tan\beta/(\mu W)$ and are plotted in Figure 2.11 and Figure 2.12.

As seen in Figure 2.11, the lateral force, F, is almost proportional to $tan \beta$ when the side-slip angle is small. After a certain value of β, the lateral force reaches saturation and does not increase further with increasing side-slip angle. From Figure 2.12, the self-aligning torque is almost proportional to $tan \beta$ when the side-slip angle is small. As β increases beyond a certain point, the self-aligning torque reaches saturation abruptly and decreases with the increase of side-slip angle. When β is small, $tan \beta \approx \beta$, the lateral force and self-aligning torque can be treated as proportional to β. When β is large, the lateral force is no longer proportional to β and has nonlinear characteristics.

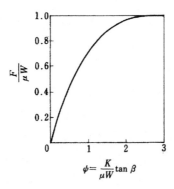

FIGURE 2.11

Lateral force to side-slip angle.

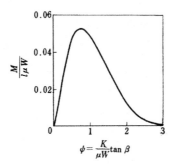

FIGURE 2.12

Self-aligning torque to side-slip angle.

In the region where the side-slip angle is small, $\tan \beta \approx \beta$, terms with more than second-orders of β can be ignored. The lateral force and self-aligning torque, in relation to β, are given by Eqns (2.15) and (2.18) as follows:

$$F \approx \frac{K_1 l^2}{2} \beta = K\beta \tag{2.28}$$

$$M \approx \frac{K_1 l^3}{12} \beta = \xi_n K\beta \tag{2.29}$$

Here, K is the generally so-called cornering stiffness, which is given by Eqn (2.19). From Eqns (2.8) and (2.16), the cornering stiffness could be written as follows:

$$K = \frac{\frac{l^2}{2} \frac{b}{d} G}{1 + \frac{\alpha^3 l^3}{12k} \frac{b}{d} G} \tag{2.30}$$

The distance of the acting point of the cornering force from the contact surface center, which is called the pneumatic trail, as shown by Figure 2.13, is defined as follows:

$$\xi_n = M/F \tag{2.31}$$

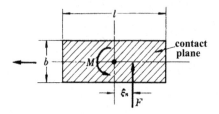

FIGURE 2.13

Pneumatic trail.

Based on Eqns (2.28) and (2.29), this value is $l/6$ when the value of β is small.

Until now, the force acting on a tire with a side-slip angle has been examined. Following this, the force acting on a tire traveling in a straight line with a camber angle will be studied. This force is called the camber thrust. Fiala analyzed the tire with camber angle as follows.

When tire camber is taken into consideration, as shown in Figure 2.14, even without any side-slip angle, the tread base centerline is not a straight line, but it becomes a part of the arc.

When referred to the same coordinate system as in Figure 2.7 and approximated to a parabola, this part of the arc will become the following:

$$y_c = -\frac{l^2\phi}{2R_0}\frac{x}{l}\left(x - \frac{x}{l}\right) \tag{2.32}$$

where R_0 is the effective tread base radius, given by the following equation:

$$R_0 = R - \frac{W}{k_e} \tag{2.33}$$

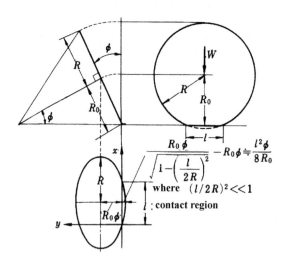

FIGURE 2.14

Tire tread base center with camber.

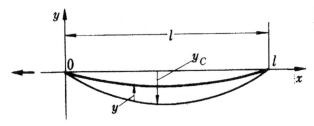

FIGURE 2.15

Tire deflection with camber.

R is the tread base radius at zero tire load, and k_e is the vertical spring constant of the tire.

This condition is shown in Figure 2.15. If the tire moves freely while having a camber angle, the contact surface center will follow the circular track as shown in Figure 2.5. However, if the tire is to travel in a straight line only, the contact surface center should follow along the x-axis. Therefore, a shear strain between the tread base and the ground is produced at the tread rubber, and a lateral force, corresponding to this displacement, acts at the contact surface.

This resultant force is the camber thrust, F_c. Assuming that the acting point of this force is concentrated to the contact surface center, the displacement of the tread base by this force is as follows:

$$y = \frac{\alpha^3 l^2 F_c}{2k} \frac{x}{l}\left(x - \frac{x}{l}\right) \tag{2.34}$$

Because the displacement from the x-axis at the contact surface is 0, the shear strain of the tread rubber at each point along the x-direction is $-(y_c + y)$. Therefore, the camber thrust that acts on the tire, F_c is as follows:

$$F_c = \int_0^l -K_0\left(y_c + y\right)dx = K_0\left[\frac{l^2\phi}{2R_0} - \frac{\alpha^3 l^2 F_c}{2k}\right]\int_0^l \frac{x}{l}\left(x - \frac{x}{l}\right)dx \tag{2.35}$$

Integrating this equation to find F_c gives the following:

$$F_c = \left(\frac{K_0}{1 + \frac{\alpha^3 l^3}{12k}K_0}\right)\frac{l^3}{12R_0}\phi = \frac{K_1 l^3}{12R_0}\phi \tag{2.36}$$

Equation (2.36) shows the camber thrust is proportional to the camber angle. The proportional constant, K_c, is called the camber thrust coefficient, and by substituting Eqn (2.8), can be expressed as follows:

$$K_c = \frac{K_1 l^3}{12R_0} = \left(\frac{\frac{l^2}{2}\frac{b}{d}G}{1 + \frac{\alpha^3 l^3}{12k}\frac{b}{d}G}\right)\frac{l}{6R_0} = \frac{l}{6R_0}K \tag{2.37}$$

From this mathematical model and theoretical discussions, the characteristics of the lateral force and moment acting on the tire can be found. This means the tire cornering characteristics are affected by E, v, I, b, d, k, l, μ and W.

E and v are dependent on the tread material and construction; I, b, and d are decided by the tire shape; W is the tire vertical load; k is mainly dependent on the air pressure and can be taken as proportional to the air pressure; l is mainly decided by the tire shape, but it also depends on the tire's vertical load and the air pressure of the tire; last, μ is dependent on the tread material and ground/road condition.

Consequently, the tire cornering characteristics are mainly affected by the following:

1. Tire material, construction, and shape
2. Tire vertical load
3. Tire air pressure
4. Ground/road condition

The theoretical calculation of the influence of 1 to 4 on the various tire parameters/properties as summarized is not easy. Neither is it straightforward to measure these effects experimentally. In contrast, the measurement of the lateral force and moment on the tire with side-slip angle or camber angle is relatively simple. Hence, it is common for the effects of 1 to 4 on tire cornering characteristics to be verified through direct experimental results.

Example 2.1

Estimate the tire side-slip angle at which the lateral force is saturated by considering the tire lateral deformation mechanism due to lateral force. Confirm that the side-slip angle estimated coincides with Eqn (2.24).

Solution

The lateral force at the saturation point is equal to μW, where μ is a friction coefficient between the tire and road surface, and W is a vertical load of the tire. Referring to Eqn (2.5), the lateral displacement of the tire tread base due to this force is expressed as follows:

$$y = \frac{\alpha^3 l^2 \mu W}{2k} \frac{x}{l} \left(1 - \frac{x}{l}\right) \tag{E2.1}$$

At the saturation point, the lateral force distribution along the contact patch of the tire is equal to the multiplication of the vertical load distribution and the friction coefficient μ:

$$\mu p = 4\mu p_m \frac{x}{l} \left(1 - \frac{x}{l}\right) \tag{E2.2}$$

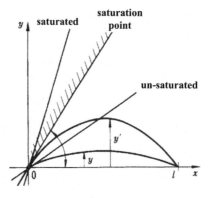

FIGURE E2.1

As shown in the above figure, shearing deformation of the tread rubber is expressed by $y'-y$, where y' is the lateral deformation of the contact patch of the tire, and the following relation is obtained:

$$\mu p b = K_0(y' - y) \tag{E2.3}$$

From the three preceding equations, y' is obtained as follows:

$$y' = \left(\frac{\alpha^3 l^2 F}{2k} + \frac{4\mu p_m b}{K_0}\right) \frac{x}{l}\left(1 - \frac{x}{l}\right) \tag{E2.4}$$

This is a maximum lateral deformation of each point of the tire contact patch. The tangential direction of y', dy'/dx, at $x = 0$ corresponds with the maximum side-slip angle at which the lateral force reaches the maximum value. Thus, the side-slip angle is described as follows:

$$\tan \beta = \left(\frac{dy'}{dx}\right)_{x=0} = \left(\frac{\alpha^3 l^2 F}{2k} + \frac{4\mu p_m b}{K_0}\right) \frac{1}{l} \tag{E2.5}$$

Using Eqns (2.8), (2.10), (2.16), and (2.19), the previous equation is transformed to the same form as Eqn (2.24):

$$\tan \beta = \frac{3\mu W}{K}.$$

2.3.2 LATERAL FORCE

In this section, the tire cornering characteristics verified experimentally so far, under given tire materials and constructions, will be investigated, and the lateral force and moment characteristics will be understood in more detail.

2.3.2.1 Common characteristics

As expected from the mathematical model, the relation between the lateral force and the side-slip angle is almost a straight line when the side-slip angle is small. After the side-slip angle exceeds a certain value, the lateral force increases at a slower rate and finally saturates at $tan \beta = 3\mu W/K$. Figure 2.16 is the typical example of the relation between cornering force and side-slip angle of the tire for passenger cars.

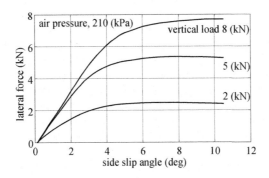

FIGURE 2.16

Tire lateral force to side-slip angle.

In this example, when β is less than around 4°, the lateral force increases in a straight line. Its increment becomes less and less after this value and finally saturates at around 8 to 10°. However, for a normal passenger car, its lateral motion usually occurs within the linear region. The slope of this line is the tire cornering stiffness, given by Eqn (2.30), and corresponds to the lateral force of the tire per unit side-slip angle. This is an important value in the evaluation of the tire cornering characteristics.

2.3.2.2 Effects of vertical load and road condition

The effect of tire vertical load on lateral force is also shown in Figure 2.16. The tire vertical load has almost no effect on lateral force at very small side-slip angles. The different saturation levels of lateral force become more obvious with larger side-slip angles. The mathematical model shows the tire load only affects lateral force in the region where there is relative slip between the tread rubber and the ground. When this region occupies the majority of the contact surface, i.e., the side-slip angle is large, the lateral force approaches the product of μ and W, and so, the effect of W is remarkable. Figure 2.17 is an example that shows the effect of the tire's vertical load on lateral force.

Next is to study the effect of the tire's vertical load on the tire cornering stiffness. From Figure 2.16, when the tire load is small, the cornering stiffness increases together with the tire load, but after a certain limit, it seems to decrease.

The cornering stiffness, divided by the corresponding tire load, is called the cornering stiffness coefficient. This cornering stiffness coefficient decreases with tire load almost linearly, as shown by Figure 2.18. Therefore, the dependence of the cornering stiffness on the tire load is written as follows:

$$K = W(c - c_1 W) \tag{2.38}$$

where, c is the cornering stiffness coefficient near zero tire load, and c_1 is the tire load dependent coefficient.

In relation to tire load, cornering stiffness could be approximated as a parabola that passes through the origin. Cornering stiffness increases with tire load to a peak value, and beyond that it decreases. Tires are normally used in the region where cornering stiffness increases with vertical load.

Figure 2.19 shows the dependence of tire cornering stiffness on tire load for a real tire. The approximation to a parabola is verified.

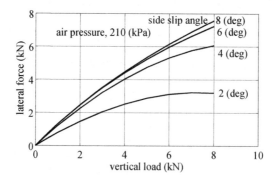

FIGURE 2.17

Lateral force to vertical load for each side slip angle.

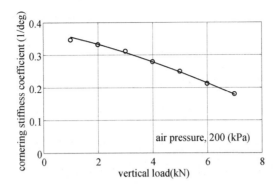

FIGURE 2.18

Effects of vertical load on cornering stiffness coefficient.

(Courtesy of Yokohama Rubber Co., Ltd.)

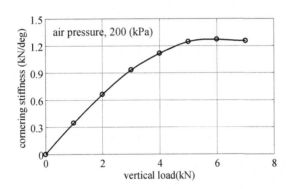

FIGURE 2.19

Effect of vertical load on cornering stiffness.

(Courtesy of Yokohama Rubber Co., Ltd.)

Usually, the effect of the tire's vertical load is expressed in the form of μW. The friction coefficient between the tread rubber and ground, μ, is expected to have an effect similar to the tire's vertical load. Figure 2.20 schematically shows the effects of the road surfaces on the cornering force as the road surface changes with side-slip angle [2].

From the figure, it can be seen that the friction coefficient has almost no effect on lateral force at small side-slip angles, but it has a more obvious effect as the side-slip angles become larger. The effect of friction coefficient on the lateral force is very similar to the effect of the vertical load.

2.3.2.3 Tire pressure effect

From analysis of the mathematical model in Section 2.3.1, it is clear that the tire's lateral force becomes larger with a smaller tread base displacement under a constant force. From Eqn (2.5), y becomes smaller with smaller $\alpha^3 l^2/2k$. If k, the spring constant of the spring support, is large and the tread base bending stiffness, EI, is also large, then the tread base displacement is small. Because spring constant, k, depends on tire air pressure, it is expected that lateral force also

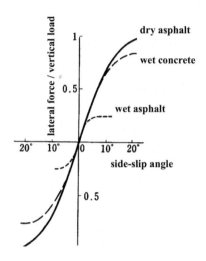

FIGURE 2.20

Effects of road surface on lateral force.

increases with tire pressure. However, the increase of tire pressure reduces the contact surface length, l, and Eqn (2.15) shows that the lateral force decreases with a decrease in contact length.

Figure 2.21 shows a good example of the relationship between lateral force and tire pressure. The increase of tire pressure contributes to the increase of k, and an increase in the lateral force is expected. However, the reduction of the contact length, l, due to the increase of the tire pressure decreases the lateral force. It is interesting to see that the lateral force is eventually almost constant to the variation in the tire pressure in this case.

The previous point is shown in more detail by the relation of the cornering stiffness to the tire pressure in Figure 2.22. If the vertical load is relatively low, the decrease in contact length

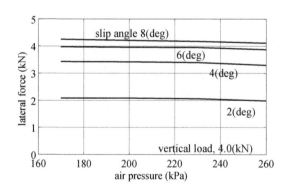

FIGURE 2.21

Effects of tire pressure on lateral force.

(Courtesy of Yokohama Rubber Co., Ltd.)

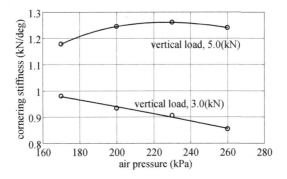

FIGURE 2.22

Effects of tire pressure on cornering stiffness.

(Courtesy of Yokohama Rubber Co., Ltd.)

contributes more than the increase of k as the tire pressure increases. In this case, the cornering stiffness decreases with an increase in tire pressure. On the other hand, if the vertical load is relatively high, the effect of the increasing k is dominant compared with the decrease in the contact length, and the cornering stiffness increases with the tire pressure. However, with even higher vertical loads, the excessive increase of the tire pressure has a greater effect on the contact length reduction, and the cornering stiffness decreases with the increase of the tire pressure. These points can be understood through Eqn (2.30).

2.3.2.4 Tire shape effect

The tread base bending stiffness, EI, is decided by the tire shape. If the tire material and the construction are given, the shape effect is dominated by the geometrical moment of inertia, I, of the tread base. This is generally larger for a larger tire. For tires with the same radius, it is larger for flatter tires with larger width. Therefore, low-profile tires are desirable for obtaining larger cornering force.

From Eqn (2.30), the cornering stiffness increases when b is larger and d is smaller. Figure 2.23 shows an example of the relationship between the cornering stiffness and the tread

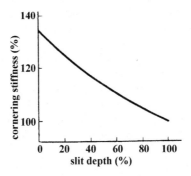

FIGURE 2.23

Effect of tire slit depth on cornering stiffness.

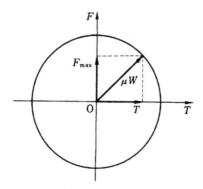

FIGURE 2.24

Friction circle.

depth, which corresponds to the tread rubber thickness. From this figure, the increase in cornering stiffness is seen for a tire with smaller d, i.e., with worn tread.

2.3.2.5 Braking and traction force effect

The effects of tire parameters and vertical load on the lateral force have already been studied. During braking or traction, the tire is acted on by the vertical load that supports the vehicle weight and the longitudinal force at the contact surface that accelerates or decelerates the vehicle. These forces also affect the lateral force.

Based on the classical Law of Friction, as shown in Figure 2.24, the lateral force F, and traction force (or braking force) T, acting on a tire, must always satisfy the following equation:

$$\sqrt{F^2 + T^2} \le \mu W \tag{2.39}$$

In other words, the result of the horizontal, in-plane forces that act on the contact surface between the tire and the ground cannot exceed the product of the tire vertical load and the friction coefficient. This means the resultant force vector is restricted to be inside a circle with radius μW. This circle is called the friction circle.

If there is a longitudinal force, either traction force or braking force, acting on the tire, the maximum cornering force, for a large side-slip angle, is given by the following equation:

$$F_{max} = \sqrt{\mu^2 W^2 - T^2} \tag{2.40}$$

and, if $T = 0$, then the previous equation is the same as Eqn (2.25).

The relation of lateral force, F_0, to side-slip angle when traction or braking force is zero is given by the line O–A$_0$ in Figure 2.25. The relation of lateral force to side-slip angle when traction (or braking) force, T, is at work is given by line O–A in the same figure. If the reduction ratio of lateral force due to traction (or braking) force is assumed to be the same at any side-slip angle, the following equation is true:

$$\frac{F}{F_0} = \frac{\sqrt{\mu^2 W^2 - T^2}}{\mu W} \tag{2.41}$$

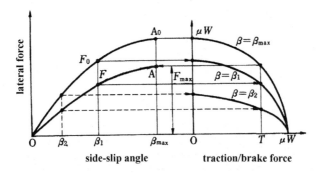

FIGURE 2.25

Effects of traction/brake force on lateral force.

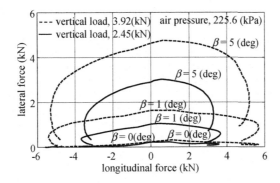

FIGURE 2.26

Lateral force with traction and braking forces.

or

$$\left(\frac{T}{\mu W}\right)^2 + \left(\frac{F}{F_0}\right)^2 = 1 \tag{2.42}$$

This means, at any given side-slip angle, the lateral force, F, and traction (or braking) force, T, will form an elliptic relationship. This ellipse coincides with the friction circle shown previously when the side-slip angle produces the maximum lateral force. Figure 2.25 shows how the relationship between F and T is satisfied as an ellipse.

Much research has been carried out on the effect of traction force and braking force on lateral force. Figure 2.26 shows an example of actual measurements that can be understood with the concept of the friction circle.

2.3.3 SELF-ALIGNING TORQUE

The theoretical analysis with the mathematical model shows that when the side-slip angle is small, the self-aligning torque increases linearly with the side-slip angle, but when side-slip angle

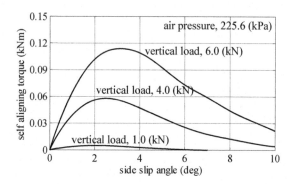

FIGURE 2.27

Self-aligning torque to side-slip angle.

is large, the self-aligning torque approaches saturation and reaches its peak at a certain value. After this point, the self-aligning torque decreases with side-slip angle.

Figure 2.27 shows a typical example of the relation between self-aligning torque and the side-slip angle of a real tire. In Figure 2.27, the effect of tire load on self-aligning torque is shown together with the side slip. The effect of tire load on lateral force is small at small side-slip angles and large at large side-slip angles. In comparison, the effect of tire load on self-aligning moment is large at any side-slip angle. One of the reasons for this is that the load effect occurs in the region where there is relative slip between the contact surface and the tread rubber, and self-aligning torque itself also has a greater contribution from the lateral force acting at both ends of the contact surface. Another reason is that the contact surface length increases with tire load, and the moment produced by the acting lateral force increases too. This can be verified from Eqn (2.18), where the third-order or higher term of l is included in the equation.

If the tire pressure is increased, as seen from the previous discussion, the lateral force increases in some case and the self-aligning torque also increases. However, on a real tire, self-aligning torque decreases if the tire pressure is increased. It is thought that while even though the lateral force increases with tire pressure, the contact surface length decreases with tire pressure. This has a large effect on self-aligning torque. Self-aligning torque increases if the tire pressure is decreased, but beyond a certain limit, the self-aligning torque does not increase with decreasing tire pressure. It is thought that if the tire pressure is too small, the decrease in lateral force has a larger effect than the effect of increasing the contact surface length.

Self-aligning torque is the moment of the lateral force around the vertical axis that passes through the contact surface center. Pneumatic trail is defined as in Figure 2.13. Using Eqns (2.22) and (2.23), Figure 2.28 shows the relationship between $\xi_n = M/F$ and ψ. The result shows that ξ_n decreases dramatically when the side-slip angle becomes larger.

2.3.4 CAMBER THRUST

From Eqn (2.36), it is expected that the camber thrust of a tire is proportional to the camber angle when the side-slip angle is zero. Figure 2.29 is an example of the relationship between the camber angle and camber thrust of a normal tire. This example shows that the camber thrust is proportional

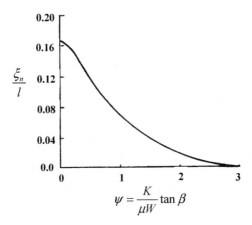

FIGURE 2.28

Pneumatic trail to side-slip angle.

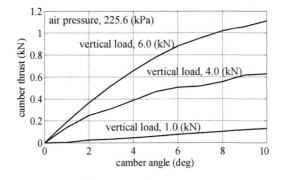

FIGURE 2.29

Camber thrust to camber angle.

to the camber angle. The effect of tire load is also noted in the same figure. As seen from the figure, the camber thrust coefficient, which is the camber thrust per unit camber angle, increases almost linearly with tire load. The tread base effective radius, R_0, is shown in Figure 2.14 and depends on tire vertical load, W, as in Eqn (2.33). The contact surface length also increases with tire load. From Eqn (2.37), it is understood that the camber thrust coefficient depends on the tire vertical load and increases with tire vertical load. Moreover, from the same equation, the camber thrust coefficient is equal to the product of the cornering stiffness and $l/6R_0$. It can now be assumed that the camber thrust coefficient has similar characteristics to the cornering stiffness. Since the ratio of contact surface length l to effective radius R_0, l/R_0, is generally about 0.3, the camber thrust coefficient is normally less than one-tenth of the cornering stiffness.

So far, the camber thrust and lateral force produced by the side-slip angle have been considered. However, the normal tire that is fitted to a real vehicle usually has both side-slip angle and camber during its travel. In this case, the tire is simultaneously acted on by the lateral force and the camber thrust, and it is regarded that they act independently.

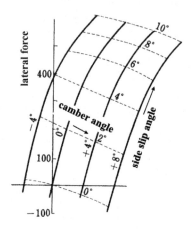

FIGURE 2.30

Effect of camber angle on tire lateral force.

Figure 2.30 shows an example of the lateral force due to both side-slip angle and camber angle, as described by Ellis [2]. The curves showing the relationship between camber angle and lateral force, at different side-slip angles, and the curves showing the relation of side-slip angle and lateral force, at different camber angles, are both parallel to each other. This shows that the lateral forces acting on the tire, produced by the camber angle and the side-slip angle, can be treated individually and independently.

In recent years, low-profile tires have become more popular, especially on passenger cars. When this kind of tire has a camber angle, the lateral distribution of the tire load could easily give larger tire loads at the inner side and smaller loads at the outer side. This kind of unequal distribution, compared to the case of equal distribution, causes a decrease in the generated lateral force at an axle. This is anticipated from the dependency of the lateral force and cornering stiffness on the tire's vertical load, which is shown by parabolic curves in Figures 2.17 and 2.19. The decrease of the lateral force due to a reduction in tire load is larger than the increase of the lateral force due to a tire load increase.

Consequently, even in cases where the camber angle is added so that the camber thrust and the lateral force caused by side-slip angle are in the same direction, the total lateral force produced by the tire with both side-slip angle and the camber angle may be reduced. This is due to the decrease of lateral force caused by tire load distribution changes, which in turn are caused by the camber angle. To fully understand this phenomenon, the experimental result using a low-profile kart tire is plotted in Figure 2.31.

2.4 TRACTION, BRAKING, AND CORNERING CHARACTERISTICS

2.4.1 MATHEMATICAL MODEL

The tire cornering characteristics from Fiala's theory have been explained in Section 2.3.1. The mathematical model assumes that with side slip, the tire tread base deforms elastically in the lateral direction toward the tire rim and, at the same time, the tread rubber deforms elastically further more toward the tread base.

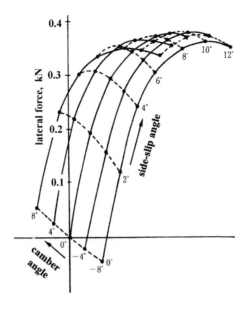

FIGURE 2.31

Effects of camber angle on lateral force of kart tire with small aspect ratio.

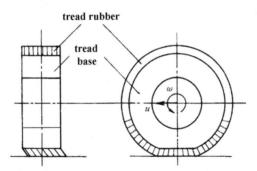

FIGURE 2.32

Tire deformable in lateral and longitudinal directions.

The effect of longitudinal force, such as traction and braking, could be considered using the same mathematical model, but the model would become too complex. Instead, the model in Figure 2.32 shows how the tread rubber is fitted circumferentially to the stiff tread base and rim, and the tread rubber is the only elastic part. This model allows elastic deformation in both the longitudinal and lateral directions. The tread rubber, similar to the previous model, is not a continuous circular body, but consists of a large number of independent springs around the tire circumference. This type of tire model is called the brush model.

This tire model will be used to understand theoretically the force generated by the tire in the longitudinal and lateral directions [3].

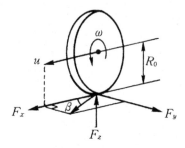

FIGURE 2.33

Tire forces in three directions.

2.4.2 TIRE LATERAL FORCE DURING TRACTION AND BRAKING

As shown in Figure 2.33, the tire is rotating with an angular velocity, ω, while traveling in a direction that forms an angle of β to the rotation plane. The velocity component in the rotation plane is taken as u. Three forces act upon this tire, namely the longitudinal force, F_x, lateral force, F_y, and vertical force, F_z.

2.4.2.1 Braking

Figure 2.34 shows how the front endpoint of the tire contact surface centerline is taken as the origin of the coordinate axes, with the x-axis in the longitudinal direction, and the y-axis in the lateral direction. The point on the tread base directly on top of point O is taken as point O'. After a fraction of time Δt, the contact surface point moves from O to P, and the point O' on the tread base moved to P'. The projected point P' on the x-axis is marked as P''.

During Δt, the distance in the x-direction of the contact point from point O is the x-coordinate of point P:

$$x = u\Delta t \tag{2.43}$$

The x-coordinate of point P' from point O' is as follows:

$$x' = R_0\omega\Delta t \tag{2.44}$$

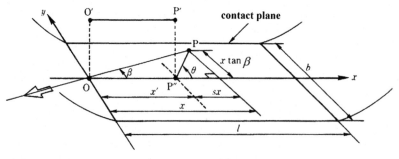

FIGURE 2.34

Tire deformation in contact plane.

Therefore, the relative displacement of point P and point P', i.e., the deformation of the tread rubber, is as follows:

$$x - x' = \frac{u - R_0\omega}{u} u\Delta t$$
$$= sx \tag{2.45}$$
$$= \frac{s}{1 - s}x'$$

whereby, s is the tire slip ratio in the longitudinal direction:

$$s = \frac{u - R_0\omega}{u} \tag{2.46}$$

and, the distance in the y-direction of the contact point, from point O, i.e., the y-coordinate of point P, is as follows:

$$y = x \tan \beta = \frac{\tan \beta}{1 - s}x' \tag{2.47}$$

Since there is no displacement of point P' in the y-direction, the previous is the deformation of tread rubber in the y-direction.

Therefore, the forces per unit length and width, acting on point P, in the x-direction and y-direction, respectively, are σ_x, σ_y:

$$\sigma_x = -K_x(x - x') = -K_x\frac{s}{1 - s}x' \tag{2.48}$$

$$\sigma_y = -K_y y = -K_y\frac{\tan \beta}{1 - s}x' \tag{2.49}$$

The sign of these forces is taken as opposite to the axes direction. Furthermore, the resultant force magnitude is as follows:

$$\sigma = \sqrt{\sigma_x^2 + \sigma_y^2}$$
$$= \sqrt{K_x^2 s^2 + K_y^2 \tan^2 \beta} \, \frac{x'}{1 - s} \tag{2.50}$$

Whereby K_x and K_y are the longitudinal and lateral tread rubber stiffness per unit width and unit length. When the tire longitudinal slip ratio and side-slip angle are produced, tire deformation occurs. As a result, a distribution of the contact surface force, proportional to x', is generated at the tire contact surface.

Assuming a tire pressure distribution that is the same as that described in Section 2.3.1 gives the following:

$$p = \frac{6F_z}{bl}\frac{x'}{l}\left(1 - \frac{x'}{l}\right) \tag{2.51}$$

As shown in Figure 2.35, the tire contact surface force is given by Eqn (2.50) in the adhesive region denoted by $0 \le x' \le x'_s$, and $x \ge x'_s$ in the slip region, the tire contact surface force is given by μp.

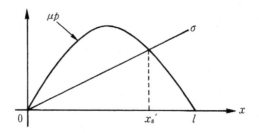

FIGURE 2.35

Force distributions in contact plane.

In the adhesive region, the forces acting at the contact surface in the x and y directions are σ_x and σ_y. In the slip region, the forces are $\mu p \cos \theta$ and $\mu p \sin \theta$. Here, θ determines the direction of the tire slip.

Substituting $\sigma = \mu p$ into Eqns (2.50) and (2.51) to find x'_s and assuming a dimensionless variable, ξ_s, gives the following:

$$\xi_s = \frac{x'_s}{l} = 1 - \frac{K_s}{3\mu F_z} \frac{\lambda}{1-s} \qquad (2.52)$$

where, if $1 - \frac{K_s}{3\mu F_z} \frac{\lambda}{1-s} < 0$ then $\xi_s = 0$, and the following is obvious:

$$\lambda = \sqrt{s^2 + \left(\frac{K_\beta}{K_s}\right)^2 \tan^2 \beta} \qquad (2.53)$$

$$K_s = \frac{bl^2}{2} K_x, K_\beta = \frac{bl^2}{2} K_y \qquad (2.54)$$

From the previous, the overall forces acting on the whole tire contact surface in the x and y directions are expressed as follows:

When $\xi_s > 0$, for a contact surface composed of adhesive and slip regions, the following results:

$$F_x = b \left(\int_0^{x'_s} \sigma_x dx' + \int_{x'_s}^{l} -\mu p \cos \theta dx' \right) \qquad (2.55)$$

$$F_y = b \left(\int_0^{x'_s} \sigma_y dx' + \int_{x'_s}^{l} -\mu p \sin \theta dx' \right) \qquad (2.56)$$

And when $\xi_s = 0$, the contact surface consists only of a slip region:

$$F_x = b \int_0^l -\mu p \cos \theta dx' \qquad (2.55)'$$

$$F_y = b \int_0^l -\mu p \sin \theta dx' \qquad (2.56)'$$

Substituting Eqns (2.50)–(2.52) into Eqns (2.55), (2.56), (2.55)′, and (2.56)′ gives F_x, F_y in the following forms:

When $\xi_s > 0$, then,

$$F_x = -\frac{K_s s}{1-s}\xi_s^2 - 6\mu F_z \cos\theta \left(\frac{1}{6} - \frac{1}{2}\xi_s^2 + \frac{1}{3}\xi_s^3\right) \tag{2.57}$$

$$F_y = -\frac{K_\beta \tan\beta}{1-s}\xi_s^2 - 6\mu F_z \sin\theta \left(\frac{1}{6} - \frac{1}{2}\xi_s^2 + \frac{1}{3}\xi_s^3\right) \tag{2.58}$$

And, when $\xi_s = 0$, then,

$$F_x = -\mu F_z \cos\theta \tag{2.57}'$$

$$F_y = -\mu F_z \sin\theta \tag{2.58}'$$

The direction of the slip force, θ, is approximated by the slip direction at the slip start point.

$$\tan\theta = \frac{K_y \frac{\tan\beta}{1-s}x'}{K_x \frac{s}{1-s}x'} = \frac{K_\beta \tan\beta}{K_s s} \tag{2.59}$$

Therefore, the following results:

$$\cos\theta = \frac{s}{\lambda} \tag{2.60}$$

$$\sin\theta = \frac{K_\beta \tan\beta}{K_s \lambda} \tag{2.61}$$

K_s, as defined by Eqn (2.54), is equivalent to the total longitudinal force per unit longitudinal slip ratio when $s \to 0$ at $\beta = 0$. K_β is equivalent to the total lateral force per unit side-slip angle when $\beta \to 0$ at $s = 0$. This can be derived from the definition of K_x and K_y by integrating the forces at the contact surface when the both slips are very small and s or $\beta = 0$. Equations (2.57) and (2.58) confirm that $(\partial F_x/\partial s)_{s=0,\beta=0} = K_s$ and $(\partial F_y/\partial \beta)_{s=0,\beta=0} = K_\beta$. Hence, if F_x and F_y are used numerically in simulations with the model described here, it is practical to determine experimentally the value of K_s, K_β, depending on the tire vertical load.

The friction coefficient, μ, is a function of F_z and the slip velocity, V_s, so an experimental equation that reflects the dependence of μ on tire load and slip velocity is desired. Here, V_s is defined as follows:

$$V_s = \sqrt{(u - R_0\omega)^2 + u^2 \tan^2\beta} = u\sqrt{s^2 + \tan^2\beta} \tag{2.62}$$

Thus, the tire longitudinal and lateral forces can be obtained numerically as functions of the longitudinal slip ratio, s, slip angle, β, tire load, F_z, and tire traveling speed, u.

$$\begin{aligned}
F_x &= F_x(s, \beta, F_z, u) \\
F_y &= F_y(s, \beta, F_z, u)
\end{aligned} \tag{2.63}$$

2.4.2.2 Accelerating

As in the case of braking, the deformation of tread rubber to the tread base at the contact surface gives the following:

$$x - x' = \frac{u - R_0\omega}{R_0\omega} R_0\omega\Delta t = sx' \tag{2.64}$$

$$y = x \tan \beta = (1+s)\tan \beta x' \qquad (2.65)$$

where s is tire longitudinal slip ratio during acceleration:

$$s = \frac{u - R_0\omega}{R_0\omega} \qquad (2.66)$$

Then, the following results:

$$\sigma_x = -K_x s x' \qquad (2.67)$$

$$\sigma_y = -K_y(1+s)\tan \beta x' \qquad (2.68)$$

$$\sigma = \sqrt{K_x^2 s^2 + K_y^2(1+s)^2 \tan^2 \beta}\, x' \qquad (2.69)$$

As with braking, the point where the contact surface changes from the adhesive region to the slip region is found from the following:

$$\xi_s = \frac{x'_s}{l} = 1 - \frac{K_s}{3\mu F_z}\lambda \qquad (2.70)$$

where, if $1 - \frac{K_s}{3\mu F_z}\lambda < 0$, then $\xi_s = 0$, and as follows:

$$\lambda = \sqrt{s^2 + \left(\frac{K_\beta}{K_s}\right)^2 (1+s)^2 \tan^2 \beta} \qquad (2.71)$$

The forces acting in the x-direction and y-direction on the tire contact surface during acceleration are derived as follows:

When $\xi_s > 0$, then,

$$F_x = -K_s s \xi_s^2 - 6\mu F_z \cos \theta \left(\frac{1}{6} - \frac{1}{2}\xi_s^2 + \frac{1}{3}\xi_s^3\right) \qquad (2.72)$$

$$F_y = -K_\beta(1+s)\tan \beta \xi_s^2 - 6\mu F_z \sin \theta \left(\frac{1}{6} - \frac{1}{2}\xi_s^2 + \frac{1}{3}\xi_s^3\right) \qquad (2.73)$$

And when $\xi_s = 0$, then,

$$F_x = -\mu F_z \cos \theta \qquad (2.72)'$$

$$F_y = -\mu F_z \sin \theta \qquad (2.73)'$$

where,

$$\tan \theta = \frac{K_y(1+s)\tan \beta x'}{K_x s x'} = \frac{K_\beta \tan \beta(1+s)}{K_s s} \qquad (2.74)$$

$$\cos \theta = \frac{s}{\lambda} \qquad (2.75)$$

$$\sin \theta = \frac{K_\beta \tan \beta(1+s)}{K_s \lambda} \qquad (2.76)$$

and the slip velocity, V_s, is as follows:

$$V_s = u\sqrt{\frac{s^2}{(1+s)^2} + \tan^2 \beta} \qquad (2.77)$$

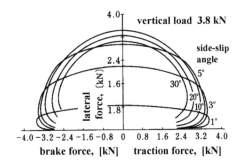

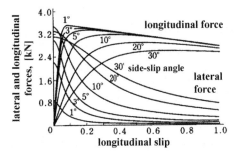

FIGURE 2.36

Interaction between lateral and longitudinal forces.

The longitudinal and lateral forces that act on a tire with side slip, while under braking or acceleration, are shown in Figure 2.36, which is plotted by using Eqns (2.57)–(2.58)′ and (2.72)–(2.73)′. It is clear from the theoretical analysis here that braking and acceleration affects the tire cornering characteristics, as explained in Section 2.3.2.

2.4.3 SELF-ALIGNING TORQUE

As described in Section 2.3.1, the lateral force acting on the contact surface is asymmetrical to the contact surface centerline, and this creates a moment around the vertical axis that passes through the tire contact center point. When the longitudinal force acts together with lateral force, the longitudinal force is offset from the tire centerline because of the lateral tire deformation. A self-aligning moment due to the longitudinal force is created in addition to that from the lateral force. The total of these moments is the self-aligning torque.

In Figure 2.37, the force per unit width and length acting at a point on the contact plane, σ_x, σ_y, is shown. The self-aligning torque described can be written as the total moment around point P as follows:

$$M = b \int \left[\left(x' - \frac{l}{2} \right) \sigma_y - y \sigma_x \right] dx' \tag{2.78}$$

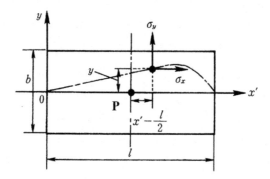

FIGURE 2.37

Lateral and longitudinal forces on contact plane.

Dividing the tire contact surface into adhesive and slip regions gives Eqn (2.78) as follows:

$$
M = b \left[\int_0^{x'_s} \left(x' - \frac{l}{2} \right) \sigma_y dx' + \int_{x'_s}^{l} \left(x' - \frac{l}{2} \right) (-\mu p \sin \theta) dx' \right]
$$

$$
- b \left[\int_0^{x'_s} y \sigma_x dx' + \int_{x'_s}^{l} \frac{\mu p \sin \theta}{K_y} (-\mu p \cos \theta) dx' \right]
$$

(2.79)

The first and second integration terms in Eqn (2.79) are the self-aligning torques due to the lateral force, while the third and fourth integration terms represent the self-aligning torques due to the longitudinal force.

Using Eqns (2.47), (2.48), and (2.49) or Eqns (2.65), (2.67), (2.68), and (2.51), the previous integration can be carried out to give the self-aligning torque.

2.4.3.1 Braking

When $\xi_s > 0$,

$$
M = b \left[\int_0^{x'_s} -\frac{K_y \tan \beta}{1 - s} \left(x' - \frac{l}{2} \right) x' dx' - \int_{x'_s}^{l} \frac{6\mu F_z \sin \theta}{bl} \left(x' - \frac{l}{2} \right) \frac{x'}{l} \left(1 - \frac{x'}{l} \right) dx' \right]
$$

$$
+ b \left[\int_0^{x'_s} \frac{K_x s \tan \beta}{(1 - s)^2} x'^2 dx' + \int_{x'_s}^{l} \left(\frac{6\mu F_z}{bl} \right)^2 \frac{\sin \theta \cos \theta}{K_y} \left(\frac{x'}{l} \right)^2 \left(1 - \frac{x'}{l} \right)^2 dx' \right]
$$

(2.80)

$$
= \frac{lK_\beta \tan \beta}{2(1 - s)} \xi_s^2 \left(1 - \frac{4}{3} \xi_s \right) - \frac{3}{2} l\mu F_z \sin \theta \xi_s^2 (1 - \xi_s)^2 + \frac{2lK_s s \tan \beta}{3(1 - s)^2} \xi_s^3
$$

$$
+ \frac{3l(\mu F_z)^2 \sin \theta \cos \theta}{5K_\beta} (1 - 10\xi_s^3 + 15\xi_s^4 - 6\xi_s^5)
$$

and when $\xi_s = 0$,

$$M = -b \int_0^l \frac{6\mu F_z \sin\theta}{bl}\left(x' - \frac{l}{2}\right)\frac{x'}{l}\left(1 - \frac{x'}{l}\right)dx' + b\int_0^l \left(\frac{6\mu F_z}{bl}\right)^2$$

$$\times \frac{\sin\theta\cos\theta}{K_y}\left(\frac{x'}{l}\right)^2\left(1 - \frac{x'}{l}\right)^2 dx' \qquad (2.81)$$

$$= \frac{3l(\mu F_z)^2 \sin\theta\cos\theta}{5K_\beta}$$

2.4.3.2 Accelerating
When $\xi_s > 0$, the following results:

$$M = b\left[\int_0^{x'_s} -K_y(1+s)\tan\beta\left(x' - \frac{l}{2}\right)x'dx' - \int_{x'_s}^l \frac{6\mu F_z \sin\theta}{bl}\left(x' - \frac{l}{2}\right)\frac{x'}{l}\left(1 - \frac{x'}{l}\right)dx'\right]$$

$$+ b\left[\int_0^{x'_s} K_x(1+s)s\tan\beta x'^2 dx' + \int_{x'_s}^l \left(\frac{6\mu F_z}{bl}\right)^2\frac{\sin\theta\cos\theta}{K_y}\left(\frac{x'}{l}\right)^2\left(1 - \frac{x'}{l}\right)^2 dx'\right]$$

$$= \frac{l}{2}K_\beta(1+s)\tan\beta\xi_s^2\left(1 - \frac{4}{3}\xi_s\right) - \frac{3}{2}l\mu F_z \sin\theta\xi_s^2(1-\xi_s)^2 + \frac{2}{3}lk_s(1+s)s\tan\beta\xi_s^3$$

$$+ \frac{3l(\mu F_z)^2 \sin\theta\cos\theta}{5K_\beta}(1 - 10\xi_s^3 + 15\xi_s^4 - 6\xi_s^5) \qquad (2.82)$$

And, when $\xi_s = 0$,

$$M = -b \int_0^l \frac{6\mu F_z \sin\theta}{bl}\left(x' - \frac{l}{2}\right)\frac{x'}{l}\left(1 - \frac{x'}{l}\right)dx' + b\int_0^l \left(\frac{6\mu F_z}{bl}\right)^2$$

$$\times \frac{\sin\theta\cos\theta}{K_y}\left(\frac{x'}{l}\right)^2\left(1 - \frac{x'}{l}\right)^2 dx' \qquad (2.83)$$

$$= \frac{3l(\mu F_z)^2 \sin\theta\cos\theta}{5K_\beta}$$

Figure 2.38 is plot of self-aligning torque against braking and traction forces, using Eqns (2.80)–(2.83). Unlike the lateral tire force, self-aligning torque changes tremendously during

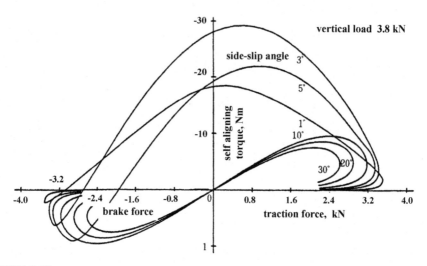

FIGURE 2.38

Effects of longitudinal forces on calculated self-aligning torque.

braking and acceleration. These changes are rather complex and depend on the braking or traction force and also the side-slip angle.

It is useful to see that instead of using Eqns (2.80)–(2.83), the self-aligning torque is sometimes calculated by the following simple equation:

$$M = \xi_n F_y + \frac{F_y}{k_y} F_x = \left(\xi_n + \frac{F_x}{k_y} \right) F_y$$

The first term is a self-aligning torque by the lateral force. The second term is a self-aligning torque produced by the longitudinal force due to the lateral offset, F_y/k_y, of the exerted point of the longitudinal force caused by the lateral force, where k_y is the tire lateral stiffness. The pneumatic trail, ξ_n, decreases with the increase of the side-slip angle as is described in Section 2.3.3, which corresponds with the increase of the slip region of the tire contact surface, and finally tends to zero at the full slip. Therefore, the pneumatic trail is expediently possible to be described by the following:

$$\xi_n = \xi_s x_{n0}$$

where, x_{n0} is the pneumatic trail at the tire slip angle almost equal to zero.

Example 2.2

Using the lateral force and the vertical load distribution diagrams along the longitudinal direction on a tire contact plane with a small side-slip angle, try to explain, schematically, how the lateral force due to small side-slip angle is proportional to side-slip angle. Show this is basically independent of the vertical load and is determined by lateral stiffness of the tire.

Solution

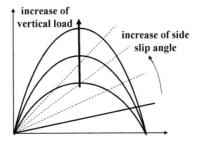

FIGURE E2.2

The above figure is a schematic diagram of the distributions with a small side-slip angle. A linear line is the lateral force due to the lateral deformation of the tread rubber. The parabolas are the distributions of the vertical loads multiplied by a friction coefficient. It is obvious from this diagram that as the side-slip angle is small, the tire lateral force expressed by the area surrounded by the linear line and one of the parabola lines is very robust to the vertical load variations. Also, as the side-slip angle is small, the surrounded plane is almost a triangle, and so, the area is almost proportional to the side-slip angle. The triangle is mostly determined by the lateral deformation of the tread rubber. The force produced by the lateral deformation depends directly on the lateral stiffness.

Example 2.3
Derive the longitudinal and lateral forces, F_x and F_y, during braking when the vertical contact pressure load is distributed uniformly along the longitudinal direction in the contact plane.

Solution
For the uniform vertical contact pressure load distribution, Eqn (2.51) is changed as follows:

$$p = \frac{F_z}{bl} \tag{E2.6}$$

As the point x' that satisfies $\sigma = \mu p$ is the point changing from adhesion to slip, putting the previous equation and Eqn (2.50) into $\sigma = \mu p$ gives us the following equations:

$$\sqrt{K_x^2 s^2 + K_y^2 \tan^2 \beta} \; \frac{x' s}{1-s} = \mu \frac{F_z}{bl} \tag{E2.7}$$

As $K_s = bl^2 K_x/2$, $K_\beta = bl^2 K_y/2$, the previous equation is transformed to the following:

$$2K_S \sqrt{s^2 + \left(\frac{K_\beta}{K_s}\right)^2 \tan^2 \beta} \; \frac{x' s}{l} \frac{1}{1-s} = 2K_S \lambda \frac{x' s}{l} \frac{1}{1-s} = \mu F_Z \tag{E2.8}$$

Thus, the following results:

$$\xi_S = \frac{x' s}{l} = \frac{\mu F_Z}{2K_S} \frac{1-s}{\lambda} \tag{E2.9}$$

Following the same procedure as in Section 2.4.2, F_x and F_y are obtained as follows. When $\xi_S \geq 1$, there is no slip region:

$$F_x = b \int_0^l \sigma_x dx' = \frac{-K_S s}{1-s} \tag{E2.10}$$

$$F_y = b \int_o^l \sigma_y dx' = \frac{-K_\beta \tan \beta}{1-s}$$ (E2.11)

When $\xi_S \leq 1$, there are adhesion and slip regions:

$$F_x = b \left(\int_0^{x'_s} \sigma_x dx' + \int_{x'_s}^l -\mu p \cos \theta dx' \right) = -\frac{K_s s}{1-s}\xi_s^2 - \mu F_z \cos \theta (1 - \xi_S)$$ (E2.12)

$$F_y = b \left(\int_o^{x'_s} \sigma_y dx' + \int_{x'_s}^l -\mu p \sin \theta dx' \right) = -\frac{K_\beta \tan \beta}{1-s}\xi_s^2 - \mu F_z \sin \theta (1 - \xi_S)$$ (E2.13)

2.5 DYNAMIC CHARACTERISTICS

The tire characteristics that have been observed until now are the steady-state characteristics of the lateral force and self-aligning torque to a fixed side-slip angle. In this section, the transient lateral force and self-aligning torque, with changing side slip, are studied. These are the dynamic tire cornering characteristics.

In Sections 2.3 and 2.4, tire cornering characteristics were dealt with theoretically, using mathematical models that considered a micro-deformation at the tire contact surface. Dynamic cornering characteristics become too complicated when analyzed with these models. Instead, the transient response of lateral force and self-aligning torque to side-slip angle will be studied using a macro tire model that deforms in the lateral direction, and it is true for only small side-slip angles.

2.5.1 LATERAL FORCE DYNAMIC CHARACTERISTICS

Figure 2.39 shows that when a sudden side-slip angle, β, occurs on a tire traveling in its rotating direction, then a lateral force is produced, and the contact surface deforms y in the lateral direction only. The lateral velocity of the contact surface is $\dot{y}$, and from Figure 2.39, the contact surface side-slip angle is $\beta - \dot{y}/V$.

Taking F as the tire lateral force and K as the cornering stiffness:

$$F = K\left(\beta - \frac{\dot{y}}{V}\right)$$ (2.84)

if the tire lateral stiffness is defined as k_y, then the following results:

$$F = k_y y$$ (2.85)

by eliminating y from these two equations, we have the following:

$$\frac{K}{k_y V}\frac{dF}{dt} + F = K\beta$$ (2.86)

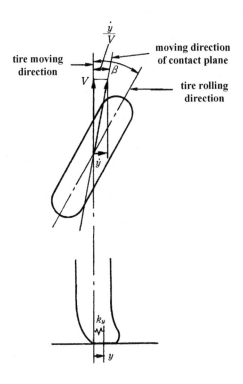

FIGURE 2.39

Tire deformation model for transient characteristics.

Applying Laplace transforms to the previous equation, the transfer function of the lateral force to side-slip angle is as follows:

$$\frac{F}{\beta}(s) = \frac{K}{1 + \dfrac{K}{k_y V}s}$$

$$= \frac{K}{1 + T_1 s} \qquad (2.87)$$

The response of lateral force to side-slip angle can be approximated by a first-order lag element with the time constant of $T_1 = K/(k_y V)$.

Using $j\omega$ instead of s in Eqn (2.87) gives the lateral force frequency response:

$$\frac{F}{\beta}\left(j\frac{\omega}{V}\right) = \frac{K}{1 + j\frac{\omega}{V}\frac{K}{k_y}} \qquad (2.88)$$

From Eqn (2.88), the frequency response of lateral force to side-slip angle is, with regard to a distance frequency, ω/V (rad/m), rather than a time frequency, ω (rad/s). Figure 2.40 is the example

FIGURE 2.40

Frequency response of lateral force to side-slip angle.

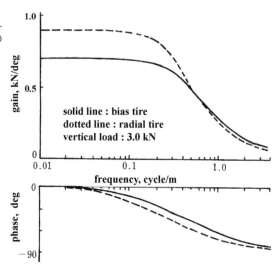

of a real measurement of such a frequency response. As seen, the response could be accurately approximated by a first-order lag element.

In this simplified case, as the steady lateral displacement of the contact surface center of the tire is $l\beta/2$, where l is the contact surface length, the lateral force is expressed by $k_y l\beta/2$. Therefore, the cornering stiffness is expressed as $K = k_y l/2$, and the time constant eventually becomes $T_1 = K/(k_y V) = l/(2V)$.

2.5.2 SELF-ALIGNING TORQUE DYNAMIC CHARACTERISTICS

The tire self-aligning torque in steady state is given by product of tire lateral force and pneumatic trail, ξ_n. As stated in Section 2.5.1, the transient response of lateral force to side-slip angle can be expressed as a first-order delay element. The response of the self-aligning torque, M_s, by this lateral force can also be approximated as a first-order lag element. The time constant is taken as T_2:

$$\frac{M_s}{\beta}(s) = \frac{\xi_n K}{1 + T_2 s} \tag{2.89}$$

In this case, when a sudden side-slip angle occurs on a tire, the tire self-torsional deformation occurs in the tire. The torque due to the torsional deformation becomes part of the transient tire self-aligning torque. The tire torsional angle is largest at the instant when the side-slip angle, β, occurs, and its magnitude is the same as the side-slip angle, but after that decreases with tire rotation and becomes zero at steady state. Therefore, the response of the torque generated by tire torsion, M_t, to the side-slip angle could be approximated by the following first-order lead element.

$$\frac{M_t}{\beta}(s) = \frac{k_t T_3 s}{1 + T_3 s} \tag{2.90}$$

here, k_t is the tire torsional stiffness.

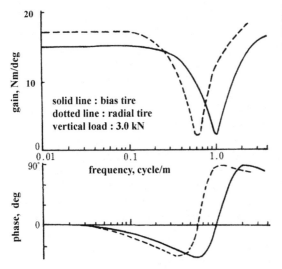

FIGURE 2.41

Frequency response of self-aligning torque to side-slip angle.

From the previous, the tire self-aligning torque, M, is the sum of M_s and M_t, given by the following equation:

$$\frac{M}{\beta}(s) = \frac{M_s}{\beta}(s) + \frac{M_t}{\beta}(s) = \frac{\xi_n K}{1 + T_2 s} + \frac{k_t T_3 s}{1 + T_3 s}$$

$$= \xi_n K \frac{1 + \left(1 + \dfrac{k_t}{\xi_n K}\right)T_3 s + \dfrac{k_t}{\xi_n K}T_2 T_3 s^2}{1 + (T_2 + T_3)s + T_2 T_3 s^2} \tag{2.91}$$

and, the frequency response is as follows:

$$\frac{M}{\beta}\left(j\frac{\omega}{V}\right) = \xi_n K \frac{1 - \dfrac{k_t V^2}{\xi_n K}T_2 T_3 \left(\dfrac{\omega}{V}\right)^2 + j\left(1 + \dfrac{k_t}{\xi_n K}\right)V T_3 \dfrac{\omega}{V}}{1 - V^2 T_2 T_3 \left(\dfrac{\omega}{V}\right)^2 + jV(T_2 + T_3)\dfrac{\omega}{V}} \tag{2.92}$$

Figure 2.41 shows the measured data of the frequency response of self-aligning torque to ω/V. The shape confirms the suitability of the approximation by Eqn (2.92).

PROBLEMS

2.1 Find the approximate value of cornering stiffness using Eqn (2.24) when the lateral force is saturated at the side-slip angle equal to $10°$ with the vertical load, 4.0 kN, on a dry road surface, $\mu = 1.0$.

2.2 Estimate the side-slip angle at which the self-aligning torque becomes maximum if the lateral force is saturated at $10°$ by referring to Eqns (2.24) and (2.26).

2.3 Confirm that the pneumatic trail is approximately equal to one-sixth of the contact length at small side-slip angles. Pay attention to the fact that the tire lateral deformation distribution along the longitudinal direction on the contact plane is almost triangular, as explained in Example 2.2.

2.4 A centrifugal force of the vehicle is balanced with the sum of the four tire lateral forces during turning. Calculate the side-slip angle of the tires when the four tires have an identical side-slip angle with the same cornering stiffness of 1.0 kN/deg. The vehicle mass, speed, and turning radius are 1500 kg, 60 km/h, and 140 m, respectively. Suppose that lateral force is proportional to side-slip angle and confirm it is less than 1.0°.

2.5 Calculate the maximum lateral force available when a braking force of 3.0 kN is applied on the tire with 5.0 kN vertical load on a dry road surface, $\mu = 1.0$.

2.6 Calculate by what percent the lateral force is reduced when the traction force, T, which is equal to the 50% of the vertical load, is applied. Refer to Eqn (2.41) on a dry road surface, $\mu = 1.0$.

2.7 Confirm schematically the nonlinear saturation property of the lateral force to the wide increase of side-slip angles using the diagram used in Example 2.2 in which the lateral force is supposed to be expressed by the surrounded area by a linear lateral force distribution line and a parabola of the vertical load distribution multiplied by the friction coefficient on the contact surface. The increase of the linear line gradient corresponds with the increase of the side-slip angle.

2.8 Confirm that the lateral force at some fixed side-slip angle has a nonlinear relation to the increase of the vertical load using the same schematic method as in Problem 2.7.

2.9 Referring to Eqn (2.87), calculate the approximate time constant for the transient response of lateral force to side-slip angle where the cornering stiffness, the tire lateral stiffness, and the vehicle speed are 1.0 kN/deg, 200 kN/m, and 10 m/s, respectively.

REFERENCES

[1] Fiala E. Seitenkrafte am rollenden Luftreifen. VDI 11, Okt., 1954;96(29).
[2] Ellis JR. Vehicle dynamics. London: London Business Book LTD; 1969 [Chapter 1].
[3] Bernard JE, Segel L, Wild RE. Tire shear force generation during combined steering and braking maneuvers, SAE Paper 770852.

FUNDAMENTALS OF VEHICLE DYNAMICS

3

3.1 PREFACE

This chapter will take the first look at the fundamentals of vehicle dynamics. As described earlier, in Chapter 1, vehicle dynamics are the motions of the vehicle generated by the steering action, through which the vehicle is capable of independent motion.

The motion of the vehicle for a given steering input is studied, and the mechanics of vehicle motion are explained. Only vehicle responses to a predetermined steering action are studied in this chapter; steering in response to vehicle motion is studied later in the book.

There is an enormous amount of previous work that has studied the vehicle response to a predetermined steering action. These studies have established the fundamentals of vehicle dynamics. This chapter describes these fundamentals, which are essential for an understanding of the independent vehicle motion due to steering, i.e., the vehicle dynamics problem [1], [2].

3.2 VEHICLE EQUATIONS OF MOTION

The first chapter identified the vehicle motion degrees-of-freedom that should be considered as lateral, yaw, and roll. However, to understand the basic characteristics of vehicle motion, more simplicity is needed, providing that the nature of the problem is not lost. The transient phenomena of the vehicle, such as sudden acceleration or deceleration are omitted, as is the case of a sudden, large steering action. With these preconditions, the vehicle can be assumed to be traveling at a constant speed, and the roll motion can be neglected. If a vehicle travels at constant speed and without roll, the vehicle's vertical height can be neglected, and only the lateral and yawing motion need to be considered. The vehicle is represented as a rigid body projected to the ground.

In describing the motion of the rigid body, the definition of a reference coordinate frame is necessary. Depending on particular body motion characteristics, there could be many ways of defining the coordinates for describing the body motion. Clever definition of a coordinate frame can simplify the description of the body motion, so selection of suitable coordinates for a particular body is important. Considering this, it is sensible to first derive the fundamental vehicle equations of motion.

3.2.1 EQUATIONS OF MOTION WITH FIXED COORDINATES ON THE VEHICLE

Consider the vehicle moving in the horizontal plane. The vehicle longitudinal and lateral directions are continuously changing with reference to a fixed coordinate frame on the ground.

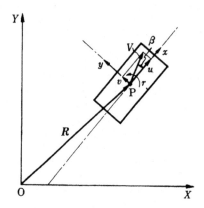

FIGURE 3.1

Coordinate axes for vehicle plane motion.

If the vehicle is examined from on board, regardless of the direction of the vehicle, the motion constraints are basically the same. It is more convenient to describe the vehicle motion by fixed coordinates on the vehicle rather than fixed coordinates on the ground.

As shown in Figure 3.1, X-Y are the fixed plane coordinates on the ground, and x-y are the fixed coordinates on the vehicle, with x in the vehicle's longitudinal direction, and y in the lateral direction. The origin of the frame is at the vehicle's center of gravity, P. The yaw angle around the vertical axis is taken as positive in the anticlockwise direction.

The vehicle is considered to be moving in plane with some speed. The position vector of point P, with reference to coordinate frame X-Y, is defined as R. The velocity vector $\dot{R}$ can be written as follows:

$$\dot{R} = ui + vj \tag{3.1}$$

Here, i and j are the respective unit vectors in the x- and y-directions. The variables u and v represent the velocity components of point P in the x- and y-directions. Differentiating Eqn (3.1) with time, the acceleration could be written as a vector of point P, as follows. Here, "$\cdot$" and "$\cdot\cdot$" mean d/dt and d^2/dt^2.

$$\ddot{R} = \dot{u}i + u\dot{i} + \dot{v}j + v\dot{j} \tag{3.2}$$

The x-y coordinate is fixed to the vehicle, and the vehicle has a yaw angular velocity of r around the vertical axis passing through point P, which is sometimes called yaw velocity or yaw rate.

The changes of i and j in time, Δt, based on Figure 3.2, are as follows:

$$\Delta i = r\Delta t j, \quad \Delta j = -r\Delta t i$$

thus,

$$i = \lim_{\Delta t \to 0} \frac{\Delta i}{\Delta t} = rj$$

$$j = \lim_{\Delta t \to 0} \frac{\Delta j}{\Delta t} = -ri$$

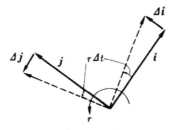

FIGURE 3.2

Time derivative of unit vectors.

Example 3.1

Show an alternative way of expressing the time derivatives of unit vectors i and j.

Solution

As shown in Figure E3.1, let i_F and j_F be the unit vectors fixed to the ground, then i and j are expressed as follows:

$$i = \cos\theta\, i_F + \sin\theta\, j_F$$

$$j = -\sin\theta\, i_F + \cos\theta\, j_F$$

where θ is the angle between the unit vectors i and i_F. Differentiating the previous equations gives the following:

$$\dot{i} = -\dot{\theta}\sin\theta\, i_F + \dot{\theta}\cos\theta\, j_F = r(-\sin\theta\, i_F + \cos\theta\, j_F) = r\,j$$

$$\dot{j} = -\dot{\theta}\cos\theta\, i_F - \dot{\theta}\sin\theta\, j_F = -r(\cos\theta\, i_F + \sin\theta\, j_F) = -r\,i$$

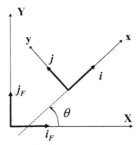

FIGURE E3.1

The final acceleration vector of point P, $\ddot{R}$, is equated next:

$$\ddot{R} = (\dot{u} - vr)i + (\dot{v} + ur)j \tag{3.3}$$

The angle between the vehicle traveling direction and longitudinal direction, β, is expressed by $\tan^{-1}(v/u)$. This is called the side-slip angle of the vehicle center of gravity. In normal cars, because $u \gg v$, $|\beta|$ is regarded to be very small. Also, if the vehicle speed is constant, $V = \sqrt{u^2 + v^2}$ is a constant as well.

It is often more convenient to describe motion using β than the motion of point P by u and v. In other words, if β is small, then the following results:

$$u = V \cos \beta \approx V, \qquad\qquad v = V \sin \beta \approx V\beta$$
$$\dot{u} = -V \sin \beta\dot{\beta} \approx -V\beta\dot{\beta} \quad \dot{v} = V \cos \beta\dot{\beta} \approx V\dot{\beta}$$

Equations (3.3) and (3.1) can be written as follows:

$$\ddot{R} = -V(\dot{\beta} + r)\beta i + V(\dot{\beta} + r)j \qquad\qquad (3.3)'$$

$$\dot{R} = Vi + V\beta j \qquad\qquad (3.1)'$$

Using Eqns (3.3)′ and (3.1)′, it can be shown that the dot product of vector $\ddot{R}$ and vector $\dot{R}$, $\ddot{R}\cdot\dot{R}$, is equal to zero. In other words, $\ddot{R}$ is perpendicular to $\dot{R}$ in the traveling direction of point P. Equation (3.4)′ shows that if β is small, the acceleration of point P has a magnitude of $V(\dot{\beta} + r)$, and it is perpendicular to the vehicle traveling direction, as shown in Figure 3.3. When β is small, it can be assumed that the direction perpendicular to the traveling direction of the vehicle almost coincides with the lateral direction, y. Therefore, a vehicle moving in plane with a constant speed, regardless of motion with reference to the X-Y coordinates, will have an acceleration of $V(\dot{\beta} + r)$ in the lateral or y-direction.

The lateral motion and yaw motion of the vehicle will generate slip angles at the tires. As described in Chapter 2, a lateral force will be produced at the tire in response to this side-slip angle. This lateral force produced by the vehicle motion becomes the force that controls the vehicle motion.

In Figure 3.4(a), the angle of the front left and right wheels with respect to the x-direction is the front wheel steering angle, δ, and the tire side-slip angles of the front and rear wheels are β_{f1}, β_{f2}, β_{r1}, and β_{r2}. The lateral forces acting on the tires are Y_{f1}, Y_{f2}, Y_{r1}, and Y_{r2}. These forces are assumed to act in the direction perpendicular to the tire heading direction, i.e., the vehicle's lateral direction, because $|\beta_{f1}|$, $|\beta_{f2}|$, $|\beta_{r1}|$, $|\beta_{r2}|$, and $|\delta| \ll 1$. The lateral motion of the vehicle is described next:

$$mV\left(\frac{d\beta}{dt} + r\right) = Y_{f1} + Y_{f2} + Y_{r1} + Y_{r2} \qquad\qquad (3.4)$$

where m is the vehicle inertia mass.

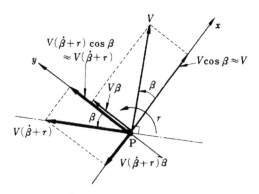

FIGURE 3.3

Acceleration and velocity at point P.

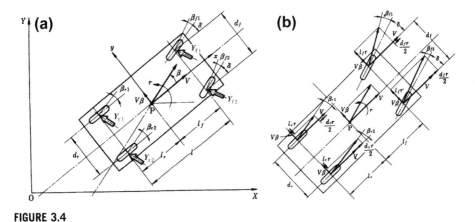

FIGURE 3.4

(a) Vehicle plane motion. (b) Side-slip angle of each tire.

The lateral forces also result in a yaw moment around the center of gravity, and the vehicle's yaw motion is described as follows:

$$I\frac{dr}{dt} = l_f(Y_{f1} + Y_{f2}) - l_r(Y_{r1} + Y_{r2}) \tag{3.5}$$

Here, I is the yaw moment of inertia, l_f and l_r are the distances of the front and rear wheel axles from the center of gravity, and the lateral forces are assumed to act along the axle. Equations (3.4) and (3.5) are the basic equations describing the plane motion of the vehicle traveling with a constant speed and without roll motion.

To study the lateral forces (Y_{f1}, Y_{f2}, Y_{r1}, and Y_{r2}) acting on the tires in more detail, it is first necessary to examine the respective tire side-slip angles: β_{f1}, β_{f2}, β_{r1}, and β_{r2}. Tire side-slip angle is defined as the angle between the tire traveling direction and the tire heading direction or the tire rotation plane. The rigid body vehicle has a velocity component of V in the longitudinal (x) direction and $V\beta$ in the lateral (y) direction. The vehicle also has an angular velocity of r around the center of gravity. Consequently, each tire will have the velocity component of the center of gravity and the velocity component due to rotation around the center of gravity.

The velocity component in x- and y-direction for each tire is shown in Figure 3.4(b). The heading direction of the front wheels has an angular displacement of δ with respect to the vehicle longitudinal direction, x. This is the actual steering angle of the front wheels. The heading direction of the rear wheels is the same as the vehicle longitudinal direction. Therefore, the side-slip angle for each tire could be written as follows:

$$\beta_{f1} \approx \frac{V\beta + l_f r}{V - d_f r/2} - \delta \approx \beta + \frac{l_f r}{V} - \delta$$

$$\beta_{f2} \approx \frac{V\beta + l_f r}{V + d_f r/2} - \delta \approx \beta + \frac{l_f r}{V} - \delta$$

$$\beta_{r1} \approx \frac{V\beta - l_r r}{V - d_r r/2} \approx \beta - \frac{l_r r}{V}$$

$$\beta_{r2} \approx \frac{V\beta - l_r r}{V + d_r r/2} \approx \beta - \frac{l_r r}{V}$$

Whereby d_f and d_r are the vehicle front and rear treads, and $|\beta|$, $|l_f r/V|$, $|l_r r/V|$, $|d_f r/2V|$, and $|d_r r/2V| \ll 1$ can be ignored as negligible. Then the left and right tire side-slip angles for both front and rear wheels are equal, and taking these as β_f and β_r, respectively, gives the following:

$$\beta_f = \beta_{f1} = \beta_{f2} = \beta + l_f r/V - \delta \tag{3.6}$$

$$\beta_r = \beta_{r1} = \beta_{r2} = \beta - l_r r/V \tag{3.7}$$

As the left and right tire side-slip angles are equal, the steering angle is small, and there is a negligible roll motion, it is suitable to consider the left and right tires of the front and rear wheels to be concentrated at the intersecting point of the vehicle's x-axis with the front and rear axles as shown in Figure 3.5. In this way, a four-wheeled vehicle could be transformed to an equivalent two-wheeled vehicle, which makes the analysis of vehicle motion simpler.

Usually, if there is no difference in the characteristics of the left and right tires, the lateral forces of the left and right tires will be equal. Taking the front and rear lateral forces as Y_f and Y_r gives the following:

$$2Y_f = Y_{f1} + Y_{f2}$$

$$2Y_r = Y_{r1} + Y_{r2}$$

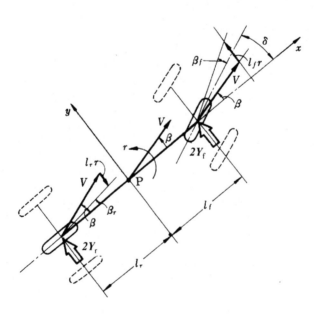

FIGURE 3.5

Equivalent bicycle model.

Because these forces act in the y-direction, Eqns (3.4) and (3.5) become the following:

$$mV\left(\frac{d\beta}{dt} + r\right) = 2Y_f + 2Y_r \tag{3.4}'$$

$$I\frac{dr}{dt} = 2l_f Y_f - 2l_r Y_r \tag{3.5}'$$

Defining the cornering stiffness of the front and rear wheels as K_f and K_r, Y_f, and Y_r are proportional to β_f and β_r, respectively. In the x-y coordinates shown in Figure 3.4(a), all angles are positive in the anticlockwise direction. When a side-slip angle is positive, Y_f and Y_r act in the negative y-direction, and can be written as shown next:

$$Y_f = -K_f \beta_f = -K_f(\beta + l_f r/V - \delta) \tag{3.8}$$

$$Y_r = -K_r \beta_r = -K_r(\beta - l_r r/V) \tag{3.9}$$

From these equations, it is clear that the forces acting on the vehicle, Y_f and Y_r, are not dominated by the vehicle's position, or attitude angle, in the fixed coordinate system. Instead, they are determined by the vehicle motion itself, namely by β, r, and steering angle, δ.

Substituting Eqns (3.8) and (3.9) into the previous Eqns (3.4)' and (3.5)' gives the following:

$$mV\left(\frac{d\beta}{dt} + r\right) = -2K_f\left(\beta + \frac{l_f}{V}r - \delta\right) - 2K_r\left(\beta - \frac{l_r}{V}r\right) \tag{3.10}$$

$$I\frac{dr}{dt} = -2K_f\left(\beta + \frac{l_f}{V}r - \delta\right)l_f + 2K_r\left(\beta - \frac{l_r}{V}r\right)l_r \tag{3.11}$$

And, rearranging the previous equations gives the following:

$$mV\frac{d\beta}{dt} + 2(K_f + K_r)\beta + \left\{mV + \frac{2}{V}(l_f K_f - l_r K_r)\right\}r = 2K_f\delta \tag{3.12}$$

$$2(l_f K_f - l_r K_r)\beta + I\frac{dr}{dt} + \frac{2\left(l_f^2 K_f + l_r^2 K_r\right)}{V}r = 2l_f K_f\delta \tag{3.13}$$

These equations now become the fundamental equations of motion describing the vehicle plane motion. The left-hand side term of Eqns (3.12) and (3.13) describes the vehicle motion characteristics in response to an arbitrary front wheel steering angle, δ, allocated at the right-hand side of the equation. From these equations, it is clear too that the vehicle's motion is not affected by the position of the vehicle or the heading direction of the vehicle with reference to the fixed coordinates on the ground.

Here, the Laplace-transformed Eqns (3.12) and (3.13) are rewritten as follows, whereby s is the Laplace transform operator and $\beta(s)$, $r(s)$, and $\delta(s)$ are the Laplace transforms for β, r, and δ:

$$\begin{bmatrix} mVs + 2(K_f + K_r) & mV + \frac{2}{V}(l_f K_f - l_r K_r) \\ 2(l_f K_f - l_r K_r) & Is + \frac{2}{V}\left(l_f^2 K_f + l_r^2 K_r\right) \end{bmatrix} \begin{bmatrix} \beta(s) \\ r(s) \end{bmatrix} = \begin{bmatrix} 2K_f \delta(s) \\ 2l_f K_f \delta(s) \end{bmatrix}$$

Hence, the characteristics equation for vehicle motion is as follows:

$$\begin{vmatrix} mVs + 2(K_f + K_r) & mV + 2(l_fK_f - l_rK_r)/V \\ 2(l_fK_f - l_rK_r) & Is + 2\left(l_f^2K_f + l_r^2K_r\right)/V \end{vmatrix} = 0$$

Expanding and arranging this equation gives the following:

$$mIV \left[s^2 + \frac{2m\left(l_f^2K_f + l_r^2K_r\right) + 2I(K_f + K_r)}{mIV} s + \frac{4K_fK_rl^2}{mIV^2} - \frac{2(l_fK_f - l_rK_r)}{I} \right] = 0 \qquad (3.14)$$

Whereby, l is the wheel base:

$$l = l_f + l_r$$

Examining the derived equations of motion, (3.12) and (3.13), it is understood that they are simultaneous first-order differential equations of β and r. In fact, this form of equation is not unique to vehicle motion only. The independent motion of ships when steered is expressed in the same form. In the case of a vehicle, front and rear lateral forces act because of the vehicle's motion and, in turn, control the vehicle motion. In the case of ships, lift force (lateral force) is generated by the attack angle between the ships longitudinal direction and traveling direction, and this force controls the ship's motion. This is how vehicle dynamics and ship dynamics have the same form of equations of motion. The aircraft gives a typical example of motion that relies on lift force, where the similarity to the motion dynamics stated previously can also be seen, particularly in the longitudinal motion. This form of motion, as expressed in Eqns (3.12) and (3.13), is shared among vehicles, ships, and aircraft as a general form of the equations of motion.

Looking at Eqns (3.12) and (3.13) in more detail, the coefficient $l_fK_f - l_rK_r$ has a big effect on vehicle motion. If this value is zero, in other words, $l_fK_f = l_rK_r$, the coupling between lateral and yaw motion is incomplete; whereby, r is no longer related to β at all, though β is still related to r. When $l_fK_f - l_rK_r \neq 0$, the sign of this term will have a big effect on the coupling of lateral and yaw motion. The relative magnitudes of l_fK_f and l_rK_r are directly related to the vehicle's basic motion characteristics, and this will be explained in detail later.

3.2.2 EQUATIONS OF MOTION WITH FIXED COORDINATES ON THE GROUND

Previously, vehicle motion has been expressed with respect to fixed coordinates on the vehicle, as this simplifies the analysis and understanding of the phenomena later on. However, in some cases, it can be convenient to express the vehicle's motion with fixed coordinates on the ground. For example, when studying the motion of a vehicle traveling at constant speed on a straight road, expressing the motion with respect to the straight road is easier and more convenient.

As shown in Figure 3.6, the X-Y coordinates fixed on the ground are considered with the straight road direction as the X-axis and the direction perpendicular to it as the Y-axis. The angle between the vehicle longitudinal direction and the X-axis, or the vehicle yaw angle, is θ. The angle between the vehicle heading direction and the X-axis is γ, and the lateral displacement of the vehicle center of gravity, P, from the X-axis is y. Normally, when considering the motion of the

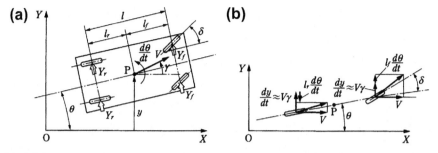

FIGURE 3.6

(a) Vehicle motion with coordinate axis fixed on the ground and (b) the simplified model.

vehicle moving in a straight line, $|\gamma| \ll 1$ and $|\theta| \ll 1$ are assumed. With these assumptions, and if the front wheel steering angle $|\delta| \ll 1$, the direction of the lateral forces, Y_f and Y_r, acting on the front and rear tires almost coincides with the Y-direction, and the vehicle motion can be expressed as follows. First, the motion of the center of gravity in the Y-direction is equated.

$$m\frac{d^2 y}{dt^2} = 2Y_f + 2Y_r \tag{3.15}$$

And, the yaw motion is as follows:

$$I\frac{d^2 \theta}{dt^2} = 2l_f Y_f - 2l_r Y_r \tag{3.16}$$

If $|\gamma|$ is small, the vehicle center of gravity has a velocity component of $V\cos\gamma \approx V$ in the X-direction, $V\sin\gamma \approx V\gamma = dy/dt$ in the Y-direction, and an angular velocity of $d\theta/dt$ around the center of gravity. The left and right wheels are considered to be concentrated to the vehicle axle centers at the front and rear, and if $|\theta|$ is small, the front and rear have additional velocity components of $l_f(d\theta/dt)$ and $-l_r(d\theta/dt)$ in the Y-direction. Consequently, from Figure 3.6(b), the angles between the front and rear wheel traveling direction and the X-axis, γ_f, and γ_r, are as follows:

$$\gamma_f = \frac{V\gamma + l_f(d\theta/dt)}{V} = \frac{1}{V}\frac{dy}{dt} + \frac{l_f}{V}\frac{d\theta}{dt}$$

$$\gamma_r = \frac{V\gamma - l_r(d\theta/dt)}{V} = \frac{1}{V}\frac{dy}{dt} - \frac{l_r}{V}\frac{d\theta}{dt}$$

The angles between the front and rear wheel heading direction and the X-axis are $\theta_f = \theta + \delta$ and $\theta_r = \theta$; thus, the front and rear tire side-slip angles, β_f and β_r, are as follows:

$$\beta_f = \gamma_f - \theta_f = \frac{1}{V}\frac{dy}{dt} + \frac{l_f}{V}\frac{d\theta}{dt} - \theta - \delta \tag{3.17}$$

$$\beta_r = \gamma_r - \theta_r = \frac{1}{V}\frac{dy}{dt} - \frac{l_r}{V}\frac{d\theta}{dt} - \theta \tag{3.18}$$

The lateral forces acting on the front and rear wheels, Y_f and Y_r, can be rewritten as follows:

$$Y_f = -K_f \beta_f = K_f \left(\delta + \theta - \frac{1}{V} \frac{dy}{dt} - \frac{l_f}{V} \frac{d\theta}{dt} \right) \tag{3.19}$$

$$Y_r = -K_r \beta_r = K_r \left(\theta - \frac{1}{V} \frac{dy}{dt} + \frac{l_r}{V} \frac{d\theta}{dt} \right) \tag{3.20}$$

Substituting these into Eqns (3.15) and (3.16) gives the following:

$$m \frac{d^2 y}{dt^2} = 2K_f \left(\delta + \theta - \frac{1}{V} \frac{dy}{dt} - \frac{l_f}{V} \frac{d\theta}{dt} \right) + 2K_r \left(\theta - \frac{1}{V} \frac{dy}{dt} + \frac{l_r}{V} \frac{d\theta}{dt} \right)$$

$$I \frac{d^2 \theta}{dt^2} = 2K_f \left(\delta + \theta - \frac{1}{V} \frac{dy}{dt} - \frac{l_f}{V} \frac{d\theta}{dt} \right) l_f - 2K_r \left(\theta - \frac{1}{V} \frac{dy}{dt} + \frac{l_r}{V} \frac{d\theta}{dt} \right) l_r$$

And, rearranging gives the following:

$$m \frac{d^2 y}{dt^2} + \frac{2(K_f + K_r)}{V} \frac{dy}{dt} + \frac{2(l_f K_f - l_r K_r)}{V} \frac{d\theta}{dt} - 2(K_f + K_r)\theta = 2K_f \delta \tag{3.21}$$

$$\frac{2(l_f K_f - l_r K_r)}{V} \frac{dy}{dt} + I \frac{d^2 \theta}{dt^2} + \frac{2\left(l_f^2 K_f + l_r^2 K_r \right)}{V} \frac{d\theta}{dt} - 2(l_f K_f - l_r K)\theta = 2l_f K_f \delta \tag{3.22}$$

These are the vehicle equations of motions with respect to the fixed coordinates on the ground.

The lateral and yaw motions of a vehicle traveling with constant speed on an almost straight road with small lateral velocity and yaw angle can be expressed in a simpler form using fixed coordinates on the ground. The analysis and understanding of the phenomena becomes easier too. The left-hand side of Eqns (3.21) and (3.22) describe the vehicle's lateral displacement, y, and yaw angle, θ, to the steering angle, δ. The term $l_f K_f - l_r K_r$ affects the coupling mode of y and θ, which means it is important in both ground and vehicle coordinate systems.

Next, the characteristic equation for Eqns (3.21) and (3.22) will be determined. The Laplace-transformed equations are written as follows:

$$\begin{bmatrix} ms^2 + \dfrac{2(K_f + K_r)}{V} s & \dfrac{2(l_f K_f - l_r K_r)}{V} s - 2(K_f + K_r) \\[3mm] \dfrac{2(l_f K_f - l_r K_r)}{V} s & Is^2 + \dfrac{2\left(l_f^2 K_f + l_r^2 K_r \right)}{V} s - 2(l_f K_f - l_r K_r) \end{bmatrix} \begin{bmatrix} y(s) \\ \theta(s) \end{bmatrix}$$

$$= \begin{bmatrix} 2K_f \delta(s) \\ 2l_f K_f \delta(s) \end{bmatrix}$$

Whereby, $y(s)$ and $\theta(s)$ are the Laplace transforms for y and θ. Hence, the characteristic equation is as follows:

$$\begin{vmatrix} ms^2 + \dfrac{2(K_f + K_r)}{V}s & \dfrac{2(l_f K_f - l_r K_r)}{V}s - 2(K_f + K_r) \\[4mm] \dfrac{2(l_f K_f - l_r K_r)}{V}s & Is^2 + \dfrac{2\left(l_f^2 K_f + l_r^2 K_r\right)}{V}s - 2(l_f K_f - l_r K_r) \end{vmatrix} = 0$$

Expanding and then rearranging these equations gives the following:

$$mIs^2 \left[s^2 + \frac{2m\left(l_f^2 K_f + l_r^2 K_r\right) + 2I(K_f + K_r)}{mIV}s + \frac{4K_f K_r l^2}{mIV^2} - \frac{2(l_f K_f - l_r K_r)}{I} \right] = 0 \qquad (3.23)$$

In this characteristic equation, if the term s^2 is omitted from the left-hand side, it matches the characteristics equation (Eqn (3.14)) for a vehicle using fixed coordinates on the vehicle. This guarantees that regardless of the fixation of the coordinates, on the vehicle or on the ground, the expressions for vehicle motion are fundamentally the same. The left-hand side of Eqn (3.23) differs from Eqn (3.14) because of the s^2 term. The existence of this independent s^2 term inside the characteristics equation shows mathematically that the vehicle, at any location on the straight road, could move freely by steering. In practice, the vehicle could make lane changes or avoid the obstacles while traveling on the road. On the other hand, this s^2 term also shows that if suitable steering is not applied when the vehicle deviates laterally from the traveling path by some disturbance, $|y|$, the deviation will get larger and larger and could result in the vehicle falling off the path (refer to Chapter 10).

Example 3.2

Describe the position of the moving vehicle center of gravity with respect to the axis fixed on the ground by using β and r in Eqns (3.12) and (3.13).

Solution

Let (X, Y) be the coordinate of vehicle center of gravity with respect to the axis fixed on the ground, and let θ be the vehicle attitude angle with respect to the X-axis as shown in Figure E3.2. Then, it is possible to have the following equations:

$$\frac{dX}{dt} = V \cos(\beta + \theta) \qquad (E3.1)$$

$$\frac{dY}{dt} = V \sin(\beta + \theta) \qquad (E3.2)$$

Integration of the previous equations gives us the equations to describe the position of the vehicle center of gravity as follows:

$$X = X_0 + V \int_0^t \cos(\beta + \theta)dt \qquad (E3.4)$$

$$Y = Y_0 + V \int_0^t \sin(\beta + \theta)dt \qquad (E3.4)$$

where

$$\theta = \theta_0 + \int_0^t r\,dt \tag{E3.5}$$

Here, X_0, Y_0, and θ_0 are the initial values of X, Y, and θ, respectively.

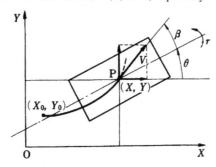

FIGURE E3.2

3.3 VEHICLE STEADY-STATE CORNERING

A theoretical understanding of a mechanical system is normally carried out by solving the equations of motion analytically, under suitable initial conditions. It may not be possible to solve the equations of motion analytically, and even if an analytical solution is possible, the solution can be extremely complicated. This can make understanding the motion characteristics difficult. Equations of motion that are difficult to solve analytically could, by use of computers, be solved numerically, but the understanding of the basic motion characteristics is almost impossible. There are several ways to understand the basic motion characteristics without the direct solution of a given equation of motion. One looks at the static characteristics by analyzing the steady-state condition of the mechanical system, and another studies the dynamic characteristics by examining the root of the characteristics equation and the response of the mechanical system to periodic external input.

Under normal conditions, a vehicle at constant speed and fixed front steering angle will make a steady circular motion with a constant radius of curvature. This is called steady-state cornering. By understanding the vehicle characteristics in steady-state cornering, the fundamental characteristics of vehicle motion can be understood.

3.3.1 DESCRIPTION OF STEADY-STATE CORNERING

3.3.1.1 Description by equations of motion

The equations of motion derived in Section 3.2 using vehicle fixed coordinates will be used to describe steady-state cornering. During steady-state cornering, there will be no changes in the side-slip angle and the yaw velocity. Here the steady-state conditions, $d\beta/dt = 0$ and $dr/dt = 0$, can be substituted into Eqns (3.12) and (3.13), giving the following:

$$2(K_f + K_r)\beta + \left\{ mV + \frac{2}{V}(l_f K_f - l_r K_r) \right\} r = 2K_f \delta \tag{3.24}$$

$$2(l_f K_f - l_r K_r)\beta + \frac{2\left(l_f^2 K_f + l_r^2 K_r\right)}{V} r = 2l_f K_f \delta \tag{3.25}$$

The solution for β and r is shown next:

$$\beta = \frac{\begin{vmatrix} 2K_f & mV + \frac{2}{V}(l_f K_f - l_r K_r) \\ 2l_f K_f & \frac{2\left(l_f^2 K_f + l_r^2 K_r\right)}{V} \end{vmatrix}}{} \frac{\delta}{\Delta} \tag{3.26}$$

$$r = \begin{vmatrix} 2(K_f + K_r) & 2K_f \\ 2(l_f K_f - l_r K_r) & 2l_f K_f \end{vmatrix} \frac{\delta}{\Delta} \tag{3.27}$$

whereby,

$$\Delta = \begin{vmatrix} 2(K_f + K_r) & mV + \frac{2}{V}(l_f K_f - l_r K_r) \\ 2(l_f K_f - l_r K_r) & \frac{2\left(l_f^2 K_f + l_r^2 K_r\right)}{V} \end{vmatrix} \tag{3.28}$$

Expanding Eqns (3.26)–(3.28) and rearranging them, gives β and r as follows:

$$\beta = \left(\frac{1 - \frac{m}{2l} \frac{l_f}{l_r K_r} V^2}{1 - \frac{m}{2l^2} \frac{l_f K_f - l_r K_r}{K_f K_r} V^2} \right) \frac{l_r}{l} \delta \tag{3.29}$$

$$r = \left(\frac{1}{1 - \frac{m}{2l^2} \frac{l_f K_f - l_r K_r}{K_f K_r} V^2} \right) \frac{V}{l} \delta \tag{3.30}$$

If the vehicle is traveling with a constant speed, V, and the yaw velocity is r, the radius of the steady-state cornering, ρ, is formulated as follows:

$$\rho = \frac{V}{r} = \left(1 - \frac{m}{2l^2} \frac{l_f K_f - l_r K_r}{K_f K_r} V^2 \right) \frac{l}{\delta} \tag{3.31}$$

Equations (3.29)–(3.31) describe the vehicle's steady-state cornering with steering angle, δ, and constant traveling speed, V. They show physically how the side-slip angle, β, yaw velocity, r, and circular radius, ρ, respond to steering angle at different traveling speeds.

Assuming that the vehicle is traveling at a very low speed ($V \approx 0$), then V^2 can be neglected in Eqns (3.29)–(3.31). β, r, and ρ can now be described as follows:

$$\beta_{(V \approx 0)} = \beta_s = \frac{l_r}{l} \delta$$

$$r_{(V \approx 0)} = r_s = \frac{V}{l} \delta \tag{3.32}$$

$$\rho_{(V \approx 0)} = \rho_s = \frac{l}{\delta}$$

Equations (3.29)–(3.31) can express the vehicle steady-state cornering as follows:

$$\frac{\beta}{\beta_s} = \frac{1 - \frac{m}{2l}\frac{l_f}{l_rK_r}V^2}{1 - \frac{m}{2l^2}\frac{l_fK_f - l_rK_r}{K_fK_r}V^2} \tag{3.29}'$$

$$\frac{r}{r_s} = \frac{1}{1 - \frac{m}{2l^2}\frac{l_fK_f - l_rK_r}{K_fK_r}V^2} \tag{3.30}'$$

$$\frac{\rho}{\rho_s} = 1 - \frac{m}{2l^2}\frac{l_fK_f - l_rK_r}{K_fK_r}V^2 \tag{3.31}'$$

The previous equations show how the conditions for vehicle steady-state cornering change with the traveling speed, V. The state of vehicle circular motion at very low speeds, $V \approx 0$, can be used as the reference.

3.3.1.2 Description by geometry

The study of vehicle steady-state cornering gave the response of β, r, and ρ to steering angle, δ, by simply putting steady-state conditions into the vehicle equations of motion derived from Section 3.2. Further derivations of these Eqns (3.29)–(3.31) or (3.29)′–(3.31)′ show how the steady state cornerings change with the traveling speed, V.

Next, the vehicle steady-state cornering will be studied geometrically to understand the vehicle motion more intuitively or in a more direct sense. Using geometry, though the steering angle and yaw velocity are positive in the anticlockwise direction in the equations of motions, they are taken as positive in the direction shown by Figure 3.7(a).

First, let us consider the steady-state cornering of the vehicle at very low speeds, $V \approx 0$. In this circumstance, a centrifugal force does not act on the vehicle, lateral forces are not needed, and no side-slip angle is produced as both front and rear wheels travel in the heading direction of the wheels, respectively, and make a circular motion around O_s, as shown in Figure 3.7(a). From this figure, the geometric relations are formulated as follows:

$$\rho_s = \frac{l}{\delta}$$

$$r_s = \frac{V}{\rho_s} = \frac{V}{l}\delta \tag{3.33}$$

$$\beta_s = \frac{l_r}{\rho_s} = \frac{l_r}{l}\delta$$

whereby $0 < \delta \ll 1$ and $l \ll \rho$. As seen from Figure 3.7(a), the actual steering angle for the left and right front wheels is not δ, but it is a little smaller for the left wheel and a little larger for the right wheel. In practice, this is achieved by a steering link mechanism, but if $\delta \ll 1$ and $\rho_s \gg d$, then the difference is very small, and the left and right wheels can be considered as having the same steering angle, δ. Equation (3.33) agrees with the steady-state cornering of Eqn (3.32) that is obtained from the equations of motion with $V \approx 0$. This geometrical relation is called the Ackermann steering geometry, and $\delta = l/\rho_s$ is called the Ackermann angle.

When the circular motion of the vehicle is considered at larger speeds, the centrifugal force becomes significant. The cornering forces at the front and rear wheels are needed to

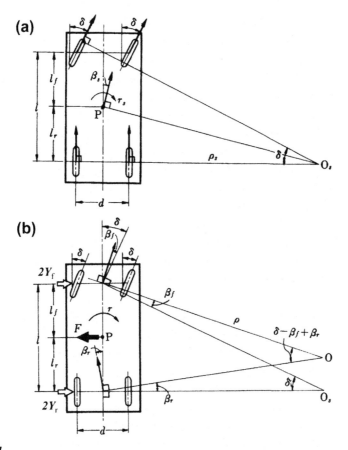

FIGURE 3.7

(a) Steady-state turning at low speed. (b) Steady-state turning with centrifugal force.

balance this centrifugal force, and the side-slip angles are produced. When the centrifugal force acts at the vehicle center of gravity, the circular motion condition shown in Figure 3.7(a) is no more accurate, and Eqn (3.33) is not accurate either. Figure 3.7(b) shows the circular motion when the front and rear wheel side-slip angles, β_f and β_r, are produced by the centrifugal force.

The center of the circular motion is the intersecting point of the two straight lines perpendicular to the front and rear wheel's traveling direction, denoted as O. The geometric relations are formulated as follows:

$$\rho = \frac{l}{\delta - \beta_f + \beta_r} \tag{3.34}$$

assuming that $0 < \delta \ll 1$, $0 < \beta_f$, $\beta_r \ll 1$, and $\rho_s \gg l, d$.

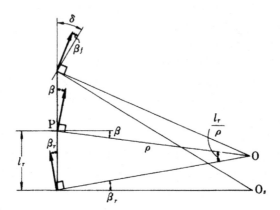

FIGURE 3.8

Side-slip angle during steady-state cornering.

Here, $r = V/\rho$. So, the following is formulated:

$$r = \frac{V(\delta - \beta_f + \beta_r)}{l} \tag{3.35}$$

Furthermore, from Figure 3.8, the following is derived:

$$\beta + \beta_r = \frac{l_r}{\rho}$$

Hence,

$$\beta = \frac{l_r}{\rho} - \beta_r = \frac{l_r}{l}\delta - \frac{l_r\beta_f + l_f\beta_r}{l} \tag{3.36}$$

Equations (3.34)–(3.36) express the vehicle circular motion and are derived from the steady-state cornering geometric relation. The front and rear wheel side-slip angles, β_f and β_r, can be found from the magnitude of the centrifugal force acting at the vehicle center of gravity. The centrifugal force is dependent on the vehicle speed, V. So, ρ, r, and β in Eqns (3.34)–(3.36) also change with V.

The steady-state cornering conditions with traveling speed have been determined by deriving Eqns (3.29)–(3.31). From the previous discussion, it is known that this occurs because the centrifugal force changes with speed. This causes the front and rear side-slip angles to change; which, in turn, changes the circular motion geometry and conditions of steady-state cornering.

Equations (3.34) and (3.36), which express the vehicle steady-state cornering, are derived from a geometric relation. They are not influenced by the relationship between lateral forces and side-slip angles, β_f and β_r, or any lateral forces acting on the front and rear wheels. As long as $|\delta| \ll 1$ and $\rho \gg l$, d, the equations are valid under any conditions. Contrary to this, it is important to note that Eqns (3.29)–(3.31) and (3.29)′–(3.31)′ are only valid when the lateral force acting on the front and rear wheels is the lateral force that is proportional to the side-slip angles β_f and β_r.

It is possible to introduce Eqns (3.29)–(3.31) from the geometric descriptions of the steady-state turning. The centrifugal force is expressed by mV^2/ρ while the lateral forces exerted on the front and rear tires are proportional to the side-slip angle and expressed by $-2K_f\beta_f$ and $-2K_r\beta_r$, respectively. If the vehicle is in steady-state turning, the following equilibrium equations arise:

$$mV^2/\rho - 2K_f\beta_f - 2K_r\beta_r = 0$$

$$-2l_f K_f\beta_f + 2l_r K_r\beta_r = 0$$

From the previous two equations, the side-slip angles are obtained:

$$\beta_f = \frac{mV^2 l_r}{2lK_f}\frac{1}{\rho}, \quad \beta_r = \frac{mV^2 l_f}{2lK_r}\frac{1}{\rho}$$

Putting β_f and β_r into Eqns (3.34)–(3.36) gives Eqns (3.29), (3.30), and (3.31), respectively.

3.3.2 STEADY-STATE CORNERING AND STEER CHARACTERISTICS

3.3.2.1 Understeer (US) and oversteer (OS) characteristics

This subsection will study in more detail how the vehicle steady-state cornering relationship with vehicle velocity is affected by the vehicle characteristics by using equations that express the vehicle steady-state cornering, as derived in Section 3.3.1.

First, look at the turning radius, ρ, given by Eqn (3.31). Taking the steering angle as δ_0, yields the following:

$$\rho = \left(1 - \frac{m}{2l^2}\frac{l_f K_f - l_r K_r}{K_f K_r}V^2\right)\frac{l}{\delta_0} \tag{3.37}$$

This equation shows how the turning radius, ρ, changes with velocity, V, for a fixed steering angle of δ_0. Figure 3.9 shows how the relationship between ρ and V is affected by the sign of $l_f K_f - l_r K_r$.

As can be seen from Eqn (3.37) or Figure 3.9, if the steering angle is constant, the radius of the vehicle path when $l_f K_f - l_r K_r = 0$ is not related to V. In other words, the radius has a constant value of l/δ_0 at any velocity. When $l_f K_f - l_r K_r < 0$, a vehicle's turning radius increases with velocity.

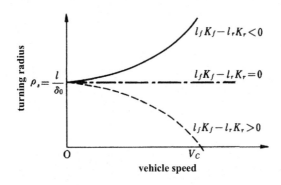

FIGURE 3.9

Relation of turning radius to vehicle speed with constant front wheel steering angle.

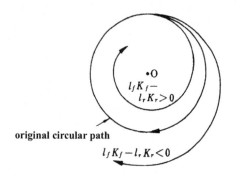

FIGURE 3.10

Change of turning radius with increase of vehicle speed.

The radius decreases with velocity if $l_f K_f - l_r K_r > 0$. In this latter case, when $V = V_c$, then $\rho = 0$ (the value and meaning of V_c will be described in the following section). This means that if the velocity increases with a fixed steering angle, the vehicle with $l_f K_f - l_r K_r < 0$ will turn out from the original circular path and make a circular path with an even larger radius. While the vehicle with $l_f K_f - l_r K_r > 0$ will, on the contrary, turn in to the inner side of the original circular path and make a circular path with an even smaller radius. These conditions are shown in Figure 3.10.

When $l_f K_f - l_r K_r < 0$, if the steering angle is maintained and the velocity increased, there is insufficient steering angle to maintain the original circular path radius. This characteristic, where steering angle is insufficient with regard to increasing velocity, is called understeer (US). When $l_f K_f - l_r K_r > 0$, if the steering angle is maintained and the velocity increased, there will be excessive steering angle to maintain the original circular path radius. This characteristic, where the steering angle is excessive with increasing velocity, is called oversteer (OS). When $l_f K_f - l_r K_r = 0$, the radius is not dependent on velocity, and the vehicle has neutralsteer characteristics (NS).

Next is to study how the steering angle, δ, should be changed to maintain steady-state cornering with a fixed radius at different velocities. Taking $\rho = \rho_0$ (constant) in Eqn (3.31) gives the following:

$$\delta = \left(1 - \frac{m}{2l^2}\frac{l_f K_f - l_r K_r}{K_f K_r}V^2\right)\frac{l}{\rho_0} \tag{3.38}$$

This equation has exactly the same form as Eqn (3.37), and a typical relation between δ and V is shown in Figure 3.11. For the vehicle to maintain a circular motion with a constant radius, a steering angle δ must be added along with velocity if $l_f K_f - l_r K_r < 0$. When $l_f K_f - l_r K_r > 0$, the steering angle must be reduced with velocity. When $V = V_c$, $\delta = 0$. Furthermore, if $l_f K_f - l_r K_r = 0$, δ is not dependent on velocity.

When the cornering forces, which are proportional to the side-slip angles, are the only lateral forces acting on the front and rear wheels, the vehicle circular motion is greatly influenced by $l_f K_f - l_r K_r$. The vehicle with $l_f K_f - l_r K_r < 0$ has US characteristics, the vehicle with $l_f K_f - l_r K_r = 0$ has NS characteristics, and the vehicle with $l_f K_f - l_r K_r > 0$ has OS characteristics. US, NS, and OS are generally called the steer characteristics (or steer properties).

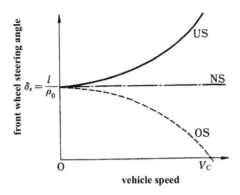

FIGURE 3.11

Front wheel steering angle to vehicle speed with constant turning radius.

Next is to study how the yaw velocity, r, of the steady-state cornering changes with vehicle steer characteristics. The yaw velocity is given by Eqn (3.30). From this equation, the relationship between r and V for steady-state cornering with constant steering angle is written as follows:

$$r = \frac{1}{1 - \frac{m}{2l^2}\frac{l_f K_f - l_r K_r}{K_f K_r}V^2}\frac{V}{l}\delta_0 \tag{3.39}$$

Using this equation, sketching the qualitative relation between r and V can be plotted, as in Figure 3.12.

The yaw velocity of the vehicle with NS characteristics increases linearly with the vehicle velocity as shown in Figure 3.12 and Eqn (3.39). If the vehicle had US characteristics, the yaw velocity also increases with the vehicle velocity, but it saturates at a certain value. In the case of OS, the yaw velocity increases rapidly with the vehicle velocity and becomes infinite at $V = V_c$.

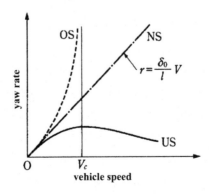

FIGURE 3.12

Steady-state yaw rate to vehicle speed.

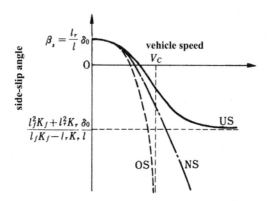

FIGURE 3.13

Steady-state side-slip angle to vehicle speed.

The next behavior to examine is how the side-slip angle, β, changes with velocity, V, for steady-state cornering. The side-slip angle β is given by Eqn (3.29). Taking $\delta = \delta_0$, the relationship between the side-slip angle and the velocity for steady-state cornering with constant steering angle is as follows:

$$\beta = \left(\frac{1 - \frac{m}{2l} \frac{l_f}{l_r K_r} V^2}{1 - \frac{m}{2l^2} \frac{l_f K_f - l_r K_r}{K_f K_r} V^2} \right) \frac{l_r}{l} \delta_0 \tag{3.40}$$

The relationship between β and V, for different vehicle steer characteristics, is shown in Figure 3.13.

Figure 3.13 and Eqn (3.40) show that β decreases with vehicle velocity, regardless of the vehicle steer characteristics. After a certain velocity, β becomes negative, and its absolute value increases continuously. If the vehicle exhibits US characteristics, β will reach a maximum value at larger velocities, and for OS characteristics, β becomes negative infinity at $V = V_c$. For vehicles with NS characteristics, the quasi-static relations between the steering angle and the path radius or yaw velocity are maintained regardless of velocity. The side-slip angle of the center of gravity, β, even with NS, or $l_f K_f - l_r K_r = 0$, is as follows:

$$\beta = \left(1 - \frac{m l_f}{2 l l_r K_r} V^2 \right) \frac{l_r}{l} \delta_0 \tag{3.40}'$$

where β does not maintain a constant value of $(l_r/l)\delta_0$ but changes proportionally to V^2, and its absolute value increases with vehicle velocity. The vehicle side-slip angle, regardless of the vehicle steer characteristics, changes with velocity due to the need of the lateral force to balance the centrifugal force. The vehicle side-slip angle is the angle between the vehicle longitudinal direction and the traveling direction of the vehicle's center of gravity; i.e., the tangent line of the circular path. It shows the attitude of the vehicle in relation to the circular path during a steady-state cornering. The side-slip angle, β, becomes negative, and its absolute value increases with vehicle speed. This means that when the vehicle increases speed, the vehicle will point into the inner side of the circular path, as shown in Figure 3.14. This tendency is even more obvious for vehicles with OS characteristics.

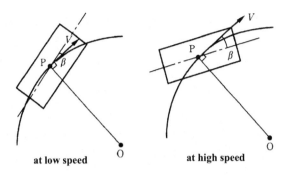

FIGURE 3.14

Vehicle relative attitude to circular path.

Example 3.3

Calculate the steady-state cornering and draw the diagrams of $\rho-V$, $r-V$, and $\beta-V$ under vehicle parameters for a normal passenger car given as $m = 1500$ kg, $l_f = 1.1$ m, $l_r = 1.6$ m, $K_f = 55$ kN/rad, $K_r = 60$ kN/rad and $\delta_0 = 0.04$ rad.

Solution

By using Eqns (3.37), (3.39) and (3.40), it is possible to draw the diagrams shown in Figure E3.3(a)–(c).

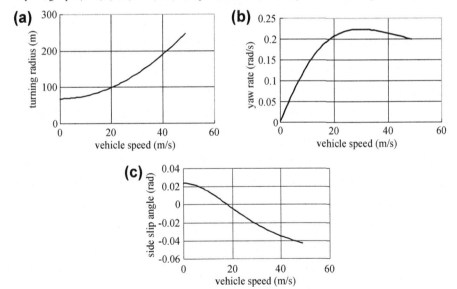

FIGURE E3.3(a)–(c)

Example 3.4

Derive the equation that describes the relation of side-slip angle to vehicle speed during steady-state turning with a constant turning radius. Draw the diagram of the relation qualitatively.

Solution

The steering angle needed to turn with the constant radius ρ_0 is expressed by Eqn (3.38). On the other hand, the side-slip angle during turning with the steering angle is given by Eqn (3.29). Eliminating δ from these two equations, the following equation is obtained:

$$\beta = \left(1 - \frac{ml_f}{2ll_rK_r}V^2\right)\frac{l_r}{\rho_0} \tag{E3.6}$$

This equation shows us how the side-slip angle changes with the vehicle speed during steady-state turning with constant turning radius, ρ_0. The qualitative relation is drawn in Figure E3.4. It is interesting to see that there is no explicit difference in the side-slip angle due to the vehicle steer characteristics (US, OS, or NS). Rather it explicitly depends on the rear cornering stiffness, K_r, itself.

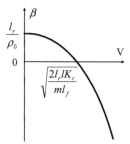

FIGURE E3.4

3.3.2.2 Stability limit velocity and stability factor

When the vehicle has an OS characteristic, the circular turning radius, ρ, with regard to a constant steering angle becomes zero when the vehicle velocity is $V = V_c$. Furthermore, the yaw angular velocity, r, and the side-slip angle, β, become infinity. When $V > V_c$, ρ, r, and β are physically meaningless. V_c can be found from the following equation:

$$1 - \frac{m}{2l^2}\frac{l_fK_f - l_rK_r}{K_fK_r}V^2 = 0 \tag{3.41}$$

If $l_fK_f - l_rK_r > 0$, the vehicle has an OS characteristic, and a real value of the velocity that satisfies Eqn (3.41) exists.

$$V_c = \sqrt{\frac{2K_fK_r}{m(l_fK_f - l_rK_r)}}l \tag{3.42}$$

Above this velocity, circular motion is no longer possible.

The critical velocity, V_c, becomes larger as $l_fK_f - l_rK_r$ reduces, as seen from Eqn (3.42). It also increases with smaller vehicle mass, m, larger front and rear tire cornering stiffness, K_f and K_r, and a larger wheelbase, l.

When the vehicle has OS characteristics, it is important to note that the vehicle motion instability at $V \geq V_c$ depends on the front steering angle being fixed. It does not mean the vehicle cannot be driven above V_c, as this depends on the driver's ability. However, because the theoretical stability limit velocity exists, vehicle designers tend to avoid the OS characteristic, and it is rare to find a vehicle that is intentionally designed to have strong OS characteristic. If the following is defined,

$$A = -\frac{m}{2l^2}\frac{l_f K_f - l_r K_r}{K_f K_r} \tag{3.43}$$

then, Eqn (3.41) becomes the following:

$$1 + AV^2 = 0 \tag{3.41}'$$

If $A < 0$, then V_c can be written as follows:

$$V_c = \sqrt{-\frac{1}{A}} \tag{3.42}'$$

Here, A is called the stability factor.

Using the stability factor, the relationships of β, r, and ρ to δ during steady-state cornering can be written as follows:

$$\beta = \frac{1 - \frac{m}{2l}\frac{l_f}{l_r K_r}V^2}{1 + AV^2}\frac{l_r}{l}\delta \tag{3.29}''$$

$$r = \frac{1}{1 + AV^2}\frac{V}{l}\delta \tag{3.30}''$$

$$\rho = (1 + AV^2)\frac{l}{\delta} \tag{3.31}''$$

The sign of the stability factor controls the vehicle steer characteristics. It is an important quantity that becomes the index of the degree of change in steady-state cornering due to vehicle velocity. In particular, the vehicle steady-state cornering is proportional to V^2 with the coefficient A.

As could be seen from Eqn (3.43), while the sign of $l_f K_f - l_r K_r$ influences the effect of velocity, a larger vehicle mass, m, smaller wheelbase, l, or smaller cornering stiffness, K_f and K_r, also increases the effect of V.

3.3.2.3 Static margin and neutral steer point

The vehicle steer characteristics, determined by the sign of $l_f K_f - l_r K_r$, have a fundamental influence on vehicle steady-state cornering. It is understood that the concept of US, NS, and OS is extremely important in the discussion of vehicle dynamic performance. More details about the physical meaning of the quantity of $l_f K_f - l_r K_r$ will be investigated.

Imagine the vehicle has the original condition $\delta = 0$, but for certain reasons, a side-slip angle at the vehicle center of gravity, β, is produced. The same side-slip angle is produced at the front and rear wheels, and lateral forces will be generated at the tires. These lateral forces produce a yaw moment around the center of gravity. The yaw motion due to this moment, based on Eqn (3.13), becomes the following:

$$I\frac{dr}{dt} + \frac{2(l_f^2 K_f + l_r^2 K_r)}{V}r = -2(l_f K_f - l_r K_r)\beta$$

If β is positive and $l_f K_f - l_r K_r$ is positive, a moment that produces negative r acts around the center of gravity. If $l_f K_f - l_r K_r = 0$, there is no moment acting, and if $l_f K_f - l_r K_r$ is negative, a moment that produces positive r acts around the center of gravity. In other words, if $l_f K_f - l_r K_r$

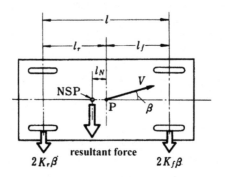

FIGURE 3.15

Resultant force of tire lateral forces due to vehicle side-slip.

is positive, the resultant force of the lateral forces at the front and rear wheels acts in front of the center of gravity. Whereas, if $l_f K_f - l_r K_r = 0$, it acts at the center of gravity, and if $l_f K_f - l_r K_r$ is negative, it acts behind the center of gravity. This acting point of the resultant force is called the neutral steer point (NSP).

If the center of gravity has a side-slip angle of β, the lateral force acting on the front and rear wheels will be $2K_f \beta$ and $2K_r \beta$. Taking the distance of NSP from the vehicle center of gravity as l_N, as shown in Figure 3.15, the moment by $2K_f \beta$ and $2K_r \beta$ around NSP must be balanced.

$$-(l_f + l_N)2K_f \beta + (l_r - l_N)2K_r \beta = 0$$

From this, the following is derived:

$$l_N = -\frac{l_f K_f - l_r K_r}{K_f + K_r} \tag{3.44}$$

NSP is in front of the center of gravity when $l_f K_f - l_r K_r$ is positive, and when $l_f K_f - l_r K_r$ is negative, it is behind the center of gravity. When $l_f K_f - l_r K_r = 0$, the NSP coincides with the center of gravity.

The dimensionless quantity of the ratio of l_N to wheelbase, l, is called the static margin (SM).

$$SM = \frac{l_N}{l} = -\frac{l_f K_f - l_r K_r}{l(K_f + K_r)} \tag{3.45}$$

Or, by transforming Eqn (3.45), it can also be written as follows:

$$SM = -\frac{l_f}{l} + \frac{K_r}{K_f + K_r} \tag{3.45}'$$

From this, the quantity of $l_f K_f - l_r K_r$, which determines the vehicle steer characteristics, can be rewritten in the form of static margin, SM. The vehicle steer characteristic is defined as follows using the SM:

When SM > 0, then US.
When SM = 0, then NS.
When SM < 0, then OS.

Moreover, if the stability limit velocity, V_c, is expressed using the term SM, the following results:

$$V_c = \sqrt{\frac{2lK_fK_r}{m(K_f + K_r)}\left(-\frac{1}{\text{SM}}\right)} \qquad (3.42)''$$

And expressing stability factor A using SM gives the following:

$$A = \frac{m}{2l}\frac{K_f + K_r}{K_fK_r}\text{SM} \qquad (3.43)'$$

3.3.2.4 Steer characteristics and geometry

The vehicle steady-state cornering characteristics have been studied by using equations that are derived from the fundamental equations of motion. These equations are introduced only for the case where the lateral tire forces are proportional to the side-slip angle.

Next, the vehicle steady-state cornering characteristics will be studied from a more practical viewpoint by using equations that express the steady-state cornering geometrically, as in Section 3.3.1, without the constraints described previously. The geometric relation between circular turning radius, ρ, to steering angle, δ has been given by Eqn (3.34):

$$\rho = \frac{l}{\delta - \beta_f + \beta_r} \qquad (3.34)$$

As seen from this equation, the relation between ρ and δ depends on the magnitude of the wheel slip angles, β_f and β_r.

$$\text{When} \quad \beta_f - \beta_r > 0, \quad \text{then} \quad \rho > \frac{l}{\delta}, \delta > \frac{l}{\rho},$$

$$\text{when} \quad \beta_f - \beta_r = 0, \quad \text{then} \quad \rho = \frac{l}{\delta}, \delta = \frac{l}{\rho}, \quad \text{and}$$

$$\text{when} \quad \beta_f - \beta_r < 0, \quad \text{then} \quad \rho < \frac{l}{\delta}, \delta < \frac{l}{\rho}.$$

In other words, if the relationship between the vehicle front and rear wheel side-slip angles is $\beta_f > \beta_r$, then the turning radius becomes larger in response to the vehicle speed with constant steering angle, and more steering angle is needed to maintain the original radius. If $\beta_f = \beta_r$, then the turning radius and steering angle do not depend on vehicle speed. If $\beta_f < \beta_r$, then the radius becomes smaller as vehicle speed increases with the steering angle constant. The steering angle must be reduced to maintain the original radius. Therefore, the vehicle steer characteristic could be defined as follows:

$$\text{When} \quad \beta_f < \beta_r, \quad \text{then US},$$

$$\text{when} \quad \beta_f = \beta_r, \quad \text{then NS and}$$

$$\text{when} \quad \beta_f < \beta_r, \quad \text{then OS}.$$

This definition is not influenced by lateral forces acting on the front and rear wheels other than the tire lateral force due to side slip, and also by whether the lateral forces are proportional to the

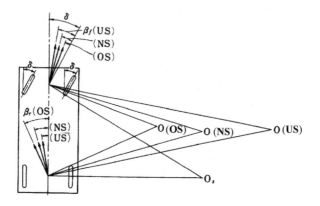

FIGURE 3.16

Side-slip angles of front and rear wheels in steady-state turning.

side-slip angle or not. Figure 3.16 shows how the vehicle circular motion, under constant steering action, changes with the relationship between β_f and β_r.

The figure also shows that the radius for a vehicle with US characteristics and $\beta_f > \beta_r$ is larger than l/δ, whereas for an NS characteristic with $\beta_f = \beta_r$, the radius is equal to l/δ, and for OS characteristics with $\beta_f < \beta_r$, it is smaller than l/δ. Furthermore, β_f and β_r increase with an increase in vehicle speed regardless of the steer characteristics of the vehicle. This results in the circular motion center moving toward the front of the vehicle. Consequently, with increasing speed, the vehicle moves inward of the circular path. This tendency is even more obvious for the vehicle with OS characteristics.

Through the geometric study of the vehicle US and OS by the sign of $\beta_f - \beta_r$, the physical meaning of vehicle steer characteristic by the sign of $l_f K_f - l_r K_r$ in Eqns (3.37) and (3.38) can be well understood.

A geometric relation described by Eqn (3.34) is rewritten as follows:

$$\delta = \frac{l}{\rho} + \beta_f - \beta_r = \left(1 + \frac{\beta_f - \beta_r}{\frac{l}{\rho}}\right) \frac{l}{\rho} \tag{3.46}$$

On the other hand, when the lateral forces exerted on both front and rear wheels are only tire cornering forces proportional to the side-slip angle, the following equation is obtained from Eqn (3.31)″:

$$\delta = \left(1 + AV^2\right) \frac{l}{\rho} \tag{3.47}$$

Then, from Eqns (3.46) and (3.47) the following is derived:

$$AV^2 = \frac{\beta_f - \beta_r}{\frac{l}{\rho}}$$

namely, the following:

$$A = \frac{\beta_f - \beta_r}{\frac{lV^2}{\rho}}$$

As V^2/ρ is the lateral acceleration during the steady-state cornering, the stability factor represents the difference in side-slip angles at front and rear tires per unit lateral acceleration during the cornering.

3.3.2.5 Steady-state cornering to lateral acceleration

If the vehicle is assumed to be in steady-state cornering with a lateral acceleration of the following:

$$\ddot{y} = \frac{V^2}{\rho g} \tag{3.48}$$

then, a centrifugal force of magnitude $mg\ddot{y}$ will act at the vehicle's center of gravity, where g is the gravitational acceleration, and $\ddot{y}$ has a gravitational unit for convenience sake. This has to be in equilibrium with the lateral forces acting at the front and rear wheels, $2Y_f$ and $2Y_r$, and the moment around the center of gravity should be zero. Thus, the following can be shown:

$$mg\ddot{y} + 2Y_f + 2Y_r = 0$$

$$2l_f Y_f - 2l_r Y_r = 0$$

which gives the following:

$$2Y_f = -m\frac{l_r}{l}g\ddot{y}$$
$$2Y_r = -m\frac{l_f}{l}g\ddot{y} \tag{3.49}$$

If the tire cornering characteristic is linear, then the following is given:

$$2Y_f = -2K_f\beta_f, \quad 2Y_r = -2K_r\beta_r$$

From Eqn (3.49), the following are obtained:

$$\beta_f = \frac{ml_r}{2K_f l}g\ddot{y}, \quad \beta_r = \frac{ml_f}{2K_r l}g\ddot{y}$$

This is the same discussion as at the end of Section 3.3.1. Substituting these into Eqn (3.46) gives the following:

$$\delta = \frac{l}{\rho} - \frac{m(l_f K_f - l_r K_r)}{2K_f K_r l}g\ddot{y} \tag{3.50}$$

Therefore, the relation between lateral acceleration, $\ddot{y}$, and the required steering angle for a given radius, $\rho = \rho_0$, of the circular motion is as follows:

$$\delta = \frac{l}{\rho_0} - \frac{m(l_f K_f - l_r K_r)}{2K_f K_r l} g\ddot{y} \tag{3.50}'$$

If the circular motion is at a constant speed of $V = V_0$, Eqn (3.48) yields the following:

$$\rho = \frac{V_0^2}{g\ddot{y}}$$

The relation between the lateral acceleration, $\ddot{y}$, and the required steering angle, δ, from Eqn (3.50), is as follows:

$$\delta = \left[\frac{l}{V_0^2} - \frac{m(l_f K_f - l_r K_r)}{2K_f K_r l}\right] g\ddot{y} \tag{3.51}$$

Using Eqns (3.50)' and (3.51), the qualitative relation between $\ddot{y}$ and δ during steady-state cornering when the tire cornering characteristic is linear will look like Figure 3.17.

Regarding the coefficient of $\ddot{y}$ in Eqn (3.50), the following are defined as the US/OS gradient:

$$U = -\frac{m(l_f K_f - l_r K_r)}{2K_f K_r l} g = glA$$

While the stability factor is the coefficient to show how the steady-state cornering depends on vehicle speed, the US/OS gradient shows its dependency on lateral acceleration.

Cornering compliances of front and rear tires are defined as side-slip angles during steady-state cornering obtained previously divided by lateral acceleration $\ddot{y}$ as follows:

$$C_f = \frac{\beta_f}{\ddot{y}} = \frac{ml_r}{2K_f l} g, \quad C_r = \frac{\beta_r}{\ddot{y}} = \frac{ml_f}{2K_r l} g$$

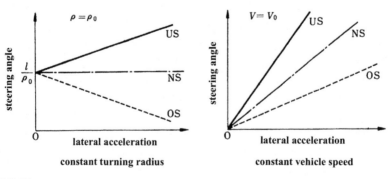

FIGURE 3.17

Required steering angle to lateral acceleration for a linear tire vehicle.

The cornering compliances of the front and rear tires are the inverse of the tire cornering stiffness per unit tire vertical load, and it is easy to see that the US/OS gradient is the difference of the front and rear cornering compliances:

$$U = C_r - C_f$$

This is the same concept as described in Subsection 3.3.2.4 that the stability factor is the difference in side-slip angles at front and rear tires per unit lateral acceleration during the steady cornering.

Table 3.1 summarizes the relation between vehicle steer characteristics with the steady-state cornering.

3.3.3 STEADY-STATE CORNERING AND TIRE NONLINEAR CHARACTERISTICS

Section 3.3.2 studied the characteristics of vehicle steady-state cornering to give a basic understanding of the vehicle dynamics. It also looked at the vehicle steer characteristics (US, NS, and OS) as an important concept for describing a basic vehicle's dynamic characteristics. Previously, it was assumed that the lateral forces acting on the front and rear wheels are the lateral forces that are proportional to the side-slip angles. However, as seen in Chapter 2, the lateral force may not necessarily be proportional to the side-slip angle. In the case of larger side-slip angles and under certain other conditions, lateral forces may show nonlinearity in their characteristics.

This subsection will examine the effect of nonlinear tire characteristics on the vehicle steady-state cornering characteristics.

As mgl_r/l and mgl_f/l are the vertical loads of the front and rear wheels, taking the ratio of the lateral force obtained by Eqn (3.49) to the vertical load at the front and rear wheels as μ_f and μ_r, respectively, the following are obtained:

$$\mu_f = \frac{|2Y_f|}{mgl_r/l} = \ddot{y}, \quad \mu_r = \frac{|2Y_r|}{mgl_f/l} = \ddot{y} \tag{3.52}$$

This means, during steady-state cornering, the lateral force at the front and rear wheels divided by their respective vertical loads is always equal to the lateral acceleration of the center of gravity, $\ddot{y}$.

Assuming the lateral force produced by the tire side-slip angles is the only lateral force, then μ_f and μ_r depend only on β_f and β_r. If the tire cornering characteristics of the front and rear wheels ($\beta_f - \mu_f$ and $\beta_r - \mu_r$) are given and the lateral acceleration of the vehicle center of gravity, $\ddot{y}$, is known, the front and rear wheel side-slip angles, β_f and β_r, at that instant can be found.

When the tire cornering characteristic is not linear, the front and rear tire cornering characteristics, $\beta_f - \mu_f$ and $\beta_r - \mu_r$, are given, for example, as in Figure 3.18. As seen previously, if the vehicle is in steady-state cornering, the lateral acceleration $\ddot{y}$ is equal to μ_f and μ_r. Thus, by knowing the lateral acceleration $\ddot{y}$, the front and rear wheel side-slip angles, β_f and β_r, at that instant are known, and $\beta_f - \beta_r$ can be determined.

For vehicle steady-state cornering, regardless of whether lateral force acting at the front and rear wheels is proportional to the side-slip angle or not, the geometric relation of Eqn (3.46) is always satisfied. From Eqn (3.46), if the constant radius of the circular motion is ρ_0, then the following results:

$$\delta = \frac{l}{\rho_0} + \beta_f - \beta_r \tag{3.46}'$$

Table 3.1 Steer Characteristics and Steady-State Turning

$A = -\dfrac{m}{2l^2}\dfrac{l_f K_f - l_r K_r}{K_f K_r}$		Relative Position of NSP and C.G.	Effects of Vehicle Speed on Steady-State Turning				
			ρ (δ = Constant)	δ (ρ = Constant)	Yaw Rate r	Side-Slip Angles	Critical Speed
US	> 0		Increase with speed	Increase with speed	Increase with speed to some extent and decrease	$\beta_f > \beta_r$	Non-critical
NS	$= 0$		Constant $\rho = \frac{l}{\delta}$	Constant $\delta = \frac{l}{\rho}$	Proportional increase with speed	$\beta_f = \beta_r$	Non-critical
OS	< 0		Decrease with speed	Decrease with speed	Rapid increase with speed	$\beta_f < \beta_r$	$V_c = \sqrt{\dfrac{2K_f K_r l^2}{m(l_f K_f - l_r K_r)}}$

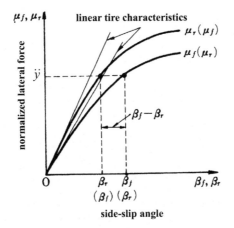

FIGURE 3.18

Nonlinear tire characteristics of front and rear.

By knowing $\ddot{y}$ and determining $\beta_f - \beta_r$ from Figure 3.18, the relation between lateral acceleration, $\ddot{y}$, and steering angle, δ, can be plotted using Eqn (3.46)$'$ for constant radius circular motion, as in Figure 3.19.

If $\mu_f < \mu_r$ with regard to tire side-slip angle, the vehicle shows US characteristics because $\beta_f > \beta_r$. If $\mu_f = \mu_r$, then $\beta_f = \beta_r$, and the vehicle characteristic is, therefore, NS. Whereas, if $\mu_f > \mu_r$, then $\beta_f < \beta_r$, and the vehicle characteristic becomes OS.

For the case when the vehicle characteristic is US, as in Figure 3.18, if $\ddot{y}$ is small and the side-slip angles are small, then $\beta_f - \beta_r$ is positive, and its value increases almost proportionally to $\ddot{y}$. Here, $\ddot{y}$ could also be assumed to be proportional to δ. When $\ddot{y}$ is greater than a certain value, $\beta_f - \beta_r$ is no longer proportional to $\ddot{y}$, and it increases rapidly with $\ddot{y}$. Consequently, δ also

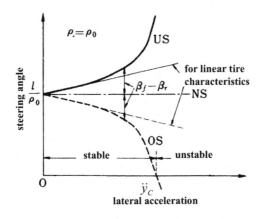

FIGURE 3.19

Steering angle to lateral acceleration under constant turning radius.

increases rapidly with $\ddot{y}$, and the vehicle shows a strong US characteristic. Whereas, for the case when the vehicle characteristic is OS, when $\ddot{y}$ is small, then $\beta_f - \beta_r$ is negative, and the value increases almost proportionally to $\ddot{y}$. Thus, δ also decreases proportionally to $\ddot{y}$. When $\ddot{y}$ is greater than a certain value, $\beta_f - \beta_r$ increases rapidly with $\ddot{y}$, and the vehicle shows a strong OS characteristic.

For the case of OS, δ decreases rapidly with large $\ddot{y}$. At $\ddot{y} = \ddot{y}_c$, $\delta = 0$. This point defines the limit where under steady-state cornering, with constant radius ρ_0, it is not possible to increase the lateral acceleration by raising the velocity. This $\ddot{y}$ is considerably smaller than the case where the tire cornering characteristic is linear. The tire nonlinear cornering characteristic further reduces the critical velocity for the OS vehicle. Chapter 2 showed that, in practice, a lateral force saturates after a certain side-slip angle. It is important to note that the vehicle with OS characteristics becomes unstable at lower speeds than that with a linear characteristic. This velocity limit for circular motion is dependent on the radius of the circular motion, which is different than that for a linear tire characteristic.

Consider the case when the vehicle is making a circular motion with a constant speed of $V = V_0$. Here, $\rho_0 = V_0^2 / g\ddot{y}$, and the relation between lateral acceleration, $\ddot{y}$, and required steering angle, δ, from Eqn (3.46), is as follows:

$$\delta = \frac{l}{V_0^2} g\ddot{y} + \beta_f - \beta_r \tag{3.53}$$

The value of $\beta_f - \beta_r$ can again be determined from Figure 3.18 by knowing $\ddot{y}$. Using Eqn (3.53), the relationship between steering angle, δ, and lateral acceleration, $\ddot{y}$, during constant-speed circular motion is shown in Figure 3.20. This figure shows that a US vehicle with $\ddot{y}$ small has δ increasing almost proportionally to $\ddot{y}$. After $\ddot{y}$ reaches a certain value, δ increases rapidly with $\ddot{y}$, and the vehicle reveals a strong US characteristic.

Whereas, for the case of OS, when $\ddot{y}$ is small, δ increases almost proportionally to $\ddot{y}$, but after $\ddot{y}$ reaches a certain value, the increase of δ with $\ddot{y}$ becomes weaker. Finally, when $\ddot{y} = \ddot{y}_c$, δ reaches its peak. When $\ddot{y}$ is greater than $\ddot{y}_c$, δ decreases. This reduction of δ decreases the turning radius at a constant speed. This means that when the vehicle is about to move to the right with regard to its

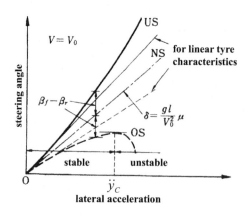

FIGURE 3.20

Steering angle to lateral acceleration under constant vehicle speed.

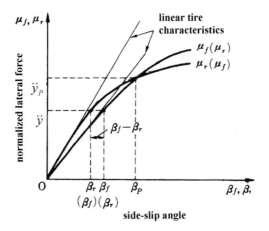

FIGURE 3.21

Nonlinear tire characteristics of front and rear.

current traveling direction, the steering should be turned to the left. This is impossible and has no physical meaning in practice. This point shows that circular motion with a radius that causes $\ddot{y} = \ddot{y}_c$ and the motion with smaller radius is not possible. For an OS vehicle, there is always a lower limit to the cornering radius possible, regardless of vehicle speed. Furthermore, $\ddot{y}_c$ at constant turning radius, as shown in Figure 3.19, has the same meaning as $\ddot{y}_c$ for the constant speed shown in Figure 3.20.

The front and rear wheel tire cornering characteristics shown in Figure 3.18 are always either $\mu_f > \mu_r$ or $\mu_f < \mu_r$ for the whole range of side-slip angles. This is not always the case and depends on circumstances.

Consider the front and rear wheel tire cornering characteristics $\beta_f - \mu_f$ and $\beta_r - \mu_r$, as shown in Figure 3.21. (A) is the case where if β_f and β_r are smaller than β_p, then $\mu_f < \mu_r$, and if β_f and β_r are larger than β_p, then $\mu_f > \mu_r$. In contrast, (B) is the case where if β_f and β_r are smaller than β_p, then $\mu_f > \mu_r$, and if β_f and β_r are larger than β_p, then $\mu_f < \mu_r$.

If the vehicle is making a circular motion of constant radius, ρ_0, Eqn (3.46)' is formed as previously. By knowing the lateral acceleration, $\ddot{y}$, $\beta_f - \beta_r$ can be determined from Figure 3.21. Using Eqn (3.46)', the relationship between steering angle, δ, and lateral acceleration, $\ddot{y}$, during constant radius circular motion is obtained as shown in Figure 3.22.

In the case of (A), when $\ddot{y}$ is small, $\beta_f - \beta_r$ increases with $\ddot{y}$; δ also increases, and the vehicle shows a US characteristic. After $\ddot{y}$ reaches a certain value, with the increase of $\ddot{y}$, $\beta_f - \beta_r$ decreases, and at $\ddot{y} = \ddot{y}_p$, the value becomes zero. As $\ddot{y} > \ddot{y}_p$, $\beta_f - \beta_r$ decreases rapidly with $\ddot{y}$, and the vehicle reveals an OS characteristic. This tendency continues to increase with $\ddot{y}$, and at $\ddot{y} = \ddot{y}_c$, $\delta = 0$. This point, as in Figure 3.19, shows that it is impossible to have circular motion with higher lateral acceleration with radius ρ_0.

In the case of (B), when $\ddot{y}$ is small, δ decreases with $\ddot{y}$, and the vehicle shows OS characteristic. After $\ddot{y}$ reaches a certain value, δ increases with $\ddot{y}$, and at $\ddot{y} = \ddot{y}_p$, the value returns to l/ρ_0. At $\ddot{y} > \ddot{y}_p$, δ increases rapidly, and the vehicle reveals a strong US characteristic.

If the vehicle is making a circular motion with a constant speed of $V = V_0$, Eqn (3.53) is formed. Therefore, by knowing the lateral acceleration $\ddot{y}$, $\beta_f - \beta_r$ can be determined from

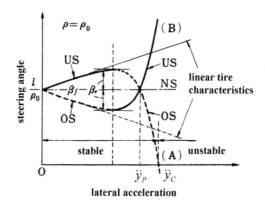

FIGURE 3.22

Steering angle to lateral acceleration under constant turning radius.

Figure 3.21 and using Eqn (3.53), the relationship between steering angle, δ, and lateral acceleration, $\ddot{y}$, during constant radius circular motion is shown in Figure 3.23.

In the case of (A), at small $\ddot{y}$, the vehicle shows a US characteristic, but at large $\ddot{y}$, the vehicle characteristic changes to OS. Circular motion with a radius that causes $\ddot{y} \geq \ddot{y}_c$ becomes impossible. (B) is opposite to (A), whereas $\ddot{y}$ becomes larger, and the vehicle shows a US characteristic.

As seen in case (A), the vehicle has a US characteristic at small lateral acceleration, $\ddot{y}$, but when $\ddot{y}$ gets large, the steer characteristic changes to OS due to the nonlinear tire characteristic, and circular motion becomes impossible after a certain velocity. While there is no critical velocity based on the concept of linear tire, the stable vehicle might fall into a statically unstable condition during circular motion at larger lateral accelerations. As shown by Figure 3.22 and Figure 3.23, the change of steer characteristic due to $\ddot{y}$ is called the reverse-steer.

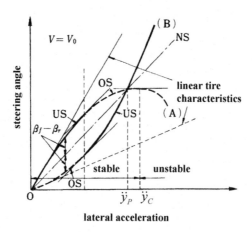

FIGURE 3.23

Steering angle to lateral acceleration under constant vehicle speed.

Figures 3.22 and 3.23 also show that in case (B), when the lateral acceleration gets larger, the vehicle exhibits a strong US characteristic, and no matter how large the steering angle is, circular motion with lateral acceleration beyond a certain point is impossible. This condition is called vehicle plow. When the front and rear wheels reach the upper limit of the lateral force at the same time, it is called vehicle drift. In the case of (A), the condition where the vehicle shows a strong OS characteristic with the rear wheels reaching the limit earlier than the front wheels and the vehicle becomes statically unstable is called vehicle spin.

3.4 VEHICLE DYNAMIC CHARACTERISTICS

Section 3.3 examined the basic characteristics of vehicle motion by looking at the vehicle steady-state cornering. The results obtained so far can only be classified as the static characteristics, in other words, the characteristics of vehicle motion in steady state. To understand the characteristics of vehicle motion in more detail, the dynamic characteristics must be examined as well.

Thus, continuing from here, the vehicle's transient response to steering input will be analyzed from different points of view to further understand the fundamental characteristics of the vehicle motion.

3.4.1 VEHICLE TRANSIENT RESPONSE TO STEERING INPUT [1]

3.4.1.1 Transient response and directional stability

The basic equations that describe the vehicle motion are defined by Eqns (3.12) and (3.13) in Section 3.2 as follows:

$$mV\frac{d\beta}{dt} + 2(K_f + K_r)\beta + \left\{mV + \frac{2}{V}(l_fK_f - l_rK_r)\right\}r = 2K_f\delta \tag{3.12}$$

$$2(l_fK_f - l_rK_r)\beta + I\frac{dr}{dt} + \frac{2\left(l_f^2K_f + l_r^2K_r\right)}{V}r = 2l_fK_f\delta \tag{3.13}$$

Once the equations of motion of a dynamic system, such as (3.12) and (3.13), are given, the vehicle response to δ can be obtained by solving the equations of motion under suitable conditions. If the system is linear, the transient behavior of the dynamic system can be understood by solving the equations of motion directly or investigating the eigenvalues of the characteristic equation.

The characteristic equation of the dynamic system that is our subject of study is given by Eqn (3.14):

$$s^2 + \left[\frac{2m\left(l_f^2K_f + l_r^2K_r\right) + 2I(K_f + K_r)}{mIV}\right]s + \left[\frac{4K_fK_rl^2}{mIV^2} - \frac{2(l_fK_f - l_rK_r)}{I}\right] = 0 \tag{3.54}$$

Or, it can be written as follows:

$$s^2 + 2Ds + P^2 = 0 \tag{3.55}$$

where

$$2D = \frac{2m\left(l_f^2 K_f + l_r^2 K_r\right) + 2I(K_f + K_r)}{mIV} \tag{3.56}$$

$$P^2 = \frac{4K_f K_r l^2}{mIV^2} - \frac{2(l_f K_f - l_r K_r)}{I} \tag{3.57}$$

And, the vehicle yaw inertia moment could be written as follows:

$$I = mk^2 \tag{3.58}$$

Here, k is called the vehicle yaw moment radius. Substituting Eqn (3.58) into Eqn (3.56), $2D$ could be written as the following:

$$2D = \frac{2}{mV}\left[(K_f + K_r)\left(\frac{1 + k^2/l_f l_r}{k^2/l_f l_r}\right) + \frac{1}{k^2}(l_f - l_r)(l_f K_f - l_r K_r)\right] \tag{3.56'}$$

And, if $l_f \approx l_r$ and $K_f \approx K_r$, then the following is derived:

$$2D = \frac{2(K_f + K_r)}{mV}\left(\frac{1 + k^2/l_f l_r}{k^2/l_f l_r}\right) \tag{3.56''}$$

By substituting Eqn (3.58) into Eqn (3.57), P^2 could be written as the following:

$$P^2 = \frac{4K_f K_r l^2}{m^2 k^2 V^2}\left(1 - \frac{m}{2l^2}\frac{l_f K_f - l_r K_r}{K_f K_r}V^2\right) \tag{3.57'}$$

Now, the response of the system with the characteristic equation given by Eqn (3.55) is expressed by $C_1 e^{\lambda_1 t} + C_2 e^{\lambda_2 t}$ with λ_1 and λ_2 as the roots of the characteristic equation:

$$\lambda_{1,2} = -D \pm \sqrt{D^2 - P^2} \tag{3.59}$$

The transient response characteristics and stability of the system is dependent on whether λ_1 and λ_2 are real numbers or complex numbers and on the sign of λ_1 and λ_2 if they are integers or the sign of the real part of λ_1 and λ_2 if they are complex numbers. Based on Eqn (3.59), the value of λ_1 and λ_2 is dependent on D and P. From Eqn (3.56), it is apparent that $D > 0$, and the transient response characteristic and motion stability of the vehicle can be classified into the following categories based on D and P:

1. When $D^2 - P^2 \geq 0$ and $P^2 > 0$, λ_1 and λ_2 are negative real numbers, and motion converges without oscillation (stable).
2. When $D^2 - P^2 < 0$, λ_1 and λ_2 are complex numbers with the negative real part, and motion converges with oscillation (stable).
3. When $P^2 \leq 0$, λ_1 and λ_2 are positive and negative real numbers, and motion diverges without oscillation (unstable).

This is under the premise that the steering angle of the vehicle is predetermined and not changeable in response to the vehicle's behavior. It should be noted that the vehicle does not always show this behavior, and the driver plays a key role in the vehicle stability. This situation not only applies to the vehicle motion, but for ships and aircrafts alike. The control of the motion by the driver (in some cases, by control actuators) on board of the moving body itself is an important matter in studying the motion of the moving bodies (refer to Chapters 8–10 for a detailed approach).

Next, the response of the vehicle to steering input, which can be divided into three categories as previously described, is investigated. Also, the motion stability will be studied in more detail, and the type of vehicle and of situation that gives motion for cases 1, 2, and 3 is investigated.

First, case 3 is examined. From Eqn (3.57), the first term of P^2 is always positive; thus, for $P^2 \leq 0$, it is only the second term that can be negative, in other words, $l_f K_f - l_r K_r > 0$. Taking V_c as the velocity where $P^2 = 0$, the following is obtained from Eqn (3.57)′:

$$1 - \frac{m}{2l^2} \frac{l_f K_f - l_r K_r}{K_f K_r} V_c^2 = 0 \tag{3.60}$$

hence,

$$V_c = \sqrt{\frac{2K_f K_r}{m(l_f K_f - l_r K_r)}} \, l = \sqrt{\frac{2l K_f K_r}{m(K_f + K_r)}\left(-\frac{1}{SM}\right)} \tag{3.61}$$

Then, for all velocities greater than V_c, $P^2 \leq 0$. This is the condition for static instability of the mechanical system as described previously in Section 3.3.2.

When the vehicle shows OS characteristics, the vehicle's lateral motion becomes unstable at velocity V_c, and it diverges without oscillation in response to a fixed steering input. As can be seen clearly from Eqn (3.61), this stability limit is greatly dependent on the SM. Figure 3.24 shows the vehicle stability limit by SM and velocity; and, it shows that the smaller the absolute value of SM and the larger the total cornering stiffness of the front and rear wheels is, the larger the stability limit velocity.

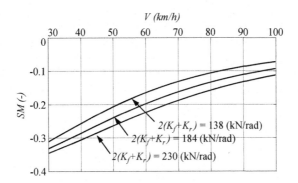

FIGURE 3.24

Relation of critical vehicle speed to SM.

Example 3.5

It is obvious that the stability condition of the vehicle is described by the following:

$$1 - \frac{m}{2l^2} \frac{l_f K_f - l_r K_r}{K_f K_r} V^2 \geq 0$$

Show that there are upper and lower limits of the front and rear cornering stiffness, K_f and K_r, respectively, for the vehicle to be stable, and draw a schematic diagram of the limits with respect to the vehicle speed.

Solution

The previous inequality is rewritten as follows:

$$\frac{2l^2}{mV_2} - \frac{l_f}{K_r} + \frac{l_r}{K_f} \geq 0$$

It turns out to be a form of upper limit of the front cornering stiffness:

$$K_f \leq \frac{l_r K_r}{l_f} \frac{V^2}{V^2 - \frac{2l^2 K_r}{m l_f}} \tag{E3.7}$$

And, as the lower limit of the rear cornering stiffness, the following inequality is obtained:

$$K_r \geq \frac{l_f K_f}{l_r} \frac{V^2}{V^2 + \frac{2l^2 K_f}{m l_r}} \tag{E3.8}$$

The schematic diagram of the upper and lower limits of the cornering stiffness with respect to vehicle speed is shown in Figure E3.5.

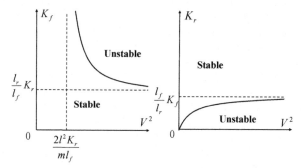

FIGURE E3.5

Next, if $l_f K_f - l_r K_r < 0$ when the vehicle shows US characteristics or if the vehicle reveals OS characteristic but $V < V_c$, then P^2 is always greater than zero, and the vehicle motion is stable. This corresponds to cases 1 and 2. Now, using Eqns (3.56) and (3.57), $D^2 - P^2$ is calculated as follows:

$$D^2 - P^2 = \frac{2(l_f K_f - l_r K_r)}{I} + \left[\left\{ \frac{m\left(l_f^2 K_f + l_r^2 K_r\right) + I(K_f + K_r)}{mI} \right\}^2 - \frac{4K_f K_r l^2}{mI} \right] \frac{1}{V^2} \tag{3.62}$$

Transforming the coefficient of $1/V^2$ gives the following:

$$\left\{ \frac{m\left(l_f^2 K_f + l_r^2 K_r\right) - I(K_f + K_r)}{mI} \right\}^2 + \frac{4(l_f K_f - l_r K_r)^2}{mI} > 0 \tag{3.63}$$

Because the coefficient of $1/V^2$ of Eqn (3.62) is always positive, if the first term $lK_f - l_rK_r$ is positive or zero, then, $D^2 - P^2$ is also positive or zero. Hence, if the vehicle steer characteristic is OS or NS, the vehicle transient steering response will always be without oscillation, stable or not. When $lK_f - l_rK_r$ is negative, the value of $D^2 - P^2$ is dependent on V, where above a certain value, $D^2 - P^2$ changes from positive to negative. In other words, when the vehicle shows US characteristics, the vehicle transient steering response is without oscillation at a vehicle speed lower than a certain value, but beyond that the response becomes oscillatory.

Using Eqns (3.45) and (3.58) and assuming that $l_f \approx l_r$ and $K_f \approx K_r$, then $D^2 - P^2$ can be calculated by transforming Eqn (3.62):

$$
\begin{aligned}
D^2 - P^2 &= \frac{2(l_fK_f - l_rK_r)}{mk^2} + \left[\frac{(K_f + K_r)^2}{m^2}\left(\frac{1 - k^2/l_fl_r}{k^2/l_fl_r}\right)^2 + \frac{4(l_fK_f - l_rK_r)^2}{m^2k^2} \right] \frac{1}{V^2} \\
&\approx \frac{(K_f + K_r)^2}{m^2}\left[-\frac{l_fl_r}{k^2}\frac{8m}{l(K_f + K_r)}\mathrm{SM} + \left\{\left(\frac{1 - k^2/l_fl_r}{k^2/l_fl_r}\right)^2 + \frac{16l_fl_r}{k^2}\mathrm{SM}^2\right\}\frac{1}{V^2} \right] \qquad (3.62)'
\end{aligned}
$$

From this equation, if SM > 0, i.e., the vehicle has a US characteristic, and the velocity, V_s, where vehicle transient steering response becomes oscillatory is as follows:

$$
V_s = \sqrt{\frac{2l(K_f + K_r)}{m}\left\{\frac{1}{16}\frac{(1 - k^2/l_fl_r)^2}{k^2/l_fl_r}\frac{1}{\mathrm{SM}} + \mathrm{SM}\right\}} \qquad (3.64)
$$

In this equation, V_s is affected by k^2/l_fl_r. From Eqn (3.64), when $k^2/l_fl_r = 1$, V_s is minimum and if k^2/l_fl_r is greater or smaller than this, V_s always becomes larger. It is interesting to see that when the vehicle yaw moment inertia is larger or smaller than a certain value, the speed where vehicle transient steering response becomes oscillatory always becomes larger. Also, an interesting thing is that V_s becomes a minimum at the following:

$$
\mathrm{SM} = \frac{|1 - k^2/l_fl_r|}{4\sqrt{k^2/l_fl_r}}
$$

This analysis has shown that the characteristics of a vehicle's transient response to steering are particularly affected by the vehicle traveling speed and steer characteristics. This is shown in Table 3.2. The stability problem of the vehicle motion, as shown in the table, is called the vehicle directional stability. The image of the vehicle response to a pulse steering input is shown in Figure 3.25. The motion response characteristic corresponding to 1, 2 and 3 in Table 3.2 is clear.

3.4.1.2 Natural frequency and damping ratio

Here, the natural frequency and damping ratio of the vehicle response to steering input will be studied. Taking ω_n as the natural frequency and ζ as the damping ratio from the characteristic equation coefficients, the following is obtained:

$$
\omega_n^2 = P^2 \qquad (3.65)
$$

$$
2\zeta\omega_n = 2D \qquad (3.66)
$$

Table 3.2 Vehicle Steer Characteristic and Transient Responses

Steer Characteristics	Transient Response	
US	$0 \le V \le V_S$	② dumping with oscillation $V \ge V_S$
NS	① dumping without oscillation	
OS	$0 \le V \le V_C$	③ diverge $V \ge V_C$

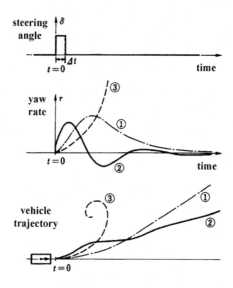

FIGURE 3.25

Qualitative understanding of vehicle transient responses.

By using Eqns (3.56), (3.57), and (3.43), ω_n and ζ are shown to be the following:

$$\omega_n = P$$

$$= \frac{2l}{V}\sqrt{\frac{K_f K_r}{mI}}\sqrt{1 + AV^2} = \frac{2\sqrt{K_f K_r l}}{mk}\frac{\sqrt{1 + AV^2}}{V}$$

$$= \frac{2\sqrt{K_f K_r l}}{mk} \frac{1}{V} \sqrt{1 + \frac{m}{2l} \left(\frac{K_f + K_r}{K_f K_r} \right) SM V^2} \tag{3.67}$$

$$\zeta = \frac{D}{P} = \frac{m\left(l_f^2 K_f + l_r^2 K_r \right) + I(K_f + K_r)}{2l\sqrt{mIK_f K_r(1 + AV^2)}}$$

$$= \frac{K_f + K_r}{2\sqrt{K_f K_r}} \frac{k}{l} \frac{\left(\frac{1 + k^2/l_f l_r}{k^2/l_f l_r} \right) + \frac{1}{k^2} \frac{(l_f - l_r)(l_f K_f - l_r K_r)}{K_f + K_r}}{\sqrt{1 + AV^2}}$$

$$= \frac{K_f + K_r}{2\sqrt{K_f K_r}} \frac{k}{l} \frac{\left(\frac{1 + k^2/l_f l_r}{k^2/l_f l_r} \right) + \frac{1}{k^2} \frac{(l_f - l_r)(l_f K_f - l_r K_r)}{K_f + K_r}}{\sqrt{1 + \frac{m}{2l} \frac{K_f + K_r}{K_f K_r} SM V^2}} \tag{3.68}$$

If $l_f \approx l_r$ and $K_f \approx K_r$, the approximation of ω_n and ζ could be written as follows:

$$\omega_n = \frac{2(K_f + K_r)}{mV} \sqrt{\frac{l_f l_r}{k^2}} \sqrt{1 + AV^2}$$

$$= \frac{2(K_f + K_r)}{mV} \sqrt{\frac{l_f l_r}{k^2}} \sqrt{1 + \frac{2m}{l(K_f + K_r)} SM V^2} \tag{3.67'}$$

$$\zeta = \frac{1 + k^2/l_f l_r}{2\sqrt{k^2/l_f l_r}} \frac{1}{\sqrt{1 + AV^2}} = \frac{1 + k^2/l_f l_r}{2\sqrt{k^2/l_f l_r}} \frac{1}{\sqrt{1 + \frac{2m}{l(K_f + K_r)} SM V^2}} \tag{3.68'}$$

Next is to study the vehicle natural frequency and damping ratio changes in relation to vehicle steer characteristics and traveling speed.

Figure 3.26 is an example of the effect of SM and traveling speed, V, on ω_n, which corresponds to Eqn (3.67)'. From this figure, the natural frequency, ω_n, in particular decreases with traveling speed V and only increases slightly with SM.

On the other hand, Figure 3.27 shows the effect of SM and traveling speed, V, on the damping ratio, ζ, which corresponds to Eqn (3.68)'. From this figure, damping ratio ζ decreases with SM, and the vehicle response to steering becomes more oscillatory. In the case of increasing V, when the vehicle exhibits US characteristics, ζ decreases, and so the vehicle motion becomes less damped and more oscillatory. When the vehicle exhibits OS characteristics, ζ increases, and the vehicle response to steering is deteriorated. Furthermore, if $k^2/l_f l_r \approx 1$ when the vehicle exhibits NS characteristics or when V is almost zero regardless of the steer characteristics, ζ is almost equal to 1.0. In other words, it could be considered as a critically damped situation.

At $\zeta < 1.0$, when the vehicle response is oscillating, if $l_f \approx l_r$ and $K_f \approx K_r$, the damped natural frequency could be determined by using Eqn (3.62)':

$$q = \sqrt{P^2 - D^2}$$

FIGURE 3.26

Effects of SM and vehicle speed on ω_n.

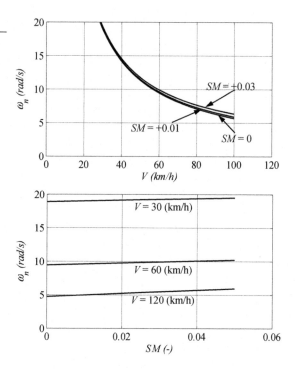

FIGURE 3.27

Effects of SM and vehicle speed on ζ.

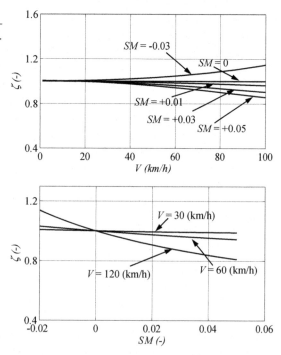

$$= \frac{4(K_f + K_r)}{m} \sqrt{\frac{l_f l_r}{k^2}} \sqrt{\frac{m}{2l(K_f + K_r)} SM - \left\{ \frac{1}{16} \frac{(1 - k^2/l_f l_r)^2}{k^2/l_f l_r} + SM^2 \right\} \frac{1}{V^2}} \qquad (3.69)$$

And, this is also equal to $\omega_n \sqrt{1 - \zeta^2}$.

3.4.1.3 Responsiveness

Until here, the focus has been on whether the vehicle transient response to steering is without or with oscillation and stability. However, the responsiveness to steering input is also an important characteristic for the vehicle discussed here, as well as for ships and aircraft.

If λ is the real part of the root of the characteristic equation for a stable linear system, the response time, as defined in the following equation, is a parameter that shows the response speed of the system:

$$t_R = -\frac{1}{\lambda} \qquad (3.70)$$

Therefore, when $D^2 - P^2 \geq 0$, the t_R for the vehicle dealt with is as follows:

$$t_R = \frac{1}{D \mp \sqrt{D^2 - P^2}} \qquad (3.71)$$

And, when $D^2 - P^2 < 0$, the following is obtained:

$$t_R = \frac{1}{D} \qquad (3.72)$$

If $l_f \approx l_r$ and $K_f \approx K_r$, using Eqn (3.56)″ and (3.62)′, the t_R values for $D^2 - P^2 \geq 0$ and $D^2 - P^2 < 0$ are expressed as follows:

$$t_R = \frac{\frac{mV}{K_f + K_r}\left(\frac{k^2/l_f l_r}{1 + k^2/l_f l_r}\right)}{1 \mp \frac{4\sqrt{k^2/l_f l_r}}{1 + k^2/l_f l_r} \sqrt{\left\{ \frac{1}{16} \frac{(1 - k^2/l_f l_r)^2}{k^2/l_f l_r} + SM^2 \right\} \frac{1}{V^2} - \frac{mSM}{2l(K_f + K_r)}}} \quad (\text{when } D^2 - P^2 \geq 0) \qquad (3.73)$$

$$t_R = \frac{mV}{K_f + K_r} \frac{k^2/l_f l_r}{1 + k^2/l_f l_r} \quad (\text{when } D^2 - P^2 < 0) \qquad (3.74)$$

Broadly speaking, it is more frequent to have the case of $D^2 - P^2 \approx 0$. In this case, the previous two equations are equal, and the standard vehicle response time, t_R, could be taken as Eqn (3.74).

From the previous discussions, the basic nature of the vehicle dynamic characteristics (such as the vehicle natural frequency, ω_n, the damping ratio, ζ, and the response time, t_R) are dependent on the traveling speed and steer characteristics, namely SM. Also, if $l_f \approx l_r$ and $K_f \approx K_r$, the transient response depends on the ratio of vehicle mass to cornering stiffness, $m/(K_f + K_r)$, wheelbase, l, and yaw inertia moment, $k^2/l_f l_r$, (which is decided by the wheelbase and center of gravity position). In particular, it should be noted that the yaw inertia moment actually affects the vehicle dynamic characteristics in the form of $k^2/l_f l_r$.

3.4.2 STEERING RESPONSE TRANSFER FUNCTION AND RESPONSE TIME HISTORY

Previously, a direct treating of Eqns (3.12) and (3.13), which are the basic equations of motion describing the vehicle motion, has been avoided. The characteristics of vehicle transient response to steering input has only been looked at by focusing on the characteristic equation and its roots.

Looking at the basic equations of motion, (3.12) and (3.13), and transforming both sides of equations gives the following:

$$\{mVs + 2(K_f + K_r)\}\beta(s) + \left\{mV + \frac{2}{V}(l_f K_f - l_r K_r)\right\}r(s) = 2K_f \delta(s) \tag{3.75}$$

$$2(l_f K_f - l_r K_r)\beta(s) + \left\{Is + \frac{2\left(l_f^2 K_f + l_r^2 K_r\right)}{V}\right\}r(s) = 2l_f K_f \delta(s) \tag{3.76}$$

where $\beta(s)$, $r(s)$, and $\delta(s)$ are the Laplace transformation of β, r, and δ. By solving the algebraic equation for $\beta(s)$ and $r(s)$, the following equations can be obtained:

$$\frac{\beta(s)}{\delta(s)} = \frac{\begin{vmatrix} 2K_f & mV + \frac{2}{V}(l_f K_f - l_r K_r) \\ 2l_f K_f & Is + \frac{2\left(l_f^2 K_f + l_r^2 K_r\right)}{V} \end{vmatrix}}{\begin{vmatrix} mVs + 2(K_f + K_r) & mV + \frac{2}{V}(l_f K_f - l_r K_r) \\ 2(l_f K_f - l_r K_r) & Is + \frac{2\left(l_f^2 K_f + l_r^2 K_r\right)}{V} \end{vmatrix}} \tag{3.77}$$

$$\frac{r(s)}{\delta(s)} = \frac{\begin{vmatrix} mVs + 2(K_f + K_r) & 2K_f \\ 2(l_f K_f - l_r K_r) & 2l_f K_f \end{vmatrix}}{\begin{vmatrix} mVs + 2(K_f + K_r) & mV + \frac{2}{V}(l_f K_f - l_r K_r) \\ 2(l_f K_f - l_r K_r) & Is + \frac{2\left(l_f^2 K_f + l_r^2 K_r\right)}{V} \end{vmatrix}} \tag{3.78}$$

Then, using the ω_n and ζ derived earlier and rewriting Eqns (3.77) and (3.78) gives the following:

$$\frac{\beta(s)}{\delta(s)} = G_\delta^\beta(0)\frac{1 + T_\beta s}{1 + \frac{2\zeta s}{\omega_n} + \frac{s^2}{\omega_n^2}} \tag{3.77}'$$

where

$$G_\delta^\beta(0) = \frac{1 - \frac{m}{2l}\frac{l_f}{l_r K_r}V^2}{1 + AV^2}\frac{l_r}{l} \tag{3.79}$$

$$T_\beta = \frac{IV}{2ll_r K_r}\frac{1}{1 - \frac{m}{2l}\frac{l_f}{l_r K_r}V^2} \tag{3.80}$$

and

$$\frac{r(s)}{\delta(s)} = G_\delta^r(0) \frac{1 + T_r s}{1 + \frac{2\zeta s}{\omega_n} + \frac{s^2}{\omega_n^2}} \tag{3.78}'$$

$$G_\delta^r(0) = \frac{1}{1 + AV^2} \frac{V}{l} \tag{3.81}$$

$$T_r = \frac{ml_f V}{2lK_r} \tag{3.82}$$

$G_\delta^\beta(0)$ is the side-slip angle gain constant, which is the value of β in response to δ during steady-state cornering, and $G_\delta^r(0)$ is the yaw rate gain constant, which is the value of r in response to δ during steady-state cornering. Equations (3.77)' and (3.78)' are the transfer functions of the response of side-slip angle, β, and yaw rate, r, to steering input, δ.

If the Laplace transformed response of β and r to δ are given, as in Eqns (3.77) and (3.78) or (3.77)' and (3.78)', the response of β and r to a given δ could be obtained by inverse Laplace transformation. When a vehicle, traveling on a straight line, is suddenly given a step steering input, the vehicle response is as follows:

$$\beta(t) = L^{-1}[\beta(s)] = L^{-1}\left[G_\delta^\beta(0)\frac{1 + T_\beta s}{1 + \frac{2\zeta s}{\omega_n} + \frac{s^2}{\omega_n^2}}\frac{\delta_0}{s}\right] \tag{3.83}$$

$$r(t) = L^{-1}[r(s)] = L^{-1}\left[G_\delta^r(0)\frac{1 + T_r s}{1 + \frac{2\zeta s}{\omega_n} + \frac{s^2}{\omega_n^2}}\frac{\delta_0}{s}\right] \tag{3.84}$$

whereby L^{-1} means inverse Laplace transformation and δ_0/s is the Laplace transformed steering angle, $\delta(s)$, a step input with a magnitude of δ_0. Applying inverse Laplace transformation formula to the previous equations, $\beta(t)$ and $r(t)$ become the following. Here, the initial value of β and r is zero. When the vehicle shows response without oscillation at $\zeta > 1$, then the following is obtained:

$$\beta(t) = G_\delta^\beta(0)\delta_0\left[1 + \frac{1 - \left(\zeta + \sqrt{\zeta^2 - 1}\right)\omega_n T_\beta}{2\left(\zeta + \sqrt{\zeta^2 - 1}\right)\sqrt{\zeta^2 - 1}}e^{\left(-\zeta - \sqrt{\zeta^2 - 1}\right)\omega_n t}\right.$$

$$\left. - \frac{1 - \left(\zeta - \sqrt{\zeta^2 - 1}\right)\omega_n T_\beta}{2\left(\zeta - \sqrt{\zeta^2 - 1}\right)\sqrt{\zeta^2 - 1}}e^{\left(-\zeta + \sqrt{\zeta^2 - 1}\right)\omega_n t}\right] \tag{3.83}'$$

$$r(t) = G_\delta^r(0)\delta_0\left[1 + \frac{1 - \left(\zeta + \sqrt{\zeta^2 - 1}\right)\omega_n T_r}{2\left(\zeta + \sqrt{\zeta^2 - 1}\right)\sqrt{\zeta^2 - 1}}e^{\left(-\zeta - \sqrt{\zeta^2 - 1}\right)\omega_n t}\right.$$

$$\left. - \frac{1 - \left(\zeta - \sqrt{\zeta^2 - 1}\right)\omega_n T_r}{2\left(\zeta - \sqrt{\zeta^2 - 1}\right)\sqrt{\zeta^2 - 1}}e^{\left(-\zeta + \sqrt{\zeta^2 - 1}\right)\omega_n t}\right] \tag{3.84}'$$

When the response is without oscillation at $\zeta = 1$, the following results:

$$\beta(t) = G_\delta^\beta(0)\delta_0\left[1 + \left\{(\omega_n^2 T_\beta - \omega_n)t - 1\right\}e^{-\omega_n t}\right] \tag{3.83}''$$

$$r(t) = G_\delta^r(0)\delta_0\left[1 + \left\{(\omega_n^2 T_r - \omega_n)t - 1\right\}e^{-\omega_n t}\right] \tag{3.84}''$$

And, when the response is with oscillation at $\zeta < 1$, the following is true:

$$\beta(t) = G_\delta^\beta(0)\delta_0\left[1 + \frac{T_\beta}{\sqrt{1-\zeta^2}}\sqrt{(1/T_\beta - \zeta\omega_n)^2 + (1-\zeta^2)\omega_n^2}\right.$$
$$\left. e^{-\zeta\omega_n t}\sin\left(\sqrt{1-\zeta^2}\omega_n t + \Psi_\beta\right)\right] \tag{3.83}'''$$

where $\Psi_\beta = \tan^{-1}\left(\dfrac{\sqrt{1-\zeta^2}\omega_n}{1/T_\beta - \zeta\omega_n}\right) - \tan^{-1}\left(\dfrac{\sqrt{1-\zeta^2}}{-\zeta}\right)$

$$r(t) = G_\delta^r(0)\delta_0\left[1 + \frac{T_r}{\sqrt{1-\zeta^2}}\sqrt{(1/T_r - \zeta\omega_n)^2 + (1-\zeta^2)\omega_n^2}\right.$$
$$\left. e^{-\zeta\omega_n t}\sin\left(\sqrt{1-\zeta^2}\omega_n t + \Psi_r\right)\right] \tag{3.84}'''$$

where $\Psi_r = \tan^{-1}\left(\dfrac{\sqrt{1-\zeta^2}\omega_n}{1/T_r - \zeta\omega_n}\right) - \tan^{-1}\left(\dfrac{\sqrt{1-\zeta^2}}{-\zeta}\right)$

The responsiveness to steering input, besides the response time that is described earlier, could also be expressed by the time to steady-state value, t_e, and time to the first peak of the response oscillation, t_p, for the yaw rate response to step steering input. This is shown in Figure 3.28.

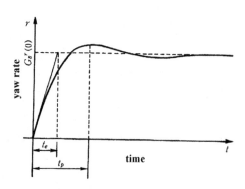

FIGURE 3.28

Responsiveness of yaw rate.

Calculating t_e and t_p gives the following terms:

$$t_e = \frac{1}{\omega_n^2 T_r} \tag{3.85}$$

$$t_p = \frac{1}{\omega_n \sqrt{1 - \zeta^2}} \left\{ \pi - \tan^{-1} \left(\frac{\sqrt{1 - \zeta^2} \omega_n T_r}{1 - \zeta \omega_n T_r} \right) \right\} \tag{3.86}$$

In particular, t_e could be defined as the approximated response time of yaw rate.

To see a time history of the vehicle responses to steering input, it is possible to use Eqn (3.83)'–(3.84)'''; however, recently, there has been an easier way to see the time history—a numerical simulation using a PC with some software, for example, Matlab-Simulink.

Equations of vehicle motion (3.12) and (3.13) are rewritten as follows:

$$\frac{d\beta}{dt} = -\frac{2(K_f + K_r)}{mV} \beta - \left\{ 1 + \frac{2}{mV^2} (l_f K_f - l_r K_r) \right\} r + \frac{2K_f}{mV} \delta \tag{3.12'}$$

$$\frac{dr}{dt} = -\frac{2(l_f K_f - l_r K_r)}{I} \beta - \frac{2\left(l_f^2 K_f + l_r^2 K_r \right)}{IV} r + \frac{2l_f K_f}{I} \delta \tag{3.13'}$$

Side-slip angle and yaw rate are obtained by integrating the right-hand side of the previous equations. It is possible to have the integral-type of block diagram of the vehicle motion to steering input as shown in Figure 3.29. This is the basis of the simulation program using Matlab-Simulink software.

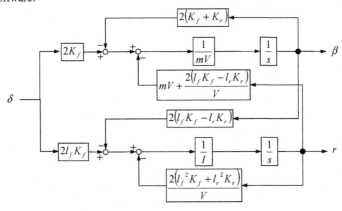

FIGURE 3.29

Integral-type block diagram of vehicle motion.

Example 3.6

Execute the simulation of the vehicle response to step steering input, $\delta = 0.04$ rad, by Matlab-Simulink at the vehicle speeds 60 km/h, 100 km/h, and 140 km/h, respectively, with the vehicle parameters as $m = 1500$ kg, $I = 2500$ kgm^2, $l_f = 1.1$ m, $l_r = 1.6$ m, $K_f = 55$ kN/rad, and $K_r = 60$ kN/rad.

Solution

The vehicle parameters for the simulation are set as in Figure E3.6(a). The simulation program is shown in Figure E3.6(b), and Figure E3.6(c) is a simulation condition. Figure E3.6(d) is a result of the simulation, and all the results are summarized in Figure E3.6(e).

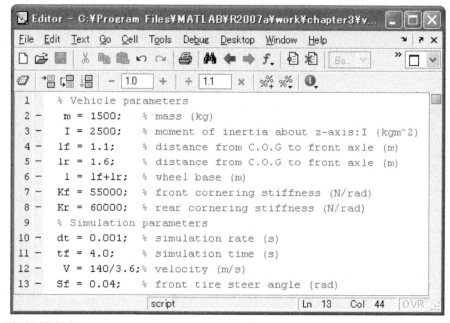

FIGURE E3.6(a)

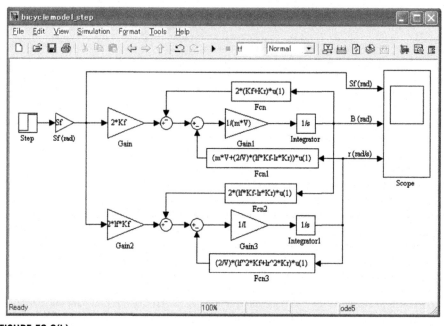

FIGURE E3.6(b)

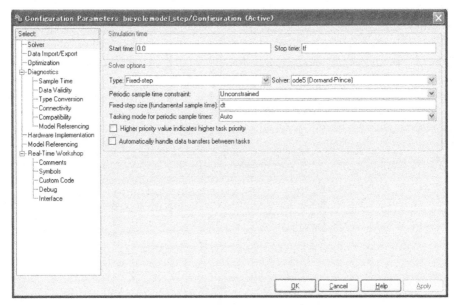

FIGURE E3.6(c)

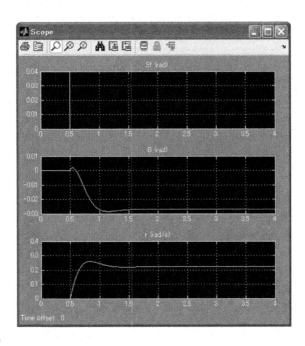

FIGURE E3.6(d)

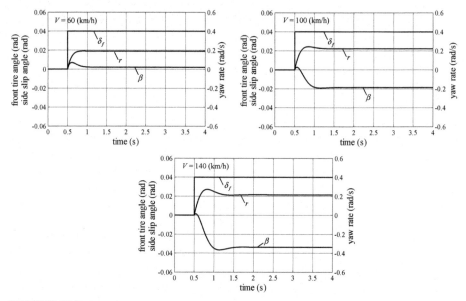

FIGURE E3.6(e)

If the vehicle motion is described with coordinates fixed on the ground, Eqns (3.21) and (3.22) will be achieved. By Laplace transforms, the following are obtained:

$$\left\{ms^2 + \frac{2(K_f + K_r)}{V}s\right\}y(s) + \left\{\frac{2(l_fK_f - l_rK_r)}{V}s - 2(K_f + K_r)\right\}\theta(s) = 2K_f\delta(s) \qquad (3.21)'$$

$$\frac{2(l_fK_f - l_rK_r)}{V}sy(s) + \left\{Is^2 + \frac{2\left(l_f^2K_f + l_r^2K_r\right)}{V}s - 2(l_fK_f - l_rK_r)\right\}\theta(s) = 2l_fK_f\delta(s) \quad (3.22)'$$

where $y(s)$ and $\theta(s)$ are the Laplace transforms of y and θ. Solving the algebraic equations of $y(s)$ and $\theta(s)$ gives the following:

$$\frac{y(s)}{\delta(s)} = \frac{\begin{vmatrix} 2K_f & \frac{2(l_fK_f - l_rK_r)}{V}s - 2(K_f + K_r) \\ 2l_fK_f & Is^2 + \frac{2\left(l_f^2K_f + l_r^2K_r\right)}{V}s - 2(l_fK_f - l_rK_r) \end{vmatrix}}{\begin{vmatrix} ms^2 + \frac{2(K_f + K_r)}{V}s & \frac{2(l_fK_f - l_rK_r)}{V}s - 2(K_f + K_r) \\ \frac{2(l_fK_f - l_rK_r)}{V}s & Is^2 + \frac{2\left(l_f^2K_f + l_r^2K_r\right)}{V}s - 2(l_fK_f - l_rK_r) \end{vmatrix}} \qquad (3.87)$$

$$\frac{\theta(s)}{\delta(s)} = \frac{\begin{vmatrix} ms^2 + \dfrac{2(K_f + K_r)}{V}s & 2K_f \\[3mm] \dfrac{2(l_f K_f - l_r K_r)}{V}s & 2l_f K_f \end{vmatrix}}{\begin{vmatrix} ms^2 + \dfrac{2(K_f + K_r)}{V}s & \dfrac{2(l_f K_f - l_r K_r)}{V}s - 2(K_f + K_r) \\[3mm] \dfrac{2(l_f K_f - l_r K_r)}{V}s & Is^2 + \dfrac{2\left(l_f^2 K_f + l_r^2 K_r\right)}{V}s - 2(l_f K_f - l_r K_r) \end{vmatrix}}$$

(3.88)

Then, using ω_n and ζ derived earlier and by rewriting equations, the equations become the following:

$$\frac{y(s)}{\delta(s)} = G_\delta^{\ddot{y}}(0)\frac{1 + T_{y1}s + T_{y2}s^2}{s^2\left(1 + \frac{2\zeta s}{\omega_n} + \frac{s^2}{\omega_n^2}\right)}$$

(3.87)'

where

$$G_\delta^{\ddot{y}}(0) = \frac{1}{1 + AV^2}\frac{V^2}{l} = VG_\delta^r(0)$$

(3.89)

$$T_{y1} = \frac{l_r}{V}$$

(3.90)

$$T_{y2} = \frac{I}{2lK_r}$$

(3.91)

$G_\delta^{\ddot{y}}(0)$ is the lateral acceleration gain constant, which is the lateral acceleration value in response to δ during steady-state cornering, and the following is derived:

$$\frac{\theta(s)}{\delta(s)} = G_\delta^r(0)\frac{1 + T_r s}{s\left(1 + \frac{2\zeta s}{\omega_n} + \frac{s^2}{\omega_n^2}\right)}$$

(3.88)'

Equations (3.87) and (3.88) or (3.87)' and (3.88)' are the transfer functions of the responses of lateral displacement and yaw angle to vehicle steering input. As previously shown, inverse Laplace transformations of $y(s)$ and $\theta(s)$ in the previous equations can obtain the lateral displacement and yaw angle responses to a given steering input.

Laplace transforms of the vehicle response to steering input and expression of the vehicle motion in the form of transfer functions are convenient when the actual response to a given steering input is desired. It is also suitable in the case of taking the vehicle as the control target in the control system and studying the control of vehicle motions and the controllability of the vehicle.

Furthermore, $\omega_n, \zeta, G_\delta^r(0), G_\delta^{\ddot{y}}(0), T_r, t_e$, and t_p, which are the coefficients in the transfer functions, are the parameters that determine the vehicle response characteristics toward steering input and are called the response parameters.

3.4.3 VEHICLE RESPONSE TO PERIODIC STEERING INPUT

Generally, among the methods for investigating dynamic characteristics of mechanical systems, the system response to periodic input is investigated. In vibration systems, this is called forced vibration; whereas, in an automatic control system, it is called the frequency response.

This method is widely used for understanding the dynamic characteristics of the vehicle, and the vehicle response to a periodic steering input is examined here. This is important, as an on-board driver can feel the vehicle's lateral acceleration and yaw rate responses to steering input very well.

First, the lateral acceleration of the vehicle's center of gravity to a periodic steering input, δ, can be written as follows by multiplying by s^2 on both sides of Eqn (3.87)′ and substituting $s = j2\pi f$:

$$G_\delta^{\ddot{y}}(j2\pi f) = G_\delta^{\ddot{y}}(0)\frac{1 - (2\pi f)^2 T_{y2} + j2\pi f T_{y1}}{1 - (2\pi f)^2/\omega_n^2 + j2\pi f 2\zeta/\omega_n} \tag{3.92}$$

where f is the frequency of the periodic steering, and $j = \sqrt{-1}$.

From this equation, the lateral acceleration gain $\left|G_\delta^{\ddot{y}}\right|$ and the phase angle $\angle G_\delta^{\ddot{y}}$ toward steering angle are as follows:

$$\left|G_\delta^{\ddot{y}}\right| = \sqrt{\frac{P_y^2 + Q_y^2}{R_y^2 + S_y^2}}\, G_\delta^{\ddot{y}}(0) \tag{3.93}$$

$$\angle G_\delta^{\ddot{y}} = \tan^{-1}\left(Q_y/P_y\right) - \tan^{-1}\left(S_y/R_y\right) \tag{3.94}$$

where

$$P_y = 1 - (2\pi f)^2 T_{y2} \quad Q_y = 2\pi f T_{y1}$$

$$R_y = 1 - (2\pi f)^2/\omega_n^2 \quad S_y = 2\pi f 2\zeta/\omega_n$$

Next, the vehicle yaw rate, r, response, $G_\delta^r(j2\pi f)$, to periodic steering, δ, can be written by substituting $s = j2\pi f$ into Eqn (3.78)′:

$$G_\delta^r(j2\pi f) = G_\delta^r(0)\frac{1 + j2\pi f T_r}{1 - (2\pi f)^2/\omega_n^2 + j2\pi f 2\zeta/\omega_n} \tag{3.95}$$

From this equation, the yaw rate gain, $\left|G_\delta^r\right|$, and the phase angle, $\angle G_\delta^r$, to steering input are as follows:

$$\left|G_\delta^r\right| = \sqrt{\frac{P_r^2 + Q_r^2}{R_r^2 + S_r^2}}\, G_\delta^r(0) \tag{3.96}$$

$$\angle G_\delta^r = \tan^{-1}(Q_r/P_r) - \tan^{-1}(S_r/R_r) \tag{3.97}$$

where

$$P_r = 1, \quad Q_r = 2\pi f T_r$$
$$R_r = 1 - (2\pi f)^2 / \omega_n^2$$
$$S_r = 2\pi f 2\zeta / \omega_n$$

In particular, the investigation of the yaw rate response to a periodic steering is very common in the study of the vehicle's inherent dynamic characteristics. The yaw rate response as expressed by Eqn (3.95) has a general form as shown in Figure 3.30. When the steering frequency is small, the yaw rate to steering gain is almost constant. As the steering frequency becomes larger, the US vehicle gain reaches a peak at a certain frequency and then decreases. The OS and NS vehicles do not have a peak, and their gain decreases with steering frequency. Furthermore, the phase lag is around zero at low steering frequencies, but it increases with frequency for all three steer characteristics. This tendency is more obvious for the vehicle with OS characteristics. For the US vehicle, the peak in the gain happens when the vehicle transient response to a fixed steering is oscillating. This peak becomes greater as the damping ratio, ζ, reduces. When the vehicle exhibits US characteristics, the peak will become larger with increasing traveling speed, V. The frequency where the peak occurs is nearly the same as the vehicle natural frequency, ω_n.

Figure 3.31 is a calculated example of the responses of yaw rate and lateral acceleration to a periodic steering for a small passenger car. From this figure, it is clear that with higher traveling speed, the motion phase lag, especially in the lateral acceleration response, becomes larger at higher frequencies. Furthermore, because the vehicle is US, a gain peak occurs at high traveling speed in the yaw rate response, and the vehicle transient response is oscillatory with insufficient damping.

Here, the relation between the frequency responses of lateral acceleration and yaw rate will be considered. The transfer function of the lateral acceleration to the steering input is derived in this section from equations of motion with fixed coordinates on the ground, Eqn (3.21)$'$ and (3.22)$'$. There is another way to derive the lateral acceleration transfer function using equations of motion with the fixed coordinates on the vehicle, Eqns (3.75) and (3.76). Based on these, the lateral

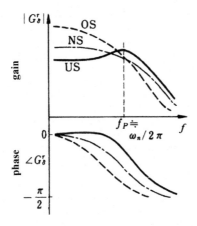

FIGURE 3.30

Conceptual diagram of yaw rate frequency response.

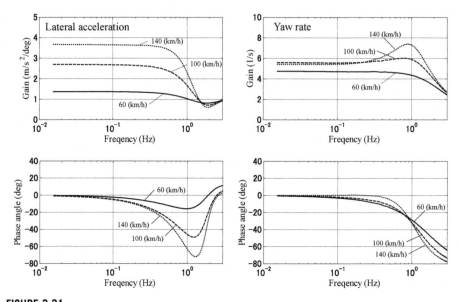

FIGURE 3.31

Yaw rate and lateral acceleration frequency response.

acceleration is expressed by $V(\dot{\beta} + r)$. A side-slip rate intervenes between the lateral acceleration response and the yaw rate.

Therefore, using Eqn (3.77)′ and (3.78)′, it is possible to describe the lateral acceleration transfer function as follows:

$$G_\delta^{\ddot{y}}(s) = V\frac{s\beta(s)}{\delta(s)} + V\frac{r(s)}{\delta(s)}$$

$$= VG_\delta^\beta(0)\frac{s(1+T_\beta s)}{1+\frac{2\zeta s}{\omega_n}+\frac{s^2}{\omega_n^2}} + VG_\delta^r(0)\frac{1+T_r s}{1+\frac{2\zeta s}{\omega_n}+\frac{s^2}{\omega_n^2}}$$

$$= \frac{1}{1+AV^2}\frac{V^2}{l}\frac{\frac{l_r}{V}s - \frac{m}{2l}\frac{l_f V}{K_r}s + \frac{I}{2lK_r}s^2}{1+\frac{2\zeta s}{\omega_n}+\frac{s^2}{\omega_n^2}} + \frac{1}{1+AV^2}\frac{V^2}{l}\frac{1+\frac{m}{2l}\frac{l_f V}{K_r}s}{1+\frac{2\zeta s}{\omega_n}+\frac{s^2}{\omega_n^2}}$$

$$= \frac{1}{1+AV^2}\frac{V^2}{l}\frac{1+\frac{l_r}{V}s + \frac{I}{2lK_r}s^2}{1+\frac{2\zeta s}{\omega_n}+\frac{s^2}{\omega_n^2}} = G_\delta^{\ddot{y}}(0)\frac{1+T_{y1}s + T_{y2}s^2}{1+\frac{2\zeta s}{\omega_n}+\frac{s^2}{\omega_n^2}}$$

The coefficient of s in the numerator of the yaw rate transfer function, $ml_f V/(2lK_r)$, is eliminated by the same term, which is a negative part of steady-state response of side-slip angle to steering input in the numerator of the side-slip transfer function. Only the term l_r/V remains in the numerator of the lateral acceleration transfer function as a coefficient of s, and this rapidly decreases with the vehicle speed.

The coefficient of s in the numerator of the transfer function, in general, has a lead effect and compensates for the response delay caused by the coefficient of s and s^2 in the denominator. The lateral acceleration response to steering has a smaller value of the coefficient of s in the numerator compared with that of the yaw rate, especially at high vehicle speed. This is partly why there is a significantly larger delay in the phase lag of the lateral acceleration compared with that of the yaw rate at high speed, as shown in Figure 3.31. The larger delay in lateral acceleration is due to the side-slip response acting in opposite to the steering angle at higher vehicle speeds. This is a very important part of the basic nature of the vehicle dynamics and is attributed to the intervention of vehicle side-slip motion between lateral acceleration and yaw rate.

Example 3.7

Calculate the yaw rate frequency responses of the vehicles with US, NS, and OS characteristics, respectively, at the vehicle speed $V = 120$ km/h using Matlab-Simulink and confirm the effects of the steer characteristics on the yaw rate frequency response shown previously in Figure 3.30.

Solution

The parameters of the US vehicle are the same as in Example 3.6, and the cornering stiffness for the NS vehicle is set as $K_f = 68.15$ kN/rad and $K_r = 46.85$ kN/rad; for the OS vehicle, it is $K_f = 72.5$ kN/rad and $K_r = 42.5$ kN/rad. The parameters of the vehicle and the sweep-type sine wave are set as in Figure E3.7(a). The simulation program of the vehicle response to the sweep-type sine wave steering input is shown in Figure E3.7(b), and Figure E3.7(c) is a result of the simulation. After finishing the vehicle response simulation to the sweep-type of sine wave steering input, the simulated data is saved as shown in Figure E3.7(d), and the yaw rate frequency response to steering input is calculated applying the Fourier Transformation to the time histories simulated, as shown in Figure E3.7(e). A result is in Figure E3.7(f), and Figure E3.7(g) shows the summarized calculation results comparing the effects of the vehicle steer characteristics on the yaw rate frequency response.

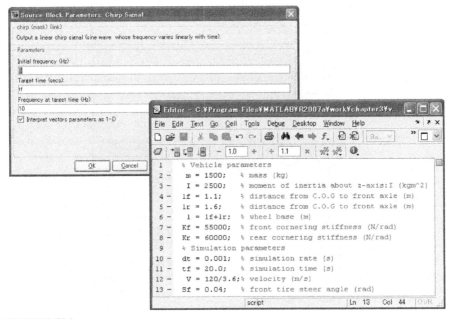

FIGURE E3.7(a)

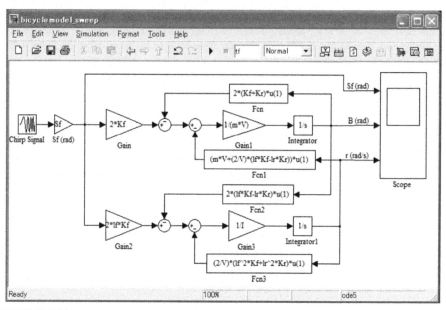

FIGURE E3.7(b)

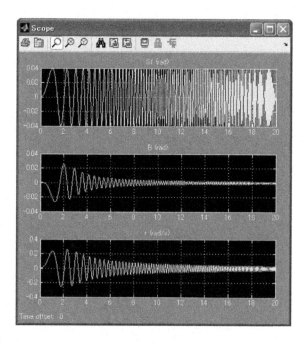

FIGURE E3.7(c)

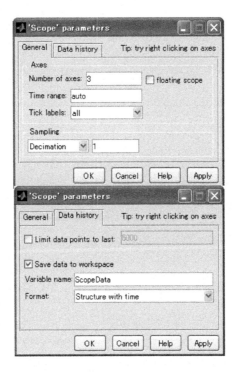

FIGURE E3.7(d)

```
 1 -   Sf = ScopeData.signals(1,1).values(:,1);% front tire steer angle (rad)
 2 -    B = ScopeData.signals(1,2).values(:,1);% side slip Angle (rad)
 3 -    r = ScopeData.signals(1,3).values(:,1);% yaw rate (rad/s)
 4      %---------------------------------------------------------------------%
 5      % The frequency response is calculated by using Fourier trasform.
 6 -   gr = etfe([r,Sf],[],2^15,0.001);
 7      % The gain and the phase angle of the frequency response are calculated.
 8 -   [amp,phase,w] = bode(gr);
 9      % An extra dimension is deleted.
10 -   amp = squeeze(amp);
11 -   phase = squeeze(phase);
12      % Drawing gain-frequency relation.
13 -   figure;
14 -   a1 = subplot(2,1,1);
15 -   graph1 = semilogx(w/(2*pi),20*log10(amp));
16 -   set(get(a1,'XLabel'),'String','Frequency (Hz)');
17 -   set(get(a1,'YLabel'),'String','Gain (dB)');
18 -   axis([0.01 3 6 18]);
19 -   grid on;
20      % Drawing phase-frequency relation.
21 -   a2 = subplot(2,1,2);
22 -   graph2 = semilogx(w/(2*pi),phase);
23 -   set(get(a2,'XLabel'),'String','Frequency (Hz)');
24 -   set(get(a2,'YLabel'),'String','Phase angle (deg)');
25 -   axis([0.01 3 -80 40]);
26 -   grid on;
```

FIGURE E3.7(e)

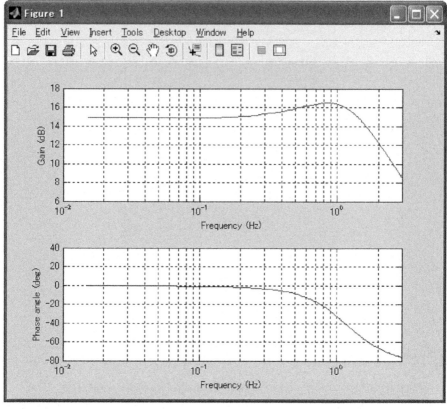

FIGURE E3.7(f)

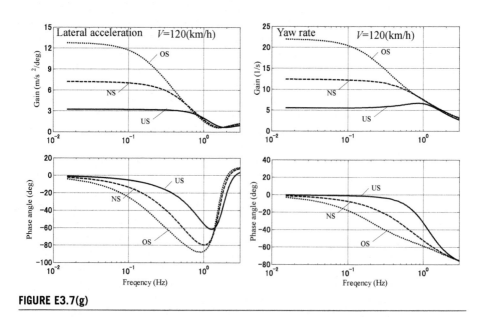

FIGURE E3.7(g)

3.4.4 EFFECT OF NONLINEAR TIRE CHARACTERISTICS

Previously, the vehicle dynamic characteristics have been studied with the assumption that the lateral force is proportional to the side-slip angles of the tires. It is important now to try to understand how the vehicle dynamics are affected when the lateral force is not proportional to the side-slip angles (e.g., at large tire slip angles).

The close relation between tire's lateral force, Y, and side-slip angles, β, has been discussed in Sections 2.3.1 and 2.4.2. For simplicity, taking K as the cornering stiffness at $\beta = 0$ and a friction force, μW, as a saturated lateral force that can be approximated as a second-order polynomial of β, the following is obtained:

$$Y = K\beta - \frac{K^2}{4\mu W}\beta^2 \qquad (3.98)$$

The relation is shown in Figure 3.32.

When a vehicle, with weight mg is making a circular motion with lateral acceleration $\ddot{y}$, as in Eqn (3.48), the lateral forces acting at the front and rear wheels are as follows:

$$2Y_f(\beta_f) = \frac{l_r mg}{l}\ddot{y} = 2K_f\beta_f - \frac{K_f^2}{\mu\frac{l_r mg}{l}}\beta_f^2 \qquad (3.99)$$

$$2Y_r(\beta_r) = \frac{l_f mg}{l}\ddot{y} = 2K_r\beta_r - \frac{K_r^2}{\mu\frac{l_f mg}{l}}\beta_r^2 \qquad (3.100)$$

where β_f and β_r are the front and rear wheel side-slip angles.

Using these equations, the equivalent cornering stiffness values, $\partial Y_f/\partial \beta_f$ and $\partial Y_r/\partial \beta_r$, are the following:

$$\frac{\partial Y_f}{\partial \beta_f} = K_f\left(1 - \frac{K_f}{\mu\frac{l_r mg}{l}}\beta_f\right) = K_f\sqrt{1 - \frac{\ddot{y}}{\mu}} \qquad (3.101)$$

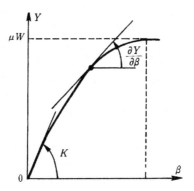

FIGURE 3.32

Approximation of tire nonlinear characteristics.

$$\frac{\partial Y_r}{\partial \beta_r} = K_r \left(1 - \frac{K_r}{\mu \frac{l_r mg}{l}} \beta_r \right) = K_r \sqrt{1 - \frac{\ddot{y}}{\mu}} \tag{3.102}$$

These are the gradients of the lateral forces to side-slip angle at the equilibrium point of circular motion with $\ddot{y}$. If $\ddot{y}/\mu \ll 1$, then the following results:

$$\frac{\partial Y_f}{\partial \beta_f} = K_f \left(1 - \frac{\ddot{y}}{2\mu} \right) \tag{3.103}$$

$$\frac{\partial Y_r}{\partial \beta_r} = K_r \left(1 - \frac{\ddot{y}}{2\mu} \right) \tag{3.104}$$

The cornering stiffness of the vehicle during circular motion decreases with the lateral acceleration when the lateral acceleration approaches the limit or the friction coefficient between the road and tire decreases abruptly. In the region where $\ddot{y}$ is small compared to μ, the cornering stiffness could be treated as decreasing linearly. The previous condition is shown in Figure 3.33.

Next, the characteristics of the vehicle motion in the region where the tire exhibits its nonlinear characteristics will be looked at. Consider the very small motion of the vehicle in response to a very small steering input from the initial condition of circular motion with the lateral acceleration $\ddot{y}$. Equations of motion at that time are expressed as follows:

$$m\left\{ g\ddot{y} + V\left(\frac{d\beta}{dt} + r \right) \right\} = 2Y_f \left(\beta_f + \delta - \beta - \frac{l_f r}{V} \right) + 2Y_r \left(\beta_r - \beta + \frac{l_r r}{V} \right) \tag{3.105}$$

$$I\frac{dr}{dt} = 2l_f Y_f \left(\beta_f + \delta - \beta - \frac{l_f r}{V} \right) - 2l_r Y_r \left(\beta_r - \beta + \frac{l_r r}{V} \right) \tag{3.106}$$

Because δ, β, and r are very small, the following results:

$$Y_f \left(\beta_f + \delta - \beta - \frac{l_f r}{V} \right) \cong Y_f (\beta_f) + \frac{\partial Y_f}{\partial \beta_f} \left(\delta - \beta - \frac{l_f r}{V} \right)$$

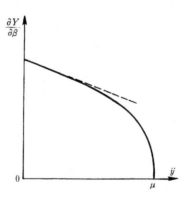

FIGURE 3.33

Change of equivalent cornering stiffness due to lateral acceleration.

$$Y_r\left(\beta_r - \beta + \frac{l_r r}{V}\right) \cong Y_r(\beta_r) + \frac{\partial Y_r}{\partial \beta_r}\left(-\beta + \frac{l_r r}{V}\right)$$

And, from the equilibrium conditions, we have the following:

$$mg\ddot{y} = 2Y_f(\beta_f) + 2Y_r(\beta_r)$$

$$2l_f Y_f(\beta_f) - 2l_r Y_r(\beta_r) = 0$$

Substituting these equations into Eqns (3.105) and (3.106) and rearranging them gives the following final equations:

$$mV\frac{d\beta}{dt} + 2\left(\frac{\partial Y_f}{\partial \beta_f} + \frac{\partial Y_r}{\partial \beta_r}\right)\beta + \left\{mV + \frac{2\left(l_f\frac{\partial Y_f}{\partial \beta_f} - l_r\frac{\partial Y_r}{\partial \beta_r}\right)}{V}\right\}r = 2\frac{\partial Y_f}{\partial \beta_f}\delta \qquad (3.107)$$

$$2\left(l_f\frac{\partial Y_f}{\partial \beta_f} - l_r\frac{\partial Y_r}{\partial \beta_r}\right)\beta + I\frac{dr}{dt} + \frac{2\left(l_f^2\frac{\partial Y_f}{\partial \beta_f} + l_r^2\frac{\partial Y_r}{\partial \beta_r}\right)}{V}r = 2l_f\frac{\partial Y_f}{\partial \beta_f}\delta \qquad (3.108)$$

These are the linearized equations of motion for the region where tire characteristics are nonlinear, based on the theory of small perturbation. In the region where tire characteristics are nonlinear, the tire cornering stiffness values of K_f and K_r are now replaced by the equivalent cornering stiffness of $\partial Y_f/\partial \beta_f$ and $\partial Y_r/\partial \beta_r$ in Eqns (3.101) and (3.102) or Eqns (3.103) and (3.104). Here, when $\ddot{y}/\mu \ll 1$, expressing the equivalent cornering stiffness as the following:

$$\frac{\partial Y_f}{\partial \beta_f} = K_f^* = K_f\left(1 - \frac{\ddot{y}}{2\mu}\right)$$

$$\frac{\partial Y_r}{\partial \beta_r} = K_r^* = K_r\left(1 - \frac{\ddot{y}}{2\mu}\right)$$

then several parameters that show the vehicle dynamic characteristics are obtained through the following equations. First of all, the stability factor is shown:

$$A^* = \frac{m}{2l^2}\frac{l_r K_r^* - l_f K_f^*}{K_f^* K_r^*} = \frac{m}{2l^2}\frac{l_r K_r - l_f K_f}{K_f K_r}\left(1 + \frac{\ddot{y}}{2\mu}\right)$$

$$= A\left(1 + \frac{\ddot{y}}{2\mu}\right) \qquad (3.109)$$

And, the natural frequency now becomes the following:

$$\omega_n^* = \frac{2\sqrt{K_f^* K_r^*}l}{mk}\frac{\sqrt{1 + A^* V^2}}{V}$$

$$= \frac{2\sqrt{K_f K_r}l}{mk}\frac{\sqrt{1 + AV^2}}{V}\left[1 - \left(1 + \frac{1}{1 + AV^2}\right)\frac{\ddot{y}}{4\mu}\right] \qquad (3.110)$$

$$= \omega_n\left[1 - \left(1 + \frac{1}{1 + AV^2}\right)\frac{\ddot{y}}{4\mu}\right]$$

Furthermore, the approximated response time of yaw rate given by Eqn (3.85) is as follows:

$$t_e^* = \frac{1}{\omega_n^{*2} T_r^*}$$

$$= \frac{1}{\omega_n^2 T_r} \left(1 + \frac{1}{1+AV^2} \frac{\ddot{y}}{2\mu} \right) \tag{3.111}$$

$$= t_e \left(1 + \frac{1}{1+AV^2} \frac{\ddot{y}}{2\mu} \right)$$

These show the change of vehicle dynamic characteristics with respect to lateral acceleration, $\ddot{y}$, in the region where the tire characteristic is nonlinear.

Figure 3.34 is an example of the vehicle yaw rate and lateral acceleration frequency response. It shows the vehicle dynamic characteristics during circular motion at different lateral acceleration values. This gives an idea of the effect of tire nonlinear characteristics. The dynamic characteristics change substantially with the lateral acceleration due to the saturation property of tire characteristic to side-slip angle.

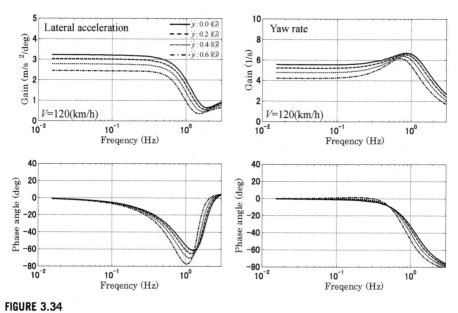

FIGURE 3.34

Effect of tire nonlinear characteristics on frequency response.

PROBLEMS

3.1 Referring to Figure 3.4(b), confirm that it is acceptable to regard the side-slip angles of right and left wheels as almost identical, and by using the bicycle vehicle model that it is reasonable when the vehicle speed is higher than 40 km/h, the yaw rate is less than 0.1 rad/s and the vehicle track is 1.4 m.

3.2 Derive Eqns (3.29) and (3.30) from Eqns (3.26)–(3.28).

3.3 Geometrically, show that the side-slip angle during steady-state cornering at low speed is described by the third equation in Eqn (3.33).

3.4 Give the geometric proof of Eqn (3.34).

3.5 Using Eqn (3.39), find the vehicle speed at which steady-state yaw rate reaches the peak value when the vehicle is understeer. This speed is called characteristic speed. Show that the peak value is half of the yaw rate value of the neutralsteer vehicle at the characteristic speed.

3.6 Find the vehicle speed at which the steady-state side-slip angle is equal to zero using Eqn (3.40), and calculate the value under $m = 1500$ kg, $l_f = 1.1$ m, $l_r = 1.6$ m, and $K_r = 60$ kN/rad.

3.7 Calculate the stability factor using Eqn (3.43) under $m = 1500$ kg, $l_f = 1.1$ m, $l_r = 1.6$ m, $K_f = 55$ kN/rad, and $K_r = 60$ kN/rad.

3.8 Calculate the static margin using Eqn (3.45) for the same vehicle parameters as used in Problem 3.7.

3.9 Calculate the critical vehicle speed for the OS vehicle with the parameters used in Example 3.7.

3.10 Confirm that for the vehicle with a static margin, SM, of almost zero, the inverse of the vehicle natural frequency, $1/\omega_n$, is nearly equal to the vehicle response time expressed by Eqn (3.74).

3.11 Using Eqn (3.110), estimate what percent of the vehicle's natural frequency is reduced due to circular turning with the lateral acceleration, 2.0 m/s^2, on a dry road surface, $\mu = 1.0$.

3.12 Execute the vehicle response simulation to a single, 0.5 Hz, sine-wave steering input with an amplitude of 0.04 rad at vehicle speeds of 60, 100, and 140 km/h, respectively, using the Matlab-Simulink simulation software. Use the same vehicle parameters as in Example 3.6.

3.13 Find the steady-state side-slip angle caused by disturbance yaw rate, Δr_C, at a steering angle equal to zero by using Eqn (3.24).

3.14 Using Eqn (3.25), find the steady-state yaw rate, Δr_R, caused by the restoring yaw moment that is produced by the side-slip angle calculated in Problem 3.13; assume the steering angle is equal to zero.

3.15 From Problem 3.14, it is possible to obtain the ratio $\Delta r_R / \Delta r_C$. A ratio larger than 1.0 means that the result is larger than the cause, and the result causes larger next results, and so on. The vehicle, eventually, becomes unstable. Find the vehicle speed that satisfies $\Delta r_R / \Delta r_C = 1.0$, and confirm that the speed found is identical to the critical vehicle speed obtained by Eqn (3.42).

REFERENCES

[1] Whitcomb DW, Milliken WF. Design implication of a general theory of automotive stability and control. Proc IMechE (AD) 1956. 367–91.
[2] Ellis JR. Vehicle Dynamics. London: London Business Book Ltd; 1969.

VEHICLE MOTION BY DISTURBANCES

4

4.1 PREFACE

The previous chapter studied the basic characteristics of vehicle motion, particularly steering input. It is clear that a vehicle moving freely in the horizontal plane, without any direct restrictions from tracks, can be disturbed laterally by some external force acting on the vehicle in the lateral direction. In other words, because the vehicle has the ability of free plane motion, it cannot avoid unwanted motion by a disturbance.

In this chapter, vehicle motion characteristics are further understood by examining the motion of the vehicle when subjected to lateral disturbances. In these studies, the vehicle steering angle is always kept as zero, and there is no controlling steering action that responds to the vehicle motion.

4.2 MOTION BY LATERAL FORCE EXERTED ON THE CENTER OF GRAVITY

When the vehicle is traveling on a banked road, for example, a component of the vehicle weight will act as a lateral force at the center of gravity, as shown in Figure 4.1. This section will look at the vehicle motion when the lateral force, Y, acts at the center of gravity.

4.2.1 VEHICLE MOTION TO A STEP CHANGE IN LATERAL FORCE

In order to study the vehicle motion due to the lateral force, Y, the vehicle response to an idealized form of lateral force will be investigated. Generally, one ideal form of lateral force in this kind of situation is a step change. Consider the lateral force shown in Figure 4.2 acting on the center of gravity of a vehicle traveling on straight line.

If this disturbance force acts a long enough time, even if the Y_0 value is small, the vehicle will eventually deviate away from its original path. The vehicle motion in this case is more conveniently expressed with coordinates fixed on the vehicle itself, as discussed in Section 3.2.1.

The vehicle equations of motion when the steering angle is zero and the lateral force acting at the center of gravity equals Y are found from Eqns (3.12) and (3.13) as follows:

$$mV\frac{d\beta}{dt} + 2(K_f + K_r)\beta + \left\{mV + \frac{2}{V}(l_f K_f - l_r K_r)\right\}r = Y \tag{4.1}$$

109

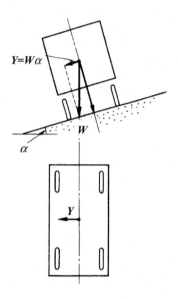

FIGURE 4.1

Lateral force exerted on the C.G.

$$2(l_f K_f - l_r K_r)\beta + I\frac{dr}{dt} + \frac{2\left(l_f^2 K_f + l_r^2 K_r\right)}{V} r = 0 \tag{4.2}$$

where the magnitude of lateral force Y is not large, and the side-slip angle condition, $|\beta| \ll 1$, is always true, even during vehicle motion.

The side slip and yaw rate response of the vehicle to Y can be obtained by Laplace transformation of Eqns (4.1) and (4.2). $\beta(s)$ and $r(s)$ are written as follows:

$$\beta(s) = \frac{\begin{vmatrix} \dfrac{Y_0}{s} & mV + \dfrac{2}{V}(l_f K_f - l_r K_r) \\[2mm] 0 & Is + \dfrac{2\left(l_f^2 K_f + l_r^2 K_r\right)}{V} \end{vmatrix}}{\begin{vmatrix} mVs + 2(K_f + K_r) & mV + \dfrac{2}{V}(l_f K_f - l_r K_r) \\[2mm] 2(l_f K_f - l_r K_r) & Is + \dfrac{2\left(l_f^2 K_f + l_r^2 K_r\right)}{V} \end{vmatrix}} \tag{4.3}$$

$$= \frac{Y_0}{mV} \frac{s + a_\beta}{s\left(s^2 + 2\zeta\omega_n s + \omega_n^2\right)}$$

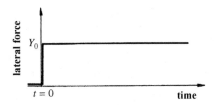

FIGURE 4.2

Step form lateral force.

$$r(s) = \frac{\begin{vmatrix} mVs + 2(K_f + K_r) & \dfrac{Y_0}{s} \\[2ex] 2(l_f K_f - l_r K_r) & 0 \end{vmatrix}}{\begin{vmatrix} mVs + 2(K_f + K_r) & mV + \dfrac{2}{V}(l_f K_f - l_r K_r) \\[2ex] 2(l_f K_f - l_r K_r) & Is + \dfrac{2\left(l_f^2 K_f + l_r^2 K_r\right)}{V} \end{vmatrix}}$$

$$= \frac{Y_0}{mIV} \frac{a_r}{s\left(s^2 + 2\zeta\omega_n s + \omega_n^2\right)}$$

(4.4)

where the following values are assigned:

$$a_\beta = \frac{2\left(l_f^2 K_f + l_r^2 K_r\right)}{IV}, \quad a_r = -2(l_f K_f - l_r K_r)$$

And, ω_n and ζ are the natural frequency and damping ratio, respectively, given by Eqns (3.67) and (3.68), and Y_0/s is the Laplace transformation of Y.

4.2.1.1 Steady-state condition

First, we will establish the final steady-state condition the vehicle reaches when acted on by a lateral force, as in Figure 4.2.

Using Laplace transforms, the steady-state values of β and r become the following:

$$\beta = \lim_{s \to 0} s\beta(s) = \frac{Y_0}{mV} \frac{a_\beta}{\omega_n^2}$$

$$= \frac{l_f^2 K_f + l_r^2 K_r}{2l^2 K_f K_r \left[1 - \dfrac{m(l_f K_f - l_r K_r)}{2l^2 K_f K_r} V^2\right]} Y_0$$

(4.5)

$$r = \lim_{s \to 0} sr(s) = \frac{Y_0}{mIV} \frac{a_r}{\omega_n^2}$$

$$= \frac{-(l_f K_f - l_r K_r)V}{2l^2 K_f K_r \left[1 - \frac{m(l_f K_f - l_r K_r)}{2l^2 K_f K_r}V^2\right]} Y_0 \qquad (4.6)$$

The previous equation shows β is always positive while r is positive when $l_f K_f < l_r K_r$ (US characteristic) and is negative when $l_f K_f > l_r K_r$ (OS characteristic) under $V < V_c$.

The physical meaning of Eqns (4.5) and (4.6) requires further discussion. For a vehicle with a US characteristic, the traveling condition and force equilibrium during steady state are shown in Figure 4.3.

If the lateral disturbance, Y_0, acts on the vehicle, the center of gravity, P, will move and produce a side-slip angle of $\beta > 0$. Due to this β, the forces of $2K_f \beta$ and $2K_r \beta$ will be exerted on the front and rear tires. The resultant force of these two forces acts at the NSP, as explained in Section 3.3.3. The magnitude of the resultant force is $2(K_f + K_r)\beta$, and it acts in the opposite direction to Y_0. If the vehicle exhibits a US characteristic, the NSP is behind the center of gravity, P, and the resultant force produces an anticlockwise yaw moment around the point P, as shown Figure 4.3.

If the vehicle motion is in steady state, a moment must act to balance this yawing moment. This moment can only be produced by a force acting on the tire, so there must be relative motion in the lateral direction, other than β, between the tire and the road surface. Here, the anticlockwise yaw motion around point P produces side-slip angles of $l_f r/V$ and $l_r r/V$ on the front and rear tires, respectively, as shown in Figure 4.3. Two forces in an opposite direction to each other with the magnitude of $l_f K_f r/V$ and $l_r K_r r/V$ are exerted on the front and rear wheels to balance the yaw moment caused by the disturbance. This is why r in Eqn (4.6) is positive and the vehicle makes an anticlockwise circular motion when it exhibits US characteristic.

The centrifugal force, mrV, also acts at the center of gravity in a direction opposite to Y_0. These forces are in the equilibrium so the vehicle is in steady-state cornering and heading outward from the circular path.

The vehicle with NS characteristic has the traveling condition and force equilibrium shown in Figure 4.4. If the characteristic is NS, the NSP coincides with point P, and the resultant force of the tire forces, $2(K_f + K_r)\beta$, acts at the same position as Y_0. This resultant force does not produce any moment around the center of gravity, and the vehicle has no yawing motion. There is no centrifugal force acting at the center of gravity. This is why r is zero in Eqn (4.6). The resultant

FIGURE 4.3

Steady state of a US vehicle.

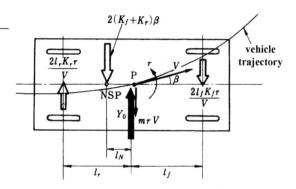

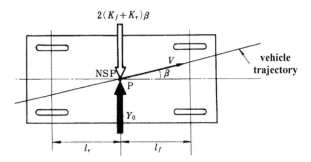

FIGURE 4.4

Steady state of a NS vehicle.

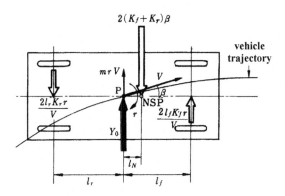

FIGURE 4.5

Steady state of an OS vehicle.

force of the tire forces is in equilibrium with the external force, Y_0. Consequently, the vehicle continues its transverse motion while producing a side-slip angle.

Figure 4.5 shows the traveling condition and force equilibrium for an OS vehicle. Here, the lateral force Y_0 moves the center of gravity to produce a side-slip angle. This side-slip angle generates forces of $2K_f\beta$ and $2K_r\beta$ at the front and rear wheels, and the resultant force acts at the NSP. If the vehicle characteristic is OS, the NSP is in front of P, and the resultant force $2(K_f + K_r)\beta$ produces a clockwise yaw moment around the vehicle's center of gravity. In steady state, there must be a moment to balance this yawing moment. This moment is obtained from the front and rear lateral forces, $2K_f l_f r/V$ and $2K_r l_r r/V$, produced by the clockwise yawing motion of the vehicle. This is why r is negative in Eqn (4.6). Then, the centrifugal force, mrV, acts at the center of gravity with the same direction as Y_0. When the vehicle exhibits an OS characteristic, the steady-state cornering is in the opposite direction to the case of US, and the vehicle heads inward of the circular path.

Figure 4.6(a) and (b) show how the vehicle side-slip angle, β, and yaw rate, r, change with traveling speed, during the steady-state cornering caused by lateral force Y_0. The figures show that when the vehicle has an OS characteristic, it becomes more sensitive to disturbances as the vehicle speed increases.

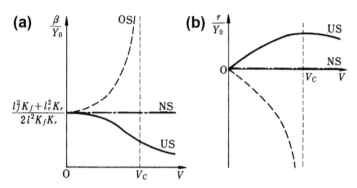

FIGURE 4.6

Steady-state cornering by lateral force Y_0, (a) side slip angle and (b) yaw rate.

4.2.1.2 Transient condition

The previous section examined the vehicle motion in steady-state condition when subjected to a lateral disturbance, Y. Next is to study the transient response of the vehicle before reaching the steady state.

The transient response of the vehicle motion, with a lateral disturbance Y, can be obtained from Laplace transforms of Eqns (4.3) and (4.4). The equations will be studied for different levels of damping.

When $\zeta > 1$, the following is obtained:

$$
\beta(t) = \frac{Y_0}{mV} \left[\frac{a_\beta}{\omega_n^2} + \frac{a_\beta - \left(\zeta + \sqrt{\zeta^2 - 1}\right)\omega_n}{2\left(\zeta + \sqrt{\zeta^2 - 1}\right)\sqrt{\zeta^2 - 1}\omega_n^2} e^{\left(-\zeta - \sqrt{\zeta^2-1}\right)\omega_n t} \right.
$$

$$
\left. - \frac{a_\beta - \left(\zeta - \sqrt{\zeta^2 - 1}\right)\omega_n}{2\left(\zeta - \sqrt{\zeta^2 - 1}\right)\sqrt{\zeta^2 - 1}\omega_n^2} e^{\left(-\zeta + \sqrt{\zeta^2-1}\right)\omega_n t} \right]
\tag{4.7}
$$

$$
r(t) = \frac{Y_0}{mIV} \left[\frac{a_r}{\omega_n^2} + \frac{a_r}{2\left(\zeta + \sqrt{\zeta^2 - 1}\right)\sqrt{\zeta^2 - 1}\omega_n^2} e^{\left(-\zeta - \sqrt{\zeta^2-1}\right)\omega_n t} \right.
$$

$$
\left. - \frac{a_r}{2\left(\zeta - \sqrt{\zeta^2 - 1}\right)\sqrt{\zeta^2 - 1}\omega_n^2} e^{\left(-\zeta + \sqrt{\zeta^2-1}\right)\omega_n t} \right]
\tag{4.8}
$$

When $\zeta = 1$, we have the following:

$$
\beta(t) = \frac{Y_0}{mV} \left[\frac{a_\beta}{\omega_n^2} + \left(\frac{\omega_n - a_\beta}{\omega_n} t - \frac{a_\beta}{\omega_n^2}\right) e^{-\omega_n t} \right]
\tag{4.7}'
$$

$$
r(t) = \frac{Y_0}{mIV} \left[\frac{a_r}{\omega_n^2} + \left(\frac{-a_r}{\omega_n} t - \frac{a_r}{\omega_n^2}\right) e^{-\omega_n t} \right]
\tag{4.8}'
$$

and, when the transient response is oscillatory at $\zeta < 1$, the following results:

$$\beta(t) = \frac{Y_0}{mV}\left[\frac{a_\beta}{\omega_n^2} + \frac{1}{\omega_n^2\sqrt{1-\zeta^2}}\sqrt{(a_\beta - \zeta\omega_n)^2 + (1-\zeta^2)\omega_n^2}\, e^{-\zeta\omega_n t}\sin\left(\sqrt{1-\zeta^2}\,\omega_n t + \Psi_\beta\right)\right]$$

$$(4.7)''$$

where the following are defined:

$$\Psi_\beta = \tan^{-1}\left(\frac{\sqrt{1-\zeta^2}\,\omega_n}{a_\beta - \zeta\omega_n}\right) - \tan^{-1}\left(\frac{\sqrt{1-\zeta^2}}{-\zeta}\right)$$

and

$$r(t) = \frac{Y_0}{mIV}\left[\frac{a_r}{\omega_n^2} + \frac{a_r}{\omega_n^2}e^{-\zeta\omega_n t}\sin\left(\sqrt{1-\zeta^2}\,\omega_n t + \Psi_r\right)\right] \qquad (4.8)''$$

$$\Psi_r = -\tan^{-1}\left(\frac{\sqrt{1-\zeta^2}}{-\zeta}\right)$$

It is rather difficult to gain a comprehensive understanding of the vehicle transient motion from the time histories of the vehicle response; so, the transient motion is considered in a more realistic manner. First, the case of a US vehicle is considered. If a step lateral force is applied at the center of gravity of a vehicle traveling in straight line, the vehicle will, initially, produce a side-slip angle, β, at the center of gravity. At this instant, the vehicle has no other motions and the front and rear side-slip angles are both equal to β. This condition is shown in Figure 4.7(a).

As the vehicle is US, the resultant force produces an anticlockwise yaw moment around the vehicle center of gravity, and the vehicle begins its yaw motion. This yaw motion acts to reduce the side-slip angle produced initially at the center of gravity. It produces new side-slip angles at the front and rear wheels, which lead to lateral forces acting in opposite directions to each other. The moment produced by these forces is in the opposite direction to the moment caused by β. The centrifugal force acts at the center of gravity in a direction opposite to Y_0. This condition is shown in Figure 4.7(b). The anticlockwise yaw motion reduces the center of gravity side-slip angle. As the centrifugal force acts in the opposite direction to the lateral force, the resultant force of the tire forces required to maintain an equilibrium condition becomes smaller compared to (a). Moreover, the acting point moves from the original NSP toward point P, and the acting moment becomes smaller compared to (a).

In this way, the transient motion of the vehicle with US characteristic when acted on by the lateral disturbance, Y, reduces the effect of the lateral force. The effect of the lateral force, Y, becomes smaller due to the vehicle motion induced by the lateral force Y. Finally, the vehicle motion reaches a steady-state condition, as shown in Figure 4.7(c).

If the vehicle exhibits an NS characteristic, the resultant force of the tire forces acts at the same point as the lateral force, P. As a result, there is no yaw moment acting on the vehicle. When $l_f K_f = l_r K_r$ in Eqns (4.1) and (4.2), r and β become uncoupled with each other. As long as the lateral force acts at the center of gravity, r is independent from any effects. In other words, for the case of NS, the vehicle response to the lateral force Y is only represented by side-slip motion.

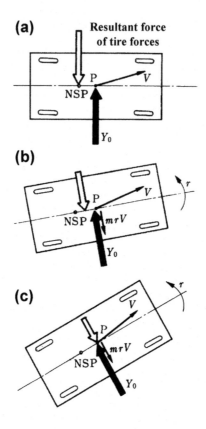

FIGURE 4.7

Transient motion of a US vehicle to lateral force Y_0, (a) initial state, (b) transient state and (c) steady state.

Consider the case of an OS vehicle, as with the two previous cases; when a lateral force is added, the vehicle will produce a side-slip angle, β, at the center of gravity. At this instance, the front and rear side-slip angles are both equal to β, and the resultant force acts at the NSP. This condition is shown in Figure 4.8(a).

As the vehicle has an OS characteristic, the resultant force produces a clockwise yaw moment around the vehicle center of gravity. This yaw motion increases the side-slip angle produced initially at the center of gravity because a centrifugal force caused by the yaw motion acts at the center of gravity in the same direction as the lateral force Y_0. This condition is shown in Figure 4.8(b). As the vehicle clockwise yaw motion increases, the center of gravity side-slip angle becomes larger, and because the centrifugal force acts in the same direction as the lateral force, the resultant force of the tire forces becomes larger than in (a). Although the position of the resultant force moves from the original NSP toward the point P, the moment acting on the vehicle may not get smaller, as in the case of US, because resultant force magnitude increases.

When the vehicle exhibits an OS characteristic, the vehicle may not always reduce the effect of Y, as in the case of US. On the contrary, the vehicle motion might possibly excite the

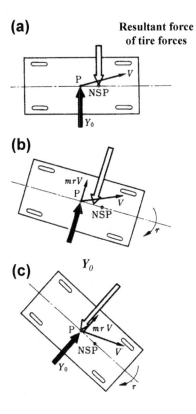

FIGURE 4.8

Transient motion of an OS vehicle to lateral force Y_0, (a) initial state, (b) transient state and (c) steady state.

effect of Y on the vehicle. In particular, the centrifugal force increases dramatically with increasing vehicle speed. Beyond a certain speed, the centrifugal force due to the vehicle yaw motion becomes too large, and the yaw moment acting on the vehicle becomes larger. Ultimately, the vehicle could fall into a spin motion. This velocity limit is the same as the stability limit velocity V_c derived in Section 3.4.1 and is the velocity that causes the denominator to be zero in Eqns (4.5) and (4.6).

If the vehicle traveling speed is less than V_c, the resultant force acts at P, and the resultant force is in equilibrium with the centrifugal force and Y_0. The vehicle reaches its steady-state condition as shown in Figure 4.8(c).

When the vehicle exhibits OS characteristics, even if there's no unstable condition, i.e., $V < V_c$, the vehicle motion caused by the lateral force acting at the center of gravity still excites the vehicle motion. When $V > V_c$, the vehicle dynamics falls into a mathematically unstable condition, and the steady condition cannot be reached.

Figure 4.9 schematically shows the vehicle motion due to the lateral force Y in correspondence to different steer characteristics.

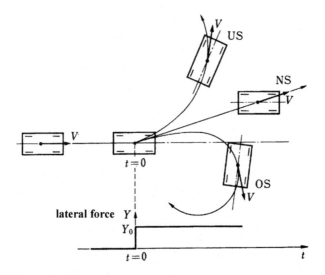

FIGURE 4.9

Vehicle motion to step form lateral force on the C.G.

Example 4.1

In order to understand the vehicle response time history to the step disturbance force exerted on the vehicle C.G. more definitely and to confirm the schematic explanation of the effects of the steer characteristics on the vehicle response in Figure 4.9, execute the vehicle response simulation to the disturbance of 4 kN exerted on the C.G. with the Matlab-Simulink software. Regard the vehicle speed as 80 km/h and the steering angle as always fixed to zero. Calculate the vehicle track after the disturbance as well as the vehicle yaw rate and side-slip responses for the US, NS, and OS vehicles with the same vehicle parameters in Example 3.7, respectively.

Solution

The simulation is based on the same equations as the ones used in Example 3.6, just putting $\delta = 0$, which are as follows:

$$\frac{d\beta}{dt} = -\frac{2(K_f + K_r)}{mV}\beta - \left\{1 + \frac{2}{mV^2}(l_f K_f - l_r K_r)\right\}r + \frac{Y_0}{mV} \tag{E4.1}$$

$$\frac{dr}{dt} = -\frac{2(l_f K_f - l_r K_r)}{I}\beta - \frac{2\left(l_f^2 K_f + l_r^2 K_r\right)}{IV}r \tag{E4.2}$$

In addition to the previous, the equations introduced in Example 3.2 are needed for the calculation of the vehicle track:

$$\frac{dX}{dt} = V\cos(\beta + \theta) \tag{E3.1}$$

$$\frac{dY}{dt} = V\sin(\beta + \theta) \tag{E3.2}$$

$$\frac{d\theta}{dt} = r \tag{E4.3}$$

From these equations, it is possible to have the integral-type of block diagram of the vehicle motion to the disturbance at the C.G., as shown in Figure E4.1(a).

The vehicle parameters used in the simulation are shown in Figure E4.1(b). Figure E4.1(c) shows the simulation program, and Figure E4.1(d) shows the results of the simulation. The summarized effects of the steer characteristics on the vehicle behaviors are shown in Figure E4.1(e).

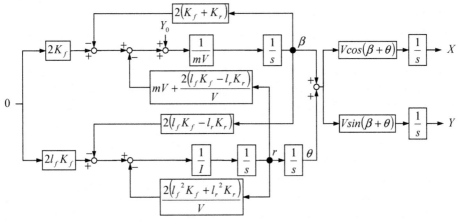

FIGURE E4.1(a)

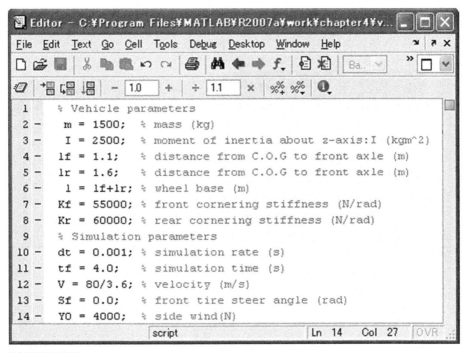

FIGURE E4.1(b)

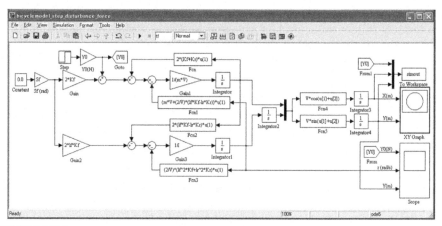

FIGURE E4.1(c)

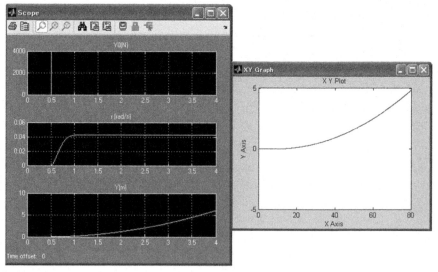

FIGURE E4.1(d)

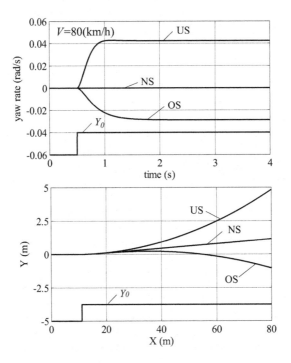

FIGURE E4.1(e)

4.2.2 **VEHICLE MOTION TO IMPULSE LATERAL FORCE**

The vehicle motion due to an idealized form of lateral force at the center of gravity was studied in the previous section. For the actual vehicle, a lateral force acting at the center of gravity for a substantially long time is rare. Instead, a condition where the vehicle travels in a straight line and part of the road banks is more practical.

In this situation, vehicle motion is acted on by a lateral force, as shown in Figure 4.10. The impulse acting on the vehicle in the lateral direction, $Y_0 \Delta t$, is not so large, and the vehicle will not deviate much from its original path. In these circumstances, the vehicle motion is more conveniently expressed by coordinates fixed on the ground.

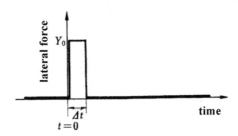

FIGURE 4.10

Pulse form lateral force on the C.G.

If the steering angle is zero, the vehicle equations of motion when acted on by the lateral force Y could be written as follows, using Eqns (3.21) and (3.22), when $|\theta| \ll 1$:

$$m\frac{d^2 y}{dt^2} + \frac{2(K_f + K_r)}{V}\frac{dy}{dt} + \frac{2(l_f K_f - l_r K_r)}{V}\frac{d\theta}{dt} - 2(K_f + K_r)\theta = Y \qquad (4.9)$$

$$\frac{2(l_f K_f - l_r K_r)}{V}\frac{dy}{dt} + I\frac{d^2\theta}{dt^2} + \frac{2\left(l_f^2 K_f + l_r^2 K_r\right)}{V}\frac{d\theta}{dt} - 2(l_f K_f - l_r K_r)\theta = 0 \qquad (4.10)$$

When Δt is very small compared to the period of the vehicle motion, the approximate Laplace transform of the lateral force Y, $Y(s)$, could be taken as $Y_0\Delta t$. Consequently, the vehicle response to Y could be obtained by carrying out Laplace transforms on the previous equations to find $y(s)$ and $\theta(s)$:

$$y(s) = \frac{\begin{vmatrix} Y_0\Delta t & \dfrac{2(l_f K_f - l_r K_r)}{V}s - 2(K_f + K_r) \\[3mm] 0 & Is^2 + \dfrac{2\left(l_f^2 K_f + l_r^2 K_r\right)}{V}s - 2(l_f K_f - l_r K_r) \end{vmatrix}}{\begin{vmatrix} ms^2 + \dfrac{2(K_f + K_r)}{V}s & \dfrac{2(l_f K_f - l_r K_r)}{V}s - 2(K_f + K_r) \\[3mm] \dfrac{2(l_f K_f - l_r K_r)}{V}s & Is^2 + \dfrac{2\left(l_f^2 K_f + l_r^2 K_r\right)}{V}s - 2(l_f K_f - l_r K_r) \end{vmatrix}} \qquad (4.11)$$

$$= \frac{Y_0\Delta t}{m}\frac{s^2 + a_{y1}s + a_{y2}}{s^2\left(s^2 + 2\zeta\omega_n s + \omega_n^2\right)}$$

$$\theta(s) = \frac{\begin{vmatrix} ms^2 + \dfrac{2(K_f + K_r)}{V}s & Y_0\Delta t \\[3mm] \dfrac{2(l_f K_f - l_r K_r)}{V}s & 0 \end{vmatrix}}{\begin{vmatrix} ms^2 + \dfrac{2(K_f + K_r)}{V}s & \dfrac{2(l_f K_f - l_r K_r)}{V}s - 2(K_f + K_r) \\[3mm] \dfrac{2(l_f K_f - l_r K_r)}{V}s & Is^2 + \dfrac{2\left(l_f^2 K_f + l_r^2 K_r\right)}{V}s - 2(l_f K_f - l_r K_r) \end{vmatrix}} \qquad (4.12)$$

$$= \frac{Y_0\Delta t}{mIV}\frac{a_r}{s\left(s^2 + 2\zeta\omega_n s + \omega_n^2\right)}$$

where

$$a_{y1} = \frac{2\left(l_f^2 K_f + l_r^2 K_r\right)}{IV}, \qquad a_{y2} = \frac{-2(l_f K_f - l_r K_r)}{I}$$

From these equations, the steady state reached by the vehicle is as follows:

$$y = \lim_{s \to 0} sy(s) = \pm\infty \quad \text{(when } l_f K_f \neq l_r K_r)$$

$$= \frac{l_f^2 K_f + l_r^2 K_r}{2l^2 K_f K_r} VY_0 \Delta t \quad \text{(when } l_f K_f = l_r K_r) \tag{4.13}$$

$$\theta = \lim_{s \to 0} s\theta(s) = \frac{-(l_f K_f - l_r K_r)VY_0 \Delta t}{2l^2 K_f K_r \left[1 - \frac{m(l_f K_f - l_r K_r)}{2l^2 K_f K_r} V^2\right]} \tag{4.14}$$

When $l_f K_f < l_r K_r$ and the vehicle has a US characteristic, $y = +\infty$; If $l_f K_f > l_r K_r$ and the vehicle has an OS characteristic, $y = -\infty$. Also, θ is positive for a US vehicle, negative for OS, and zero for NS.

The vehicle has the same transient response, for $Y = Y_0$ during $0 \leq t \leq \Delta t$, as the response to a step lateral force. At $t = \Delta t$, the lateral force Y becomes zero, and the vehicle center of gravity side-slip angle becomes small immediately. The lateral force acting at the tires due to the yaw motion becomes more predominant. These forces reduce the yaw motion, and the vehicle maintains a constant yaw angle relative to the original traveling direction. This is positive for a US vehicle and negative for an OS vehicle. Consequently, the vehicle lateral displacement, y, relative to the coordinates fixed on the ground, increases in the positive direction for US and negative direction for OS. When the vehicle exhibits an NS characteristic, the yaw angle is not produced, and the vehicle continues to travel straight, maintaining the original traveling direction after a constant lateral displacement.

Figure 4.11 shows the vehicle response to the lateral force, Y, for different steer characteristics.

4.2.3 VEHICLE MOTION DUE TO EXTERNAL DISTURBANCES AND STEER CHARACTERISTICS

As explained previously, the vehicle motion due to the lateral forces acting at the center of gravity is greatly affected by the vehicle steer characteristics. Based on the studies until now,

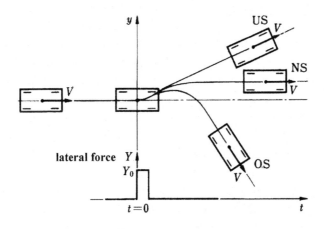

FIGURE 4.11

Vehicle motion to pulse form lateral force on the C.G.

concerning the vehicle motions due to disturbances acting at the center of gravity in the lateral direction, the concept of steer characteristics could be rearranged as follows.

When a traveling vehicle suddenly has a yaw velocity of r, then a centrifugal force of mrV acts at the vehicle center of gravity. As seen in the previous sections, lateral forces will act at the front and rear tires due to the side-slip angle of the center of gravity caused by the external force. The acting point of the resultant force is at the NSP. If the NSP is at the rear of the center of gravity, the resultant force will control the yawing motion caused by the yaw velocity, r. The US characteristic controls motions produced by the disturbances.

If the NSP is in front of the center of gravity, the resultant force will excite the yaw motion. Here, the OS characteristic has a mechanism of exciting the motions produced by the disturbance through its own motion. From the motion mechanics point of view, this is why a vehicle with an OS characteristic is termed to be inferior in terms of directional stability. A NS characteristic means that the forces acting at the tires due to the side-slip angle of the center of gravity have no effect on the yawing motion.

These findings are portrayed in Figure 4.12.

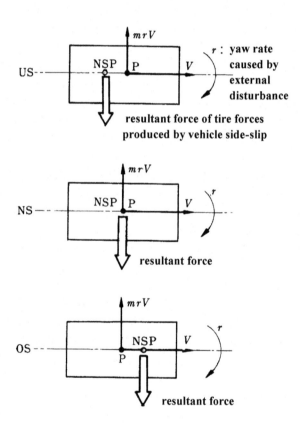

FIGURE 4.12

Vehicle motion to external disturbance and steer characteristics.

The vehicle steer characteristic not only describes the vehicle steady-state cornering characteristic, but it also gives a strong indication of the vehicle's motion characteristics under external disturbances. It is understood here that the cross-coupling of vehicle side-slip and yaw motions plays a principle role in determining the vehicle motion due to the disturbances.

From Section 4.2.1, if a constant lateral force, Y_0, acts at the center of gravity of a vehicle traveling with constant velocity, V, a yaw velocity of r given by Eqn (4.6) will be produced. The center of gravity acceleration, $\ddot{y}$, is as follows:

$$\ddot{y} = Vr = \frac{-\frac{m(l_f K_f - l_r K_r)}{2l^2 K_f K_r}V^2}{1 - \frac{m(l_f K_f - l_r K_r)}{2l^2 K_f K_r}V^2}\frac{Y_0}{m} \tag{4.15}$$

The lateral force of magnitude, Y_0, could be expressed as a unit of acceleration as follows:

$$\ddot{y}_0 = \frac{Y_0}{m} \tag{4.16}$$

The ratio of $\ddot{y}$ and $\ddot{y}_0$ is called the understeer rate [1], U_R. The vehicle steer characteristic could also be based on this value.

$$U_R = \frac{\ddot{y}}{\ddot{y}_0} \tag{4.17}$$

A vehicle with U_R of 100% has a center of gravity acceleration, Y_0/m, and the circular motion radius is mV^2/Y_0. The vehicle with U_R of 0% is NS, and if U_R is negative, the vehicle is OS.

The vehicle steer characteristic is not only a characteristic of the vehicle steady-state cornering, but it is also a characteristic to the disturbance at its center of gravity when the steering angle is zero. Substituting Eqns (4.15) and (4.16) into Eqn (4.17), the understeer rate, U_R, could be written using the stability factor A from Eqn (3.43):

$$U_R = \frac{AV^2}{1 + AV^2} \tag{4.18}$$

When $A > 0$, U_R is positive and less than 1 for all velocities. In other words, when the vehicle has US characteristics and the external force of magnitude, Y_0, is applied to the center of gravity, the lateral acceleration is always less than Y_0/m. On the contrary, when $A < 0$, U_R becomes negative, and at the velocity which satisfy $AV^2 < -0.5$, the absolute value of U_R is greater than 1. When the vehicle has OS characteristics, the lateral acceleration is less than Y_0/m when the vehicle speed is lower than a certain velocity. It becomes greater than Y_0/m at higher speeds.

The vehicle response to a sudden yaw moment disturbance, as shown in Figure 4.13, helps to better understand that the OS vehicle is more sensitive to the external disturbance and less stable in terms of directional stability. The vehicle motion caused by the yaw moment is described by the following equations:

$$m\frac{d^2y}{dt^2} + \frac{2(K_f + K_r)}{V}\frac{dy}{dt} + \frac{2(l_f K_f - l_r K_r)}{V}\frac{d\theta}{dt} - 2(K_f + K_r)\theta = 0 \tag{4.9}'$$

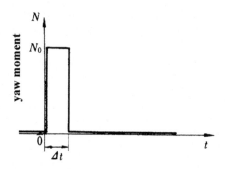

FIGURE 4.13

Disturbance yaw moment.

$$\frac{2(l_f K_f - l_r K_r)}{V}\frac{dy}{dt} + I\frac{d^2\theta}{dt^2} + \frac{2\left(l_f^2 K_f + l_r^2 K_r\right)}{V}\frac{d\theta}{dt} - 2(l_f K_f - l_r K_r)\theta = N \qquad (4.10)'$$

Applying the Laplace transform to the previous equations, the steady-state yaw angle to the yaw moment input is obtained as follows:

$$\theta = \lim_{s \to 0} s\theta(s) = \frac{(K_f + K_r)VN_0\Delta t}{2l^2 K_f K_r\left[1 - \frac{m(l_f K_f - l_r K_r)}{2l^2 K_f K_r}V^2\right]}$$

Here, Δt is small enough that the yaw moment, N, is a pulse form with the magnitude N_0 of which the Laplace transformation is $N_0\Delta t$.

This equation shows the change of the vehicle attitude angle caused by the disturbance yaw moment with respect to the vehicle speed, as in Figure 4.14. It is obvious that the stronger the vehicle OS aspect is, the larger the change of the attitude angle and the more sensitive it is to the disturbance.

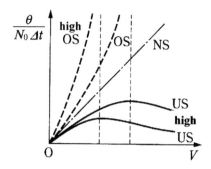

FIGURE 4.14

Change of vehicle attitude angle to disturbance yaw moment.

4.3 VEHICLE MOTION DUE TO A LATERAL WIND DISTURBANCE

Vehicles traveling at high speeds may encounter lateral wind disturbances, which result in lateral motion. This section will look at the basic motion characteristics of the vehicle traveling in a straight line when it is acted on by a lateral wind disturbance.

4.3.1 LATERAL WIND DISTURBANCE FORCE

If the vehicle traveling in a straight line at velocity, V, as shown in Figure 4.15, is subjected to a lateral wind of velocity w, the lateral force, Y_w, and yaw moment, N_w, acting on the vehicle are expressed as follows:

$$Y_w = C_y \frac{\rho}{2} S(V^2 + w^2) \tag{4.19}$$

$$N_w = C_n \frac{\rho}{2} lS(V^2 + w^2) \tag{4.20}$$

Here, C_y is the lateral force coefficient, C_n is the yawing moment coefficient, and both are functions of the relative side-slip angle due to the airflow, β_w. C_n is positive in the anticlockwise direction, ρ is the air density, S is the vehicle frontal area, and l is the vehicle dimension normally taken as the wheelbase.

Figure 4.16 shows the lateral force coefficient and yawing moment coefficient related to the relative side-slip angle of the airflow for a normal passenger car [2]. This figure shows C_y and C_n increase with β_w and are significantly affected by the vehicle shape.

The acting point of the lateral force, Y_w, is called the aerodynamic center (AC). The distance between the AC and the vehicle center of gravity is l_w, and it is positive if the AC is behind the center of gravity. The yaw moment acting on the vehicle N_w can be written as follows:

$$N_w = -l_w Y_w$$

The vehicle motion due to the lateral wind is caused by a lateral force, Y_w, acting at the AC, a distance l_w from the center of gravity, as shown in Figure 4.17.

Strictly speaking, C_y and C_n change with β_w, and β_w changes with vehicle motion. For example, if the vehicle is subjected to a constant lateral wind, Y_w and N_w are not constant but

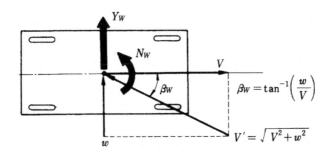

FIGURE 4.15

Disturbance force and yaw moment caused by side wind.

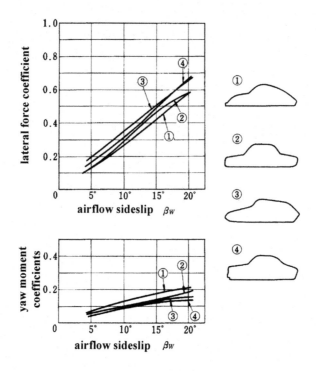

FIGURE 4.16

Lateral force and yaw moment coefficients of side wind.

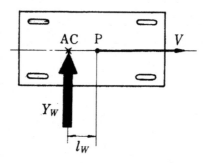

FIGURE 4.17

Lateral force by side wind.

change with the vehicle motion. However, if the transient vehicle motion is not so large, Y_w and N_w can be viewed independently from the vehicle motion, which simplifies the analysis.

4.3.2 VEHICLE MOTION TO LATERAL WIND WITH CONSTANT SPEED

The vehicle motion when subjected to a constant speed lateral wind is shown in Figure 4.18. At this time, the lateral force, Y_w, is assumed to be a step force that acts at the vehicle AC.

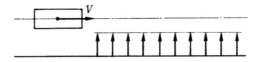

FIGURE 4.18

Side wind with constant speed.

It is convenient to express the vehicle motion with coordinates fixed on the vehicle, as in Section 4.2.1. By using Eqns (3.12) and (3.13), the equations of motion are derived as follows:

$$mV\frac{d\beta}{dt} + 2(K_f + K_r)\beta + \left\{mV + \frac{2}{V}(l_f K_f - l_r K_r)\right\}r = Y_w \tag{4.21}$$

$$2(l_f K_f - l_r K_r)\beta + I\frac{dr}{dt} + \frac{2\left(l_f^2 K_f + l_r^2 K_r\right)}{V}r = -l_w Y_w \tag{4.22}$$

The vehicle response to Y_w can be obtained by applying Laplace transforms to Eqns (4.21) and (4.22) as in Section 4.2.1. $\beta(s)$ and $r(s)$ are obtained as follows:

$$\beta(s) = \frac{Y_{w0}}{mV} \frac{s + b_\beta}{s\left(s^2 + 2\zeta\omega_n s + \omega_n^2\right)} \tag{4.23}$$

$$r(s) = \frac{-l_w Y_{w0}}{I} \frac{s + b_r}{s\left(s^2 + 2\zeta\omega_n s + \omega_n^2\right)} \tag{4.24}$$

where

$$b_\beta = \frac{2\left(l_f^2 K_f + l_r^2 K_r\right) + 2l_w(l_f K_f - l_r K_r)}{IV} + \frac{ml_w V}{I}$$

$$= \frac{2}{IV}\left[l_f^2 K_f + l_r^2 K_r - l_w l_N(K_f + K_r)\right] + \frac{ml_w V}{I}$$

$$b_r = \frac{2(K_f + K_r)}{mV} + \frac{2(l_f K_f - l_r K_r)}{ml_w V}$$

$$= \frac{2(l_w - l_N)}{ml_w V}(K_f + K_r)$$

Y_{w0} is the magnitude of the step lateral force. From these equations, the steady-state value of β and r are as follows:

$$\beta = \frac{\left[\left(l_f^2 K_f + l_r^2 K_r\right) - l_w l_N(K_f + K_r)\right] + \frac{ml_w}{2}V^2}{2l^2 K_f K_r\left[1 - \frac{m(l_f K_f - l_r K_r)}{2l^2 K_f K_r}V^2\right]}Y_{w0} \tag{4.25}$$

$$r = \frac{(l_N - l_w)(K_f + K_r)V}{2l^2 K_f K_r\left[1 - \frac{m(l_f K_f - l_r K_r)}{2l^2 K_f K_r}V^2\right]}Y_{w0} \tag{4.26}$$

where l_N is the distance between NSP and the center of gravity.

$$l_N = -\frac{(l_f K_f - l_r K_r)}{K_f + K_r} \tag{3.44}$$

From Eqn (4.26), when $l_N > l_w$, then $r > 0$; if $l_N = l_w$, $r = 0$; and when $l_N < l_w$, $r < 0$. l_N and l_w are taken as positive if the NSP and AC are behind the center of gravity. When $l_f K_f - l_r K_r > 0$, it is assumed that $V < V_c$.

The vehicle subjected continuously to lateral wind with constant speed, traveling in a straight line, will ultimately enter into anticlockwise circular motion if the AC is in front of the NSP. It produces no circular motion except at the transient period when the AC coincides with the NSP, and it produces clockwise circular motion when the AC is behind the NSP. The relative position of the vehicle center of gravity to the AC and NSP has no direct effect. Figure 4.19 shows the results of a computer simulation of the motion of a passenger car when it is acted on by the lateral wind disturbance.

The steady-state value of the lateral acceleration is Vr, and from Eqn (4.26), the steady-state value of the lateral acceleration per unit lateral wind force could be expressed as the following:

$$S_w = \frac{Vr}{Y_{w0}} = \frac{(l_N - l_w)(K_f + K_r)V^2}{2l^2 K_f K_r \left[1 - \frac{m(l_f K_f - l_r K_r)}{2l^2 K_f K_r}V^2\right]} \tag{4.27}$$

This is termed the sensitivity coefficient of lateral wind and is suggested to be the index that shows the vehicle sensitivity to a lateral wind disturbance.

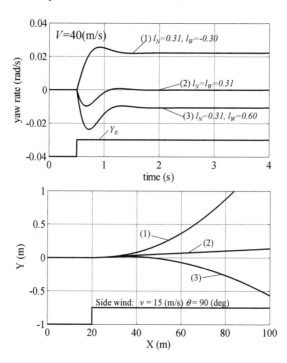

FIGURE 4.19

Vehicle motion subjected to side wind with constant speed.

Example 4.2

Find the front wheel steering angle needed to keep the vehicle running on a straight path having a side wind with constant wind speed. Also, calculate the attitude angle of the vehicle with respect to the straight path during this period.

Solution

The steady-state yaw rate response to the constant side wind force is expressed by Eqn (4.26). On the other hand, the steady-state yaw rate response to the steering input is given by Eqn (3.39). If the sum of the both yaw rates is zero, it is regarded that the vehicle yaw motion caused by the side wind is eliminated by the yaw motion due to the front wheel steering, i.e., the vehicle runs on a straight line having the side wind, which is described by the following:

$$\frac{(l_N - l_w)(K_f + K_r)V}{2l^2 K_f K_r \left[1 - \frac{m(l_f K_f - l_r K_r)}{2l^2 K_f K_r}V^2\right]} Y_{w0} + \frac{1}{\left[1 - \frac{m(l_f K_f - l_r K_r)}{2l^2 K_f K_r}V^2\right]} \frac{V}{l}\delta_0 = 0 \qquad (E4.4)$$

From this equation, the steering angle is obtained as follows:

$$\delta_0 = \frac{l_w - l_N}{2l}\left(\frac{1}{K_f} + \frac{1}{K_r}\right)Y_{w0} \qquad (E4.5)$$

It is understood that if l_N is greater than l_w, the steering angle in the counter direction to the wind is needed; on the other hand, greater l_w than l_N requires the steering angle in the following direction to the wind in order for the vehicle to keep the straight path under the side wind.

The attitude angle of the vehicle to the straight path is identical to the side-slip angle, β, of the vehicle. The side wind force is balanced by the total lateral force of the front and rear tires:

$$-2K_f(\beta - \delta_0) - 2K_r\beta + Y_{w0} = 0 \qquad (E4.6)$$

Putting the previous Eqn (E4.5) into this equation, the attitude angle is obtained as follows:

$$\beta = \frac{(l_f + l_w)Y_{w0}}{2lK_r} = \frac{(l_f + l_w)Y_{w0}}{2(l_f + l_N)(K_f + K_r)} \qquad (E4.7)$$

In this process, pay attention to the relation $l_N = -(l_f K_f - l_r K_r)/(K_f + K_r)$.

4.3.3 VEHICLE MOTION DUE TO A LATERAL WIND GUST

Consider the situation where the vehicle is subjected to a sudden lateral gust with the front steering angle fixed to zero. It is assumed that lateral force Y_w as shown in Figure 4.20 acts at the vehicle AC.

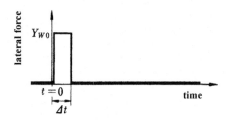

FIGURE 4.20

Lateral force by gust wind.

The vehicle motion is expressed in terms of absolute coordinates fixed on the ground for the same reason as in Section 4.2.2. From Eqns (3.21) and (3.22), the equations of motion are derived as follows:

$$m\frac{d^2y}{dt^2} + \frac{2(K_f + K_r)}{V}\frac{dy}{dt} + \frac{2(l_f K_f - l_r K_r)}{V}\frac{d\theta}{dt} - 2(K_f + K_r)\theta = Y_w \quad (4.28)$$

$$\frac{2(l_f K_f - l_r K_r)}{V}\frac{dy}{dt} + I\frac{d^2\theta}{dt^2} + \frac{2\left(l_f^2 K_f + l_r^2 K_r\right)}{V}\frac{d\theta}{dt} - 2(l_f K_f - l_r K_r)\theta = -l_w Y_w \quad (4.29)$$

If Δt is small enough, then applying Laplace transforms to Eqns (4.28) and (4.29) as in Section 4.2.2, $y(s)$ and $\theta(s)$ are as follows:

$$y(s) = \frac{Y_{w0}\Delta t}{m}\frac{s^2 + b_{y1}s + b_{y2}}{s^2\left(s^2 + 2\zeta\omega_n s + \omega_n^2\right)} \quad (4.30)$$

$$\theta(s) = \frac{-l_w Y_{w0}\Delta t}{I}\frac{s + b_r}{s\left(s^2 + 2\zeta\omega_n s + \omega_n^2\right)} \quad (4.31)$$

where

$$b_{y1} = \frac{2\left(l_f^2 K_f + l_r^2 K_r\right) - 2l_w(l_f K_f - l_r K_r)}{IV}$$

$$= \frac{2\left(l_f^2 K_f + l_r^2 K_r\right) + 2l_N l_w(K_f + K_r)}{IV}$$

$$b_{y2} = -\frac{2(l_f K_f - l_r K_r) + 2l_w(K_f + K_r)}{I}$$

$$= \frac{2(l_N - l_w)(K_f + K_r)}{I}$$

From these equations, the steady-state values are as follows:

$$y = \pm\infty \quad \text{(when } l_N \neq l_w)$$

$$= \frac{l_f^2 K_f + l_r^2 K_r + l_N L_W(K_f + K_r)}{2l^2 K_f K_r\left[1 - \frac{m(l_f K_f - l_r K_r)}{2l^2 K_f K}V^2\right]}VY_{w0}\Delta t \quad \text{(when } l_N = l_w) \quad (4.32)$$

$$\theta = \frac{(l_N - l_w)(K_f + K_r)V}{2l^2 K_f K_r\left[1 - \frac{m(l_f K_f - l_r K_r)}{2l^2 K_f K_r}V^2\right]}Y_{w0}\Delta t \quad (4.33)$$

The equations for vehicle steady-state motion are arranged as follows, assuming $V < V_c$ when $l_f K_f - l_r K_r > 0$:

when $l_N - l_w > 0$, then $y = +\infty$, $\theta = $ a positive constant value
when $l_N - l_w = 0$, then $y = $ a positive constant value, $\theta = 0$
when $l_N - l_w < 0$, then $y = -\infty$, and $\theta = $ a negative constant value

The vehicle traveling in a straight line and subjected to a lateral gust will change its direction toward the leeward and travel forward following the wind gust if the AC is in front of the NSP. It will drift to the leeward temporarily and then regain its traveling direction if the AC and NSP coincide. If the AC is in front of the NSP, the vehicle will drift to the leeward temporarily and then change its direction to the windward and travel forward against the wind gust direction.

Example 4.3

Execute a Matlab-Simulink simulation of the vehicle motion subjected to a wind gust of 25 m/s for 0.4 s. The vehicle is running at $V = 40$ m/s and has the same parameters as in Example 3.6. Set the positions of NSP and AC the same as in Figure 4.19.

Solution

The wind force can be calculated by Eqn (4.19), where, C_y is estimated using Figure 4.16, $S = 1.5$ m^2, and $\rho = 1.29$ kg/m^3. As a result of the calculation, $Y_w = 2.07$ kN.

The equations of motion of the vehicle subjected to the side wind gust are described by Eqns (4.28) and (4.29). The equations are rewritten in the following forms:

$$\frac{dv}{dt} = -\frac{2(K_f + K_r)}{mV}v - \frac{2(l_fK_f - l_rK_r)}{mV}r + \frac{2(K_f + K_r)}{m}\theta + \frac{Y_w}{m}$$

$$\frac{dr}{dt} = -\frac{2(l_fK_f - l_rK_r)}{IV}v - \frac{2\left(l_f^2K_f + l_r^2K_r\right)}{IV}r + \frac{2(l_fK_f - l_rK_r)}{I}\theta + \frac{l_wY_w}{I}$$

$$\frac{dy}{dt} = v \tag{E4.8}$$

$$\frac{d\theta}{dt} = r \tag{E4.9}$$

It is possible to have the integral-type of block diagram of the vehicle motion to the side wind gust as shown in Figure E4.3(a). The vehicle parameters are set as in Figure E4.3(b). Based on the previous block diagram, the simulation program is obtained as in Figure E4.3(c). A simulation result is shown in Figure E4.3(d), and the summarized results are in Figure E4.3(e).

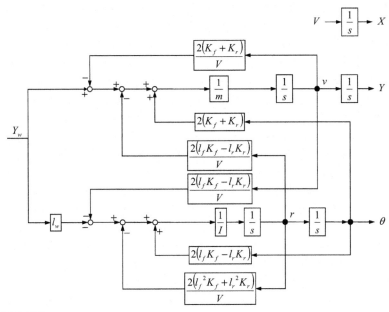

FIGURE E4.3(a)

```
1      % Vehicle parameters
2 -     m = 1500;    % mass (kg)
3 -     I = 2500;    % moment of inertia about z-axis:I (kgm^2)
4 -    lf = 1.1;     % distance from C.O.G to front axle (m)
5 -    lr = 1.6;     % distance from C.O.G to front axle (m)
6 -     l = lf+lr;   % wheel base (m)
7 -    Kf = 55000;   % front cornering stiffness (N/rad)
8 -    Kr = 60000;   % rear cornering stiffness (N/rad)
9      % Simulation parameters
10 -   dt = 0.001;   % simulation rate (s)
11 -   tf = 4.0;     % simulation time (s)
12 -    V = 40;      % velocity (m/s)
13 -   Sf = 0.0;     % front tire steer angle (rad)
14 -    v = 25;      % side wind velocity (m/s)
15 -   cy = 0.96;
16 -    p = 1.29;
17 -    s = 1.5;
18 -   Yw = cy*0.5*p*s*(V^2+v^2);
19 -   lw = -0.30;
```

FIGURE E4.3(b)

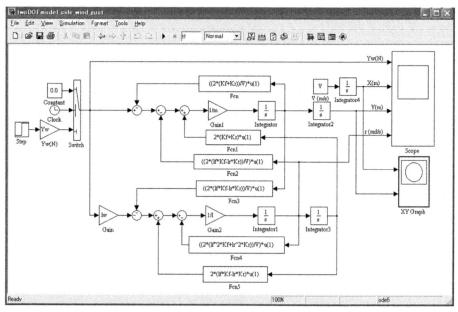

FIGURE E4.3(c)

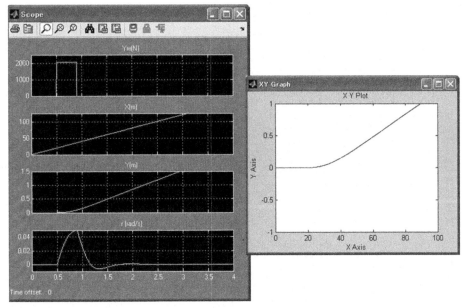

FIGURE E4.3(d)

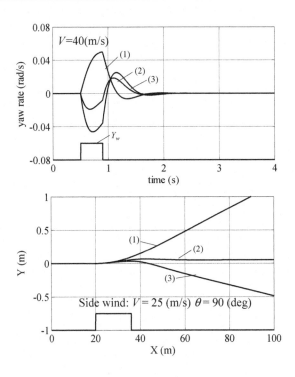

FIGURE E4.3(e)

Table 4.1 Vehicle Motion Subjected to Lateral Disturbance

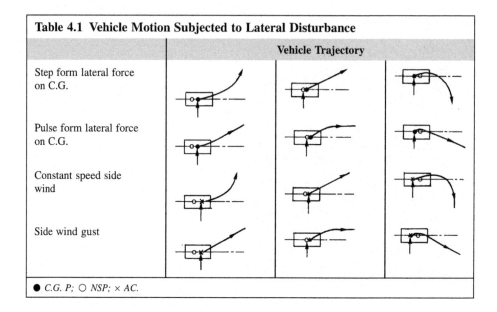

	Vehicle Trajectory		
Step form lateral force on C.G.			
Pulse form lateral force on C.G.			
Constant speed side wind			
Side wind gust			

● *C.G. P;* ○ *NSP;* × *AC.*

4.4 CONCLUSION OF VEHICLE MOTION BY DISTURBANCES

The vehicle motions when subjected to lateral external forces at the center of gravity are greatly affected by the positions of the center of gravity, P, and NSP; whereas, the vehicle motions when subjected to lateral wind are greatly affected by the positions of the AC and NSP.

This is summarized and shown in Table 4.1.

The yaw rate reached by the vehicle, regardless of the type of disturbance, is either proportional to $1/(2l^2 K_f K_r)$ or $(K_f + K_r)/(2l^2 K_f K_r)$. Usually, vehicles with high cornering stiffness values and large wheelbases are less sensitive, i.e., more robust to disturbances.

PROBLEMS

4.1 Calculate the steady-state side-slip angle and yaw rate for the vehicle running with the constant lateral force, $Y_0 = 2.0$ kN, exerted on the C.G., and then draw the $\beta - V$ and $r - V$ diagram, respectively. Use the vehicle parameters in Example 3.3.

4.2 Find the vehicle speed at which the yaw rate response in Problem 4.1 reaches the peak value.

4.3 Find the steering angle needed to keep straight running under a constant side force disturbance at the C.G.

4.4 Find the attitude angle of the vehicle relative to the straight path in Problem 4.3.

4.5 Execute a Matlab-Simulink simulation of the vehicle response subjected to the disturbance force of 2.0 kN for 1.0 s at the C.G. Compare the effects of the vehicle speeds on the responses with each other. Use the same vehicle parameters as in Example 3.6.

4.6 Calculate the lateral force and yaw moment exerted on the vehicle running at 100 km/h with a the side wind of 10 m/s. Refer to Figure 4.16 for the lateral force and yaw moment coefficients of a specific vehicle, and let ρ, l, and S be equal to 1.25 kg/m^3, 2.7 m, and 1.5 m^2, respectively.

4.7 Confirm that in order to avoid the vehicle heading against the wind direction as a result of the side wind, stronger US characteristics are required for the vehicle with the AC shifted to the rearward direction relative to the C.G. position.

REFERENCES

[1] Bergman W. Bergman gives new meaning to under-steer and over-steer. SAE J 1965;73(12).
[2] Ichimura H. Aerodynamic characteristics of passenger cars. JSAE J 1978;32(4) [In Japanese].

STEERING SYSTEM AND VEHICLE DYNAMICS

5

5.1 PREFACE

The previous chapters looked at vehicle motion with the front steering angles fixed to a certain value. In reality, the front steering angle is controlled through the steering wheel, not directly. The mechanism from the front wheel to the steering wheel has a motion freedom, and the front wheel steering angle is viewed as one of the motion degree-of-freedom. This mechanism, from the front wheel to the steering wheel, is called the steering system. In this chapter, the effect of the steering system characteristics on the vehicle's dynamic performance will be studied. In order to do so, the equations of motion are derived for the steering system. These are used to examine the effect of the steering system characteristics on the vehicle motion. Under normal traveling conditions of the vehicle, the steering wheel is controlled by the driver's hand. Thus, the effect of the driver's hand will be studied from a theoretical approach.

5.2 STEERING SYSTEM MODEL AND EQUATIONS OF MOTION

The steering system of normal vehicles is assumed to be constructed as in Figure 5.1. The steering wheel rotation is transferred through the steering wheel shaft and gearbox to the tie rod, and then through the knuckle arm, which allows the front wheel rotation around the kingpin.

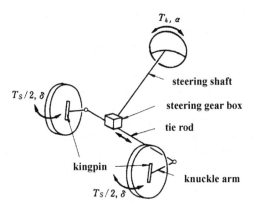

FIGURE 5.1

Vehicle steering system.

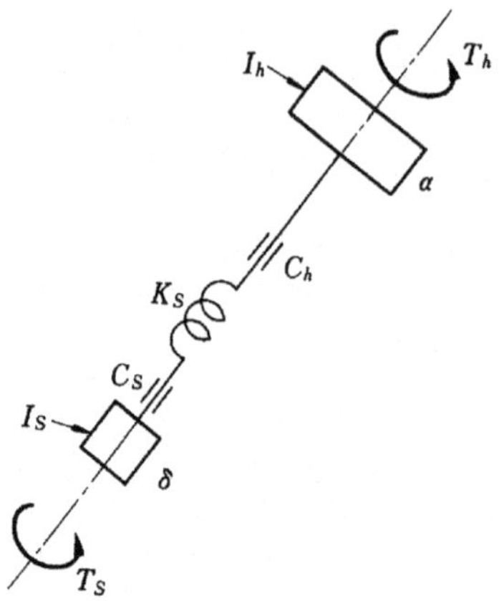

FIGURE 5.2

Steering system model.

If all the steering system motion is converted to rotating motion around the kingpin, an equivalent modeling of the steering system can be assumed, as shown in Figure 5.2. A rotating body equivalent to the steering wheel with moment of inertia I_h is connected to another rotating body equivalent to the front wheels, with moment of inertia I_s, through a rotating shaft equivalent to the steering wheel shaft and gearbox with a spring constant of K_s (I_h, I_s, and K_s are the moment of inertias and spring constant converted around the kingpin). It is also considered that there is a damping friction at the steering wheel shaft and kingpin, and these are defined as the damping coefficients of C_h and C_s. If the rotational angle of the steering wheel, converted around the kingpin, is taken as α and actual front wheel steering angle is δ, the above equivalent steering system becomes a two-degree-of-freedom torsional vibration system. Here, I_s and C_s are the totals of the left and right wheels.

The torque, T_h, provided by the human driver at the steering wheel can be seen as an external moment. On the other hand, when the vehicle is moving with a certain steering wheel angle, there is a torque acting to bring the steering wheel back to its original position. This is due to the moment around the kingpin produced by the lateral force that acts at the front wheels. This must also be taken into account as an external moment to the steering system. Chapter 2 showed that the acting point of the lateral force on the front wheel is shifted slightly behind the contact surface centerline. It is common that the intersecting point of the line extending from the kingpin shaft and the ground is always at the front of the contact surface centerline. This condition is shown by Figure 5.3.

This figure shows that if a side-slip angle in the anticlockwise direction, β_f, is produced at the front wheel, the lateral force will act as a moment around the kingpin in the anticlockwise direction. If the cornering stiffness of the front tire is K_f, the moment acting at the tire around the kingpin, $T_s/2$, is as follows:

$$\frac{T_s}{2} = (\xi_n + \xi_c)K_f\beta_f = \xi K_f\beta_f \tag{5.1}$$

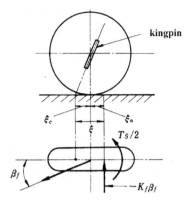

FIGURE 5.3

Self-aligning torque on front wheel.

whereby

$$\xi = \xi_n + \xi_c$$

ξ_n is the pneumatic trail, and ξ_c is called the castor trail. $2\xi K_f$ is called the restoring moment coefficient of the steering system.

If the front wheel side-slip angle is expressed by the vehicle motion variables, β_f becomes $\beta + l_f r/V - \delta$, as in Chapter 3, and the moment T_s can be written as follows:

$$T_s = 2\xi K_f \left(\beta + \frac{l_f r}{V} - \delta \right) \tag{5.2}$$

This moment is an external one acting at the front wheel to the steering system.

Taking all these into account, the steering wheel rotation motion transferred around the kingpin can be expressed as follows:

$$I_h \left(\frac{d^2\alpha}{dt^2} + \frac{dr}{dt} \right) + C_h \frac{d\alpha}{dt} + K_s(\alpha - \delta) = T_h \tag{5.3}$$

and similarly, the front wheel rotational motion can be expressed as follows:

$$I_s \left(\frac{d^2\delta}{dt^2} + \frac{dr}{dt} \right) + C_s \frac{d\delta}{dt} - K_s(\alpha - \delta) = 2\xi K_f \left(\beta + \frac{l_f r}{V} - \delta \right) \tag{5.4}$$

The reason why dr/dt is included in Eqns (5.3) and (5.4) is because the angular acceleration relative to the absolute space is respectively $d^2\alpha/dt^2 + dr/dt$ and $d^2\delta/dt^2 + dr/dt$, which is due to the steering system being attached to the traveling vehicle. In normal vehicle motion, it is assumed that $d^2\alpha/dt^2 \gg dr/dt$ and $d^2\delta/dt^2 \gg dr/dt$. Hence, the above steering system equations of motion could be written as follows:

$$I_h \frac{d^2\alpha}{dt^2} + C_h \frac{d\alpha}{dt} + K_s(\alpha - \delta) = T_h \tag{5.5}$$

$$I_s \frac{d^2\delta}{dt^2} + C_s \frac{d\delta}{dt} + K_s(\delta - \alpha) = 2\xi K_f \left(\beta + \frac{l_f r}{V} - \delta \right) \tag{5.6}$$

5.3 EFFECTS OF STEERING SYSTEM CHARACTERISTICS ON VEHICLE MOTION

5.3.1 EFFECTS OF STEERING SYSTEM CHARACTERISTICS ON VEHICLE MOTION WITH FIXED STEERING ANGLE

The previous section showed the steering system could be treated as a mechanical system with two-degree-of-freedom, α and δ. This section will study the effect of steering system characteristics on the vehicle motion by treating the steering angle, α, as fixed and not a state of motion of the mechanical system. In reality, this is when the steering wheel angle is maintained purposely at a fixed value regardless of the vehicle motion or when the steering wheel angle is set with a fixed pattern regardless of steering wheel inertia, damping force, or restoring force. As mentioned earlier, in the previous chapters, the front wheel steering angle is given freely and fixed. When the front wheel steering angle or the steering wheel angle is fixed, this is called fixed control.

Here, α is no longer treated as a state of motion, Eqn (5.5) becomes meaningless, and the steering system motion could be expressed by Eqn (5.6).

$$I_s \frac{d^2\delta}{dt^2} + C_s \frac{d\delta}{dt} + K_s(\delta - \alpha) = 2\xi K_f \left(\beta + \frac{l_f r}{V} - \delta \right) \tag{5.6}$$

And, α, here, is not a variable but something known as an input. In relation to α, the front wheel steering angle, δ, is known from Eqn (5.6). The vehicle motion in relation to steering angle, δ, had been derived in Chapter 3.

$$mV \frac{d\beta}{dt} + 2(K_f + K_r)\beta + \left\{ mV + \frac{2}{V}(l_f K_f - l_r K_r) \right\} r = 2K_f \delta \tag{3.12}$$

$$2(l_f K_f - l_r K_r)\beta + I \frac{dr}{dt} + \frac{2\left(l_f^2 K_f + l_r^2 K_r \right)}{V} r = 2l_f K_f \delta \tag{3.13}$$

These cross-coupled Eqns (5.6), (3.12), and (3.13) are the equations of the vehicle motion due to a steering angle, α, when taking the steering system into account.

To study the effect of static characteristics of the steering system, either the steering angle is fixed or rapid operation of the steering wheel is omitted, i.e., $d^2\delta/dt^2$ and $d\delta/dt$ are small. I_s and C_s are small too; thus, $I_s(d^2\delta/dt^2)$ and $C_s(d\delta/dt)$ can be neglected. The front wheel steering angle is determined by the following equation:

$$K_s(\delta - \alpha) = 2\xi K_f \left(\beta + \frac{l_f r}{V} - \delta \right) \tag{5.7}$$

δ is derived as follows:

$$\delta = \frac{K_s}{K_s + 2\xi K_f} \alpha + \left(1 - \frac{K_s}{K_s + 2\xi K_f} \right) \left(\beta + \frac{l_f r}{V} \right)$$

By taking the following,

$$e = \frac{K_s}{K_s + 2\xi K_f} = \frac{1}{1 + \frac{2\xi K_f}{K_s}} \tag{5.8}$$

$$\delta = e\alpha + (1 - e)\left(\beta + \frac{l_f}{V}r\right) \tag{5.9}$$

and putting δ into Eqns (3.12) and (3.13) gives the following:

$$mV\frac{d\beta}{dt} + 2(eK_f + K_r)\beta + \left\{mV + \frac{2}{V}(l_f eK_f - l_r K_r)\right\}r = 2eK_f\alpha \tag{5.10}$$

$$2(l_f eK_f - l_r K_r)\beta + I\frac{dr}{dt} + \frac{2(l_f^2 eK_f + l_r^2 K_r)}{V}r = 2l_f eK_f\alpha \tag{5.11}$$

Equations (5.10) and (5.11) express the vehicle motion to steering angle, α, when considering the static characteristic of the steering system. Comparing these equations with Eqns (3.12) and (3.13) shows that the front wheel cornering stiffness, K_f, is replaced by eK_f in Eqns (5.10) and (5.11). In other words, the vehicle response to steering wheel angle, α, has the characteristics of the vehicle response to front wheel steering angle with K_f replaced by eK_f. From Eqn (5.8), the value of e is smaller than one, and the equivalent front wheel cornering stiffness is decreased. Larger restoring moment coefficients and smaller steering system stiffness values increase this effect. If the front wheel cornering stiffness becomes smaller, the vehicle steer characteristics change toward US, and the vehicle exhibits a larger tendency to US. It is true to say that a larger restoring moment coefficient and smaller steering system stiffness gives stronger US characteristics and better directional stability of the vehicle.

It is impractical to assume that the front wheel steering angle is possible to be fixed to a prior given value. Even if the vehicle shows some OS tendency from the theoretical study as in Chapter 3, the vehicle might actually have a US tendency when the steering system restoring moment coefficient and stiffness are considered. In other words, the vehicle's actual US and OS steer characteristics not only depend on the tire characteristics and longitudinal positions of front and rear wheels, but they also greatly depend on the steering system stiffness. The front wheel steering angle, $\alpha - \delta$, produced by the steering system restoring moment, T_s, is called the steering system compliance steer.

Example 5.1
Calculate the effect of the compliance steer on the equivalent reduction of the cornering stiffness when the cornering stiffness of the front tire itself, K_f, pneumatic trail + caster trail, ξ, and rigidity of the steering system, K_s, are equal to 80 kN/rad, 0.035 m, and 10.0 kNm/rad, respectively. Also confirm the effect of the equivalent reduction of the cornering stiffness on the steer characteristics of the vehicle with $m = 1500$ kg, $l_f = 1.1$ m, $l_r = 1.6$ m, and $K_r = 60.0$ kN/rad.

Solution
Using Eqn (5.8), the following is obtained:

$$e = \frac{1}{1 + \frac{2\xi K_f}{K_s}} = \frac{1}{1 + \frac{2.0 \times 0.035 \times 80000}{10000}} = 0.64$$

So, the equivalent cornering stiffness of the front wheel is $0.64 \times 80.0 = 51.2$ kN/rad.

Using Eqn (3.43), the stability factor of the vehicle with the compliance steer taken into consideration is as follows:

$$A = -\frac{m}{2l^2}\frac{l_f K_f - l_r K_r}{K_f K_r} = -\frac{1500}{2 \times 2.7^2}\frac{1.1 \times 51000 - 1.6 \times 60000}{51000 \times 60000} = 0.00134$$

and, the stability factor of the vehicle with no consideration of the compliance steer is as follows:

$$A = -\frac{m}{2l^2}\frac{l_f K_f - l_r K_r}{K_f K_r} = -\frac{1500}{2 \times 2.7^2}\frac{1.1 \times 80000 - 1.6 \times 60000}{80000 \times 60000} = 0.00017$$

As shown in Example 5.1, it is very important to understand that the compliance steer has a great effect on reducing the equivalent cornering stiffness of the front wheel and eventually the vehicle steer characteristics. A normal value of the coefficient e is around 0.5 to 0.7, which means that once the tire is put on the front axle of the vehicle, the cornering stiffness of the front wheel is 50–70% of that of the original tire itself. It should be understood throughout this book that the symbol K_f, used for the cornering stiffness of the tire, includes the effect of the compliance steer unless otherwise stated.

Previously, the steering angle, α, was either fixed or prevented from rapid changes so that $I_s(d^2\delta/dt^2)$ and $C_s(d\delta/dt)$ in Eqn (5.6) could be neglected. This allowed study of the effect of static characteristics of the steering system. Strictly speaking, if the steering wheel can be operated more quickly, this should be considered. The equations of motion for the vehicle motion to steering angle, α, can be obtained by slightly modifying Eqns (5.6), (3.12), and (3.13).

$$mV\frac{d\beta}{dt} + 2(K_f + K_r)\beta + \left\{mV + \frac{2}{V}(l_f K_f - l_r K_r)\right\}r - 2K_f\delta = 0 \tag{5.12}$$

$$2(l_f K_f - l_r K_r)\beta + I\frac{dr}{dt} + \frac{2(l_f^2 K_f + l_r^2 K_r)}{V}r - 2l_f K_f\delta = 0 \tag{5.13}$$

$$-2\xi K_f\beta - \frac{2l_f\xi K_f}{V}r + I_s\frac{d^2\delta}{dt^2} + C_s\frac{d\delta}{dt} + (K_s + 2\xi K_f)\delta = K_s\alpha \tag{5.14}$$

Based on these equations, it is expected that the front wheel inertia and damping friction around the kingpin will cause a delay in the response of the front wheel steering angle, δ, to steering angle, α. The delay is also in the vehicle response to steering angle. The smaller the restoring moment coefficient and the stiffness of the steering system are, the greater this effect is. Figure 5.4 is an example of the response of vehicle yaw rate, r, to periodical steering angle, α, with different steering system stiffness values [1]. As expected from Eqns (5.12)–(5.14), with the decrease of the steering system stiffness, the vehicle response lag to steering angle becomes larger.

The previous discussion showed that the smaller the steering system stiffness is, the more liable the vehicle is to a US characteristic. Chapter 3 described how a vehicle with a US characteristic has less delay in the response to actual steering angle. However, the US vehicle with low steering system stiffness has a larger delay to steering angle, particularly when the front wheel inertia and damping friction cannot be neglected. Therefore, excessively reducing the steering system stiffness is not something that is desired in the vehicle transient response to somewhat higher speed steering input. In the case of the restoring moment coefficient, the larger the coefficient, the more likely the vehicle will show US characteristics. Equations (5.12)–(5.14)

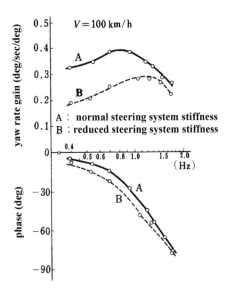

FIGURE 5.4

Effect of steering system stiffness.

show a larger restoring moment coefficient will not cause any delay in the vehicle response to steering angle. Therefore, the restoring moment coefficient, $2\xi K_f$, is desired to be as large as possible provided the steering force is not too heavy.

Example 5.2

Section 3.4.4 showed that it is possible to analyze the vehicle dynamics in the tire's nonlinear region by making the equivalent cornering stiffness decrease with increasing the lateral acceleration. In this section, however, only the saturation property of the tire to side-slip angle is dealt with, and nothing is taken into consideration for the fact that the pneumatic trail decreases with increasing side-slip angle. As the pneumatic trail decreases with the side-slip angle, namely with the lateral acceleration, the aligning torque decreases, and the compliance steer is decreased, which has an effect on the loss of the cornering stiffness due to the compliance steer.

Referring to Sections 2.3.3 and 3.4.4, the pneumatic trail, ξ', decreases proportionally to the side-slip angle and tends to be zero at the saturation point of the lateral force. The following equation is obtained:

$$\xi' = \xi\left(1 - \frac{K_f}{\mu\frac{L}{T}W}\beta_f\right) \tag{E5.1}$$

Taking the compliance steer in the steering system into account, find the equivalent cornering stiffness of the front tire at the large tire side-slip region.

Solution

Equation (3.101) suggests the following description for the pneumatic trail, ξ':

$$\xi' = \xi\left(1 - \frac{K_f}{\mu\frac{L}{T}W}\beta_f\right) = \xi\sqrt{1 - \frac{\ddot{y}}{\mu}} \tag{E5.2}$$

Also the cornering stiffness of the front wheel at a large side-slip angle, K_f', is described as follows:

$$K_f' = K_f \sqrt{1 - \frac{\ddot{y}}{\mu}} \tag{E5.3}$$

So, referring to Eqn (5.8), the cornering stiffness reduction coefficient at a large side-slip angle, e', is the following:

$$e' = \frac{1}{1 + \frac{2\xi' K_f'}{K_s}} = \frac{1}{1 + \frac{2\xi K_f}{K_s}\left(1 - \frac{\ddot{y}}{\mu}\right)} = \frac{e}{1 - (1 - e)\frac{\ddot{y}}{\mu}} \tag{E5.4}$$

From the previous equation, the equivalent cornering stiffness at a large side-slip angle, considering the compliance steer, is as follows:

$$e' K_f' = \frac{e}{1 - (1 - e)\frac{\ddot{y}}{\mu}} K_f \sqrt{1 - \frac{\ddot{y}}{\mu}} \approx e K_f \left\{ 1 - \left(e - \frac{1}{2}\right)\frac{\ddot{y}}{\mu} \right\} \tag{E5.5}$$

It is interesting in Example 5.2 that the reduction rate of the cornering stiffness due to the side-slip angle, namely to the lateral acceleration, in Eqn (E5.5) decreases compared with Eqns (3.101) and (3.102) or (3.103) and (3.104). This means that the equivalent cornering stiffness reduction due to the side-slip angle at the front wheel is smaller than that at the rear wheel, especially under large tire side-slip angle. It is interesting to see that though the steering system compliance steer makes the vehicle US, there is a possibility to weaken this aspect at larger side slips and make the vehicle tend to OS. This is achieved by reducing the effects of compliance steer on the effective cornering stiffness of the front tire at large side slips.

5.3.2 EFFECTS OF STEERING SYSTEM CHARACTERISTICS ON VEHICLE MOTION WITH NON-FIXED STEERING ANGLE

The previous section studied the effect of the steering system characteristics on the vehicle motion when the steering angle is assumed fixed at a given value. This section will look at the effect the steering system characteristics have on the vehicle motion when the steering angle, α, is not fixed but has a motion degree-of-freedom. This is typical when drivers take their hands off the steering wheel completely or when the driver gives a torque to the steering wheel regardless of the steering angle. This situation is called free control.

The vehicle and steering system equations of motion during the free control are derived as follows:

$$mV\frac{d\beta}{dt} + 2(K_f + K_r)\beta + \left\{ mV + \frac{2}{V}(l_f K_f - l_r K_r) \right\} r = 2K_f \delta \tag{3.12}$$

$$2(l_f K_f - l_r K_r)\beta + I\frac{dr}{dt} + \frac{2(l_f^2 K_f + l_r^2 K_r)}{V} r = 2l_f K_f \delta \tag{3.13}$$

$$I_h\frac{d^2\alpha}{dt^2} + C_h\frac{d\alpha}{dt} + K_s(\alpha - \delta) = T_h \tag{5.5}$$

$$I_s\frac{d^2\delta}{dt^2} + C_s\frac{d\delta}{dt} + K_s(\delta - \alpha) = 2\xi K_f\left(\beta + \frac{l_f r}{V} - \delta\right) \tag{5.6}$$

Here, the front wheel moment of inertia around the kingpin, I_s, is small compared to the steering wheel moment of inertia, I_h, and can be neglected. Also, if the damping coefficients, C_h and C_s, are neglected as small values, Eqns (5.5) and (5.6) become the following:

$$I_h \frac{d^2\alpha}{dt^2} + K_s(\alpha - \delta) = T_h \tag{5.5'}$$

$$K_s(\delta - \alpha) = 2\xi K_f \left(\beta + \frac{l_f r}{V} - \delta \right) \tag{5.6'}$$

By substituting Eqn (5.6)' into Eqn (5.5)', the following is derived:

$$I_h \frac{d^2\alpha}{dt^2} - 2\xi K_f \left(\beta + \frac{l_f r}{V} - \delta \right) = T_h \tag{5.5''}$$

Here, Eqn (5.6)' is same as Eqn (5.7), and δ is obtained from this equation as Eqn (5.9):

$$\delta = e\alpha + (1-e)\left(\beta + \frac{l_f}{V} r \right)$$

If it is assumed that the steering system stiffness is large, which means $K_s = \infty$, Eqn (5.8) gives $e = 1$, and the following is obtained:

$$\delta = \alpha \tag{5.6''}$$

The vehicle equations of motion with the steering system during free control could be written in a simpler form, based on Eqns (3.12), (3.13), (5.5)", and (5.6)":

$$mV \frac{d\beta}{dt} + 2(K_f + K_r)\beta + \left\{ mV + \frac{2}{V}(l_f K_f - l_r K_r) \right\} r - 2K_f \alpha = 0 \tag{5.15}$$

$$2(l_f K_f - l_r K_r)\beta + I \frac{dr}{dt} + \frac{2(l_f^2 K_f + l_r^2 K_r)}{V} r - 2l_f K_f \alpha = 0 \tag{5.16}$$

$$-2\xi K_f \beta - \frac{2l_f \xi K_f}{V} r + I_h \frac{d^2\alpha}{dt^2} + 2\xi K_f \alpha = T_h \tag{5.17}$$

To further simplify the analysis in order to examine the effect of the steering system characteristics on the vehicle motion, the following assumptions are made:

$$K_f = K_r = K$$

$$l_f = l_r = \frac{l}{2}$$

$$I = mk^2 = ml_f l_r = m\left(\frac{l}{2} \right)^2$$

Substituting these equations into Eqns (5.15)–(5.17) gives the following:

$$mV \frac{d\beta}{dt} + 4K\beta + mVr - 2K\alpha = 0 \tag{5.15'}$$

$$m\left(\frac{l}{2}\right)^2\frac{dr}{dt}+\frac{l^2K}{V}r-lK\alpha=0 \tag{5.16$'$}$$

$$-\frac{2\xi K_f}{I_h}\beta-\frac{\xi lK_f}{I_hV}r+\frac{d^2\alpha}{dt^2}+\omega_s^2\alpha=\frac{T_h}{I_h} \tag{5.17$'$}$$

whereby,

$$\omega_s^2=\frac{2\xi K_f}{I_h}=\frac{2\xi K}{I_h} \tag{5.18}$$

And, ω_s^2 is the natural frequency of the steering system.

Applying Laplace transforms to the simplified Eqns (5.15)$'$–(5.17)$'$, gives the characteristic equations as follows:

$$\frac{4}{m^2l^2V}\begin{vmatrix} mVs+4K & mV & -2K \\ 0 & m\left(\frac{l}{2}\right)^2s+\frac{l^2K}{V} & -lK \\ -\omega_s^2 & -\frac{l\omega_s^2}{2V} & s^2+\omega_s^2 \end{vmatrix}=A_4s^4+A_3s^3+A_2s^2+A_1s+A_0=0 \tag{5.19}$$

where

$$A_0=\frac{4K}{ml}\omega_s^2$$

$$A_1=\frac{4K}{mV}\omega_s^2$$

$$A_2=\omega_s^2+\frac{16K^2}{m^2V^2} \tag{5.20}$$

$$A_3=\frac{8K}{mV}$$

$$A_4=1$$

For the motion to be stable, the coefficients of the characteristic Eqn (5.19) have to fulfill the Routh stability conditions as follows:

$$A_0,\ A_1,\ A_2,\ A_3,\ A_4>0 \tag{5.21}$$

$$\begin{vmatrix} A_1 & A_0 & 0 \\ A_3 & A_2 & A_1 \\ 0 & A_4 & A_3 \end{vmatrix}=A_1A_2A_3-A_0A_3^2-A_1^2A_4>0 \tag{5.22}$$

A_0 and A_1 are positive when $\xi>0$ from Eqn (5.18), while A_2, A_3, and A_4 are always positive. Therefore, for the vehicle motion to be stable, trail ξ has to be always positive.

Substituting Eqn (5.20) into Eqn (5.22) and rearranging it gives the following:

$$\frac{32K^2}{m^2V^2} + \omega_s^2 - \frac{16K}{ml} > 0 \tag{5.23}$$

By deriving this condition so that it is not related to traveling speed, V, then the following is obtained:

$$\omega_s^2 - \frac{16K}{ml} > 0 \tag{5.24}$$

Defining the following:

$$\omega_y^2 = \frac{16K}{ml} = \frac{4lK}{m\left(\frac{l}{2}\right)^2} \tag{5.25}$$

The denominator at the right-hand side of Eqn (5.25) is equivalent to the yaw moment of inertia, and the numerator is equivalent to the moment per unit yaw angle by the front and rear wheels cornering stiffness (lateral stiffness). Hence, ω_y is a vehicle constant that could be viewed as the vehicle yaw natural frequency. Using this ω_y, Eqn (5.24) can be rewritten as follows:

$$\omega_s^2 - \omega_y^2 > 0 \tag{5.26}$$

For the vehicle motion to be stable during free control, regardless of the vehicle speed, the steering system natural frequency, ω_s, must be greater than the vehicle's yaw natural frequency ω_y defined by Eqn (5.25).

If $\omega_s < \omega_y$, a stability limit velocity, V_{cr}, that causes the vehicle motion to be unstable will exist. This is obtained by finding the V that makes the left-hand side of Eqn (5.23) equal to zero.

$$V_{cr} = \sqrt{\frac{2lK}{m} \frac{1}{1 - \left(\omega_s / \omega_y\right)^2}} \tag{5.27}$$

If the vehicle traveling speed V is less than V_{cr}, then the vehicle is stable, but when $V > V_{cr}$ then the vehicle motion will be unstable [2].

Equation (5.18) shows that ω_s depends greatly on the trail ξ. Using Eqn (5.27), the relation between the trail ξ and the stability limit velocity, V_{cr}, for a normal passenger car is shown in Figure 5.5. It is clear that ξ is desired to be as large as possible within the region of smaller steering torques. As the steering system natural frequency, ω_s, is also desired to be as big as possible, a smaller I_h is better. Furthermore, the larger K_s is, the larger ω_s will be.

5.3.3 EFFECTS OF DRIVER'S HANDS AND ARMS

In the previous chapters, the effect of the steering system characteristics on the vehicle motion was carried out with the assumptions that the steering wheel angle is either completely fixed or totally free. In a real vehicle situation, even when a driver is not intentionally maneuvering the vehicle, the steering wheel is lightly supported by the driver's hand and arm. The steering wheel is neither completely fixed nor totally free. The effect of the driver's hand and arm on

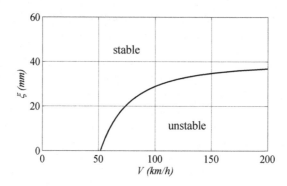

FIGURE 5.5

Relation of critical vehicle speed to trail ξ.

the vehicle's inherent motion will be looked at with some simple assumptions. The driver is not assumed to be a mechanism in controlling the vehicle motion (this matter is dealt with in Chapter 10).

The steering wheel is supported by the driver's hand and arm, which is assumed to play the role as an equivalent spring. Taking α_h as the steering wheel angle given by the human's hand and α as the real steering wheel angle, a torque of T_h will act at the steering wheel by the driver's hand.

$$T_h = K_h(\alpha_h - \alpha) \qquad (5.28)$$

where K_h is the equivalent spring constant of the human's hand and arm.

As in Section 5.3.2, the equations of motion of the vehicle and steering wheel, inclusive of the human's hand and arm effect, can be written as follows by substituting Eqn (5.28) into Eqn (5.17):

$$mV\frac{d\beta}{dt} + 2(K_f + K_r)\beta + \left\{ mV + \frac{2}{V}(l_f K_f - l_r K_r) \right\}r - 2K_f\alpha = 0 \quad (5.15)$$

$$2(l_f K_f - l_r K_r)\beta + I\frac{dr}{dt} + \frac{2(l_f^2 K_f + l_r^2 K_r)}{V}r - 2l_f K_f\alpha = 0 \quad (5.16)$$

$$-2\xi K_f\beta - \frac{2l_f\xi K_f}{V}r + I_h\frac{d^2\alpha}{dt^2} + (2\xi K_f + K_h)\alpha = K_h\alpha_h \qquad (5.29)$$

From Eqn (5.29), the natural frequency of the steering system now becomes the following:

$$\omega_s = \sqrt{\frac{2\xi K_f + K_h}{I_h}} \qquad (5.30)$$

Consequently, a human's hand and arm has the effect of increasing the steering system natural frequency, ω_s. In Section 5.3.2, it is shown that ω_s is desired to be as large as possible for the vehicle motion to be stable, and from this point of view, K_h is desired to be large.

For the vehicle motion where the steering system inertia could be neglected, Eqn (5.29) becomes the following:

$$-2\xi K_f \beta - \frac{2l_f \xi K_f}{V} r + (2\xi K_f + K_h)\alpha = K_h \alpha_h \tag{5.31}$$

From Eqn (5.31),

$$\alpha = \frac{K_h}{K_h + 2\xi K_f} \alpha_h + \left(1 - \frac{K_h}{K_h + 2\xi K_f}\right)\left(\beta + \frac{l_f}{V} r\right)$$

$$e_h = \frac{K_h}{K_h + 2\xi K_f} = \frac{1}{1 + \frac{2\xi K_f}{K_h}} \tag{5.32}$$

$$\alpha = e_h \alpha_h + (1 - e_h)\left(\beta + \frac{l_f}{V} r\right) \tag{5.33}$$

Substituting Eqn (5.33) into Eqns (5.15) and (5.16) and rearranging gives the following:

$$mV \frac{d\beta}{dt} + 2(e_h K_f + K_r)\beta + \left\{mV + \frac{2}{V}(l_f e_h K_f - l_r K_r)\right\} r = 2e_h K_f \alpha_h \tag{5.34}$$

$$2(e_h l_f K_f - l_r K_r)\beta + I\frac{dr}{dt} + \frac{2\left(l_f^2 e_h K_f + l_r^2 K_r\right)}{V} r = 2l_f e_h K_f \alpha_h \tag{5.35}$$

Comparing these to the original Eqns (5.15) and (5.16), the front wheel cornering stiffness has changed from K_f to $e_h K_f$ as before. In other words, when the effect of the human's hand and arm are taken into consideration, the front wheel cornering stiffness decreases because e_h is less than one. The more the front wheel cornering stiffness decreases, the more the vehicle will be US and the better the directional stability. Therefore, from this point of view, K_h is desired to be as small as possible.

There are now two contradictory requirements for K_h from the point of view of vehicle stability. This theoretical argument coincides with the reality. A human driver neither holds the steering wheel strongly (large K_h) such that it is fixed when driving at high speed, nor does the driver let the steering wheel free (K_h is zero) by leaving his hand off the wheel. The driver holds the steering wheel just lightly enough for practical driving.

Consequently, it could be said that the driver's hand and arm, supporting the steering wheel lightly, further stabilizes the vehicle motion.

PROBLEMS

5.1 In order to understand how the compliance steer in the steer system has an equivalent effect on reducing the front tire cornering stiffness from K_f to eK_f in the fixed control of the vehicle, derive Eqns (5.10) and (5.11) by yourself by putting Eqn (5.9) into Eqns (3.12) and (3.13).

5.2 Calculate the natural frequency of the steering system when ξ, K, and I_h are 0.04 m, 60.0 kN/rad, and 20.0 kg/m², respectively.

5.3 Calculate the upper limit of the inertia moment of the steering wheel equivalently around the kingpin for the free control vehicle to be stable at any vehicle speed under the given parameters: $\xi = 0.04$ m, $m = 1500$ kg, $l = 2.7$ m, and $K = 60$ kN/rad.

5.4 Show that the stability condition of the free control vehicle at any speed is approximately written as $I_h/I \leq \xi/2l$.

REFERENCES

[1] Nishii K, Higuchi T. Effect of component factors on controllability and stability of vehicles. JSAE J 1972;26(7) [In Japanese].

[2] Okada T. Fundamental theory on automobile stability and control. JSAE J 1972;26(7) [In Japanese].

VEHICLE BODY ROLL AND VEHICLE DYNAMICS

6

6.1 PREFACE

The previous chapters have dealt with vehicle motion under the assumptions that the wheel is rigidly attached to the vehicle body and there is no relative displacement between the body and the wheels. These are good assumptions, proven from practical experience, for understanding the basic vehicle motion characteristics. However, in the case of normal passenger cars, the vehicle body and wheels are connected to each other by soft and elastic connections to improve the vehicle ride comfort. This mechanism is generally called the suspension system. The vehicle body is called the sprung mass, and the wheels are called the unsprung mass. The suspension system between the vehicle body and the wheels allows a relative up–down displacement between the vehicle body and the wheels. When the vehicle moves laterally, a centrifugal force acts at the vehicle's center of gravity, causing the vehicle to tilt to the direction of the centrifugal force. This tilt is called the vehicle roll. If the suspension system is considered, the vehicle will have a roll degree-of-freedom that is produced together with vehicle lateral motion.

This chapter will look into the roll mechanism, derive the vehicle equations of motion including the roll motion, and study the effects of suspension system characteristics and vehicle roll on the vehicle motion.

6.2 ROLL GEOMETRY

R. Eberan's hypothesis of the roll center as the vehicle's geometric instantaneous rotation center and his assumption that this roll center is always fixed [1] have long been taken as the standard approach. This hypothesis is generally used due to its simplicity. Based on this hypothesis, the roll mechanism of the vehicle will be studied with a constant lateral acceleration, which is caused by a constant centrifugal force.

6.2.1 ROLL CENTER AND ROLL AXIS

In general, there are various types of suspension systems, from the simple rigid-axle type to the independent suspension that is common in passenger cars. The relative vertical displacement or angular displacement between the sprung and unsprung masses is dependent on the structure of the suspension system.

The front- and rear-wheel roll centers are also determined by the suspension system configuration. The line that connects the front and rear roll centers is called the roll axis. The roll center

153

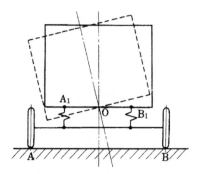

FIGURE 6.1

Roll center for rigid-axle suspension.

is the vehicle's instantaneous rotation center in the plane perpendicular to the vehicle's longitudinal direction, which contains the left and right wheels' ground contact point. The wheels are considered rigid in both up–down and left–right directions, and the ground contact point is fixed.

Figure 6.1 shows the axle-type suspension system. The vehicle body at points A_1 and B_1 can only have vertical displacement relative to the unsprung mass due to the springs. Even if the sprung mass rolls, the unsprung mass including the wheels is assumed rigid and, thus, does not move; the roll center is at point O. In other words, when a rolling moment acts on the vehicle, the vehicle body will produce a roll angle, ϕ, relative to the wheels with respect to point O.

Figure 6.2 shows a typical independent-type suspension—often called the double-wishbone suspension. As its name implies, each wheel can move independently relative to the vehicle body. If the vehicle body is fixed, the instantaneous rotation centers of the left and right unsprung mass relative to the vehicle body are the points O_1 and O_2, respectively. Point O_1 is the intersecting point of the extended lines of A_1–A_2 and A_3–A_4, and point O_2 is the intersecting point of the extended lines of B_1–B_2 and B_3–B_4. Here, when the vehicle body rolls during cornering, the wheel contact points with the ground (A and B) are fixed, and the unsprung masses must roll around them. Points O_1 and O_2 move in the direction perpendicular to O_1A and O_2B. O_1 and O_2 are the virtual points on the vehicle body as well as on the unsprung masses. Consequently, the

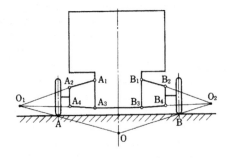

FIGURE 6.2

Roll center for wishbone suspension.

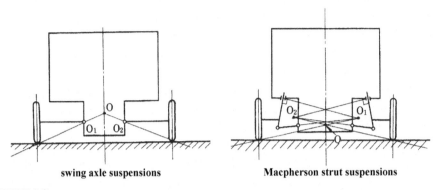

swing axle suspensions   **Macpherson strut suspensions**

FIGURE 6.3

Roll center for independent suspensions.

vehicle body's instantaneous rotating center, or the roll center, is the intersection of the extended lines of O_1A and O_2B, which is point O.

Based on this, the roll center for other types of suspension systems is shown in Figure 6.3.

It is clear that the vehicle roll center position is dependent on the structure of the suspension system. Usually, the suspension system and the vehicle are symmetrical on the left and right, and the roll center is always on the symmetric axis. In this case, it is the height of the roll center that is dependent on the suspension system structure.

The roll center is the vehicle's instantaneous rotation center, and its position can move during suspension movement. Point O shown here is the roll center when the roll angle is zero; if the vehicle rolls, the roll center will also move. To understand this, the roll center O' during body roll is shown for two types of suspension system—the wishbone and the swing axle suspension system—in Figure 6.4.

If the roll angle is not large, the movement of the roll center is small, and it is possible to assume the roll centers are fixed at point O. It is still possible to understand the vehicle roll mechanism, even with a moving roll center. But, the fixed roll center concept is easier to understand and gives a good understanding of the basic vehicle dynamics. Based on Eberan's

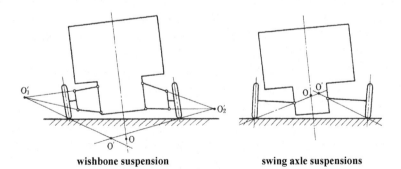

wishbone suspension   **swing axle suspensions**

FIGURE 6.4

Change of roll center position due to body roll.

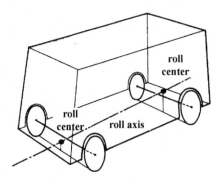

FIGURE 6.5

Roll axis.

roll center hypothesis, the front and rear roll centers are determined, and if the vehicle body is rigid, the vehicle's fixed roll axis is determined as shown in Figure 6.5. The roll center at the front and rear may not have the same height above the ground, and the roll axis is not necessarily parallel to the vehicle longitudinal axis.

Furthermore, when vehicle motion is accompanied by large roll angles, the fixed roll center and roll axis concept is not suitable anymore. In such cases, vehicle roll is usually dealt with as the indeterminate problem of the vehicle's four wheels.

6.2.2 ROLL STIFFNESS AND LOAD TRANSFER

Now, the vehicle is assumed to have a constant lateral acceleration and centrifugal force acting at the vehicle's center of gravity. The center of gravity does not normally coincide with the vehicle roll axis, but is usually above the roll axis, as shown in Figure 6.6. The centrifugal force acting at the center of gravity produces a rolling moment around the roll axis resulting in a constant roll angle. If the vehicle body rolls, the left and right vertical springs of the suspension system will be stretched at one side and be compressed on the other side. This produces an equilibrium moment to the rolling moment due to the centrifugal force. The magnitude of the moment produced by the stretch and the compression of the spring per unit roll angle is called the roll stiffness.

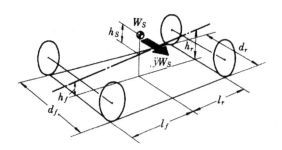

FIGURE 6.6

Roll center and C.G. heights.

Here, the respective roll stiffness for the front and rear suspension system is defined as $k_{\phi f}$ and $k_{\phi r}$, the roll center height from the ground as h_f and h_r, the front and rear treads as d_f and d_r, the distance between the vehicle's center of gravity and the roll axis as h_s, and the distance between the front and rear axle to the center of gravity as l_f and l_r. The weight of the unsprung mass is small compared to the weight of the sprung mass and could be neglected. In this case, the vehicle weight is taken to be equal to the vehicle's body weight, and it is written as W_s. The vehicle's lateral acceleration is taken as $\ddot{y}$ (the same as in Section 3.3.3), and the centrifugal force acting on the vehicle is $\ddot{y}W_s$. Assuming that the vehicle is rigid and the roll angle is small, the rolling moment by the centrifugal force is $\ddot{y}W_sh_s$, and the roll moment by the vehicle weight due to tilting of the vehicle body is $W_sh_s\phi$, then the vehicle roll angle becomes the following:

$$\left(K_{\phi f} + K_{\phi r}\right)\phi = \ddot{y}W_sh_s + W_sh_s\phi \quad \text{or} \quad \phi = \frac{\ddot{y}W_sh_s}{K_{\phi f} + K_{\phi r} - W_sh_s} \tag{6.1}$$

The centrifugal force, $\ddot{y}W_s$, acting on the vehicle requires tire cornering forces to achieve equilibrium. Distributing the $\ddot{y}W_s$ force acting at the center of gravity to the front and rear wheels, the forces $\ddot{y}W_sl_r/l$ and $\ddot{y}W_sl_f/l$ could be considered to act on the front and rear wheel, respectively, where $l = l_f + l_r$. These forces are equal to the front- and rear-wheel lateral forces.

If the vehicle body rolls, the left and right wheels at both the front and rear axles will increase in load at one side and decrease at the other side. This is called the load transfer due to roll. Defining the load transfer for the front and rear as ΔW_f and ΔW_r, respectively, the roll moment around the roll center at the front and rear wheels in the plane perpendicular to the vehicle longitudinal direction has to be in equilibrium, as shown in Figure 6.7. The following equations are derived:

$$K_{\phi f}\phi = \Delta W_f d_f - \frac{\ddot{y}W_sl_r}{l}h_f \tag{6.2}$$

$$K_{\phi r}\phi = \Delta W_r d_r - \frac{\ddot{y}W_sl_f}{l}h_r \tag{6.3}$$

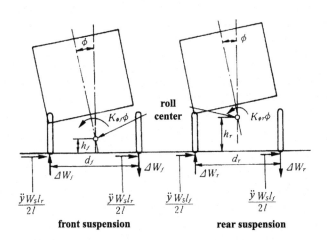

FIGURE 6.7

Transversal load transfer due to body roll.

At this time, it is assumed that there is no load transfer between the front and the rear. Substituting Eqn (6.1) into Eqns (6.2) and (6.3) to find ΔW_f and ΔW_r gives the following equations:

$$\Delta W_f = \frac{\ddot{y} W_s}{d_f} \left[\frac{K_{\phi f} h_s}{K_{\phi f} + K_{\phi r} - W_s h_s} + \frac{l_r}{l} h_f \right] \tag{6.4}$$

$$\Delta W_r = \frac{\ddot{y} W_s}{d_r} \left[\frac{K_{\phi r} h_s}{K_{\phi f} + K_{\phi r} - W_s h_s} + \frac{l_f}{l} h_r \right] \tag{6.5}$$

These equations give the load transfer between the left and right wheels due to a constant lateral acceleration. The equations show that a higher vehicle center of gravity distance from the roll axis, h_s, results in a larger load transfer at the front and rear wheels. Furthermore, a load transfer at the front and rear wheels is basically proportional to the front and rear roll stiffness ratios to the total roll stiffness, respectively. The last term in Eqns (6.4) and (6.5) depends on the height of the roll axis from the ground, h_f and h_r.

6.2.3 CAMBER CHANGE AND ROLL STEER

If the ground contact points of the wheels are fixed, as the vehicle body rolls, the unsprung mass, including the wheels, tilts relative to the ground. This gives the camber change of the wheel, which is measured relative to the ground and is due to body roll. The vehicle roll also gives the wheels an up-and-down displacement relative to the vehicle body. At such time, depending on the structure of the suspension system, the wheels may produce some angular displacement in the horizontal plane along with the up-and-down movement relative to the vehicle body. This is called roll steer.

The camber change and roll steer are dependent on the structure of the suspension system. The suspension system is designed with keen consideration of these characteristics, often using them to affect the vehicle dynamics or sometimes trying to avoid them completely. This chapter will skip the detailed explanation of camber change and roll steer mechanism for various suspension systems, and it will only look at the basic characteristics of camber change and roll steer. The collective term for camber change and roll steer is sometimes called alignment change due to roll.

In axle-type suspensions, the wheel does not produce any camber change due to vehicle roll. The camber change due to roll only occurs for independent suspension systems, where depending on the suspension structure, there could be one of two cases: camber change in the same direction as roll, which is called positive camber, or in the opposite direction, negative camber. This is shown in Figure 6.8.

Independent suspension systems are constructed by a linkage mechanism, and the vehicle roll angle and camber change can be determined from geometric analysis of the linkage. Figure 6.9 shows the actual measured value and calculated value of the camber change for a wishbone-type suspension system. This relationship varies substantially with the arrangement of the links, even for suspension systems of the same type. From the figure, if the roll angle is not large, the camber change can be considered as nearly proportional to the roll angle. As the roll angle becomes large, this linear relation is lost, and nonlinearity appears. This is generally true for other types of suspension systems. The nonlinear characteristic of the camber change is one of the main factors that influences the vehicle motion at large lateral accelerations. For

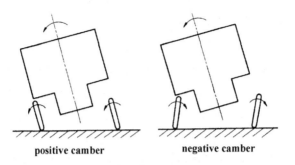

positive camber   negative camber

FIGURE 6.8

Camber change due to body roll.

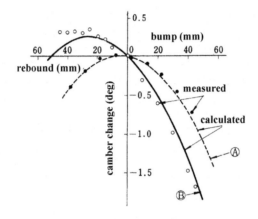

FIGURE 6.9

Chamber changes to suspension stroke.

lower lateral accelerations, the camber change could be considered to be proportional to the roll angle, in either direction shown in Figure 6.8.

Example 6.1

Investigate the geometric condition of the suspension mechanism, such as in Figures 6.2 or 6.3, which basically determines the positive or negative camber to the body roll.

Solution

Any part of the left half of the vehicle body is going downward and the right half upward due to the positive body roll. The unsprung mass rolls around the tire contact point, for example point A in Figure 6.2, which is a camber change. So any point of the left side of the unsprung mass relative to the centerline of the left wheel is going downward and the right side goes upward due to the positive camber of the unsprung mass. Due to the negative camber, the left side is going upward and the right side is going downward.

As the virtual point, O_1, is the common point fixed to the vehicle body and the left-side unsprung mass, it must move in the same direction as the body roll and the camber change during roll motion. If point O_1 is to the left of the body center and to the right of the wheel centerline, then the suspension has negative camber. If point O_1 is to the left of the body center and to the left of the wheel centerline, or to the right of the body center and the wheel centerline, then the suspension has to show the positive camber.

The roll steer due to the roll of the vehicle body is also dependent on the suspension system structure. For an axle-type suspension, the sprung mass and unsprung mass are often connected using leaf springs. The mounting point of the spring at the vehicle axle moves in the rear-and-forward direction and causes the axle to produce an angular displacement relative to the vehicle body in the horizontal plane. This is roll steer for an axle-type suspension, which is sometimes called axle steer due to roll.

For independent suspension systems, the amount of roll steer can be determined from geometric analysis of the linkage. Similar to camber change, the roll steer direction and magnitude, relative to the roll angle, can vary substantially with the arrangement of the links. The suspension systems are usually designed to control the amount of steer by careful arrangement of the links. For independent suspensions, if the roll angle is small, the roll steer can be considered to be proportional to the roll angle. The direction of the roll steer can sometimes be in either the positive or negative direction, depending on the suspension system structure.

The roll steer for an independent-type suspension is sometimes called the toe change due to the vertical stroke of the suspension. Figure 6.10 shows an example of toe change to suspension stroke. Roll steer in the direction toward the inner side of the vehicle is called toe-in, and that in the opposite direction is called toe-out.

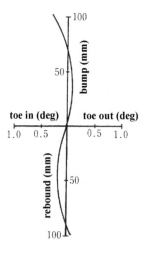

FIGURE 6.10

Roll steer (toe change to suspension stroke).

6.3 BODY ROLL AND VEHICLE STEER CHARACTERISTICS

By examining the vehicle in steady-state cornering, the vehicle steer characteristics that fundamentally influence the vehicle dynamic performance have been understood. During steady-state cornering, a constant centrifugal force acts at the vehicle's center of gravity, and if the suspension system is considered, the vehicle will produce a constant roll angle. The previous sections explained body roll geometrically. This section will try to study the effect of vehicle roll on the steer characteristics by considering vehicle steady-state cornering with body roll. The effect of suspension lateral stiffness on vehicle steer characteristics will also be investigated.

6.3.1 LOAD TRANSFER EFFECT

As described in Chapter 2, the tire lateral force changes with tire load in the form of a saturating curve. When there is a load transfer between the left and the right wheels, the sum of their lateral forces will be lower than when load transfer is not considered. The larger the load transfer is, the greater the reduction in total lateral force is.

Figure 6.11 shows a typical relationship between the load transfer and the lateral force. For an axle with two wheels of load W, a load transfer of ΔW occurs between the left and right side. This yields lateral forces of $\overline{P_1A_1}$ and $\overline{P_2A_2}$, and the sum of them is $2\overline{BA}$. In contrast, the lateral force when there is no load transfer is $2\overline{PA}$. The reduction of the lateral force at this axle, due to roll, is exactly equal to $2\overline{PB}$.

During vehicle body roll, the lateral load transfer occurs at the front and rear axles, as described by Eqns (6.4) and (6.5). Consequently, the front- and rear-wheel cornering stiffness will decrease according to the magnitude of ΔW_f and ΔW_r. To undergo steady-state cornering at the same radius as without load transfer, with the same magnitude of centrifugal force, the front- and rear-wheel side-slip angles must increase according to the magnitude of ΔW_f and ΔW_r. This will produce the lateral forces for equilibrium.

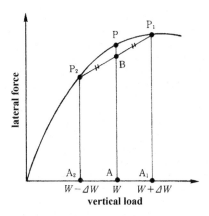

FIGURE 6.11

Load transfer and cornering force.

The vehicle steer characteristic is determined by the relative magnitude of the front- and rear-wheel side-slip angles. If $\Delta W_f > \Delta W_r$, the vehicle steer characteristic will tend to US, whereas if $\Delta W_f < \Delta W_r$, the steer characteristic will change to OS.

The load transfer for the front and rear wheels, as shown by Eqns (6.4) and (6.5), is dependent on the following:

- The front and rear roll center heights, h_f and h_r.
- The ratio of front and rear roll stiffness, $K_{\phi f}/K_{\phi r}$.
- The front and rear tracks, d_f and d_r.

If the following conditions are true: $h_f \rightarrow$ large, $h_r \rightarrow$ small, $K_{\phi f}/K_{\phi r} \rightarrow$ large, $d_f \rightarrow$ small, and $d_r \rightarrow$ large, then ΔW_f will become larger than ΔW_r, and the vehicle steer characteristic changes to US. In contrast, the vehicle steer characteristic will change to OS if $h_f \rightarrow$ small, $h_r \rightarrow$ large, $K_{\phi f}/K_{\phi r} \rightarrow$ small, $d_f \rightarrow$ large, and $d_r \rightarrow$ small.

Following is a further examination of the reduction in the axle equivalent cornering characteristics due to the load transfer between the left and right wheels.

Curve OP in Figure 6.12 is the relationship between the lateral force and tire side-slip angle for a tire with the load W. The lateral force for the tire with the extra load of ΔW is shown by curve OP_1, and the lateral force for the tire with less load of ΔW is shown by curve OP_2. The sum of these two curves is the lateral force for the axle when there is a load difference of ΔW between the left and right wheels. This is equal to two times curve OB.

If there is no load difference, the lateral force for that axle is shown by curve OP, which is larger than curve OB. The shapes of these two curves, OP_1 and OP_2, are not changed if the axle lateral force is divided by the axle load.

At lateral accelerations of $\ddot{y}_1$, $\ddot{y}_2$, and $\ddot{y}_3$, there are load transfers of ΔW_1, ΔW_2, and ΔW_3, respectively, at the vehicle axle. The curves of the axle force divided by the axle load to side-slip angle for each load transfer are seen in Figure 6.13. As described in Section 3.3.3, the vertical axis in Figure 6.13 is identical to the vehicle's lateral acceleration. Projecting points $\ddot{y}_1$, $\ddot{y}_2$, and $\ddot{y}_3$ from the vertical axis of Figure 6.13 onto the curve for load transfers ΔW_1, ΔW_2, and ΔW_3, respectively, and connecting all the points on the curves gives a new

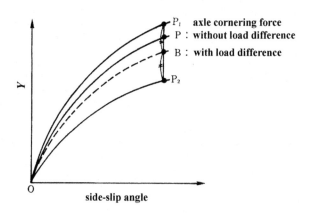

FIGURE 6.12

Axle cornering force and vertical load.

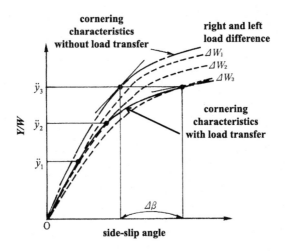

FIGURE 6.13

Effect of load transfer on axle cornering force.

curve of equivalent cornering force of the axle when lateral load transfer occurs according to the lateral acceleration.

The previous investigation gives how a change of tire characteristics due to load transfer is, and the change of vehicle motion characteristics due to this will become more obvious at large lateral accelerations because of the nonlinear tire characteristics. The tire characteristics obtained above would be used in studying vehicle steady state cornering in Section 3.3.3.

6.3.2 CAMBER CHANGE EFFECT

The wheel camber could occur in either the same or the opposite direction to the roll direction, as described in Section 6.2.3. Here, it is assumed the first case gives positive camber change, and the second gives a negative camber change. In either case, the camber change results in a force that acts in the lateral direction (camber thrust). This is proportional to the camber angle, as described in Chapter 2. In steady-state cornering, the camber thrust becomes one of the forces that balances the centrifugal force at the C.G. Positive camber angles produce a camber thrust that acts in the same direction as the centrifugal force. In this case, larger wheel side-slip angles are needed to achieve steady-state cornering at the same radius and speed as when camber change is not considered. In contrast, negative camber angles produce a camber thrust that acts in the opposite direction to the centrifugal force. Here, the cornering force and the wheel side-slip angles can become smaller.

The vehicle steer characteristics are determined by the relative magnitude of the front- and rear-wheel side-slip angles. Consequently, the positive camber change alters the vehicle steer characteristic to US at the front wheels and OS at the rear wheels. Negative camber changes have an opposite effect and change the vehicle steer characteristic to OS at the front wheels and US at the rear wheels.

6.3.3 ROLL STEER EFFECT

The angular displacement of the wheel due to roll is defined as roll steer. A positive roll steer acts in the same direction as the front steering angle, steered by steering wheel, whereas a negative roll steer acts in the opposite direction.

Figure 6.14 shows the geometry of a steady-state cornering vehicle with roll steer. Here, α_f and α_r are the front and rear roll steers. The geometric relation of steady-state cornering excluding roll steer is given by Eqn (3.34). With roll steer as in Figure 6.14, the equation becomes the following:

$$\rho = \frac{l}{\delta - \beta_f + \beta_r + \alpha_f - \alpha_r} \tag{6.6}$$

Equation (6.6) shows that the vehicle steer characteristic is determined by the front and rear roll steer angles, α_f and α_r, as well as the front- and rear-wheel side-slip angles, β_f and β_r. When a cornering radius at a constant steering angle increases with speed or lateral acceleration, the steer characteristics is termed US. If the radius decreases, the steer characteristic is called OS. If the front roll steer, α_f, is positive, it will change the vehicle steer characteristics to OS, and if it is negative, the vehicle will tend to US. On the contrary, if the rear roll steer, α_r, is positive, it will change the vehicle steer characteristics to US, and a negative roll steer will result in an OS vehicle.

Figure 6.15 schematically shows the effect of roll steer on the vehicle steer characteristic. Rewriting Eqn (6.6) gives the following:

$$\delta = \frac{l}{\rho} + \beta_f - \beta_r + \alpha_r - \alpha_f \tag{6.6'}$$

As described in Section 3.3.3, $\beta_f - \beta_r$ is determined in relation to the lateral acceleration, $\ddot{y}$, during cornering. Also, the front and rear roll steers, α_f and α_r, are found from the roll angle (or the suspension stroke), which is proportional to the lateral acceleration, $\ddot{y}$. Based upon the previous, Eqn (6.6)' can be used to investigate the vehicle steer characteristics in the relationship between the steady-state steering angle, δ, and the lateral acceleration $\ddot{y}$, when roll steer is considered.

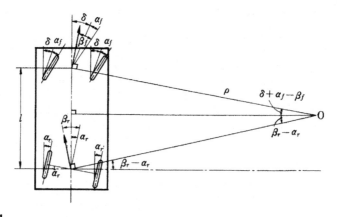

FIGURE 6.14

Steady-state cornering with roll steer.

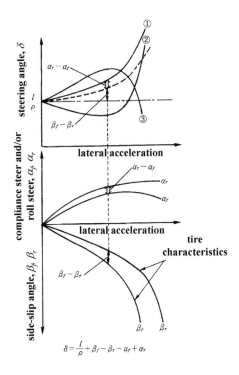

FIGURE 6.15

Vehicle steer characteristics with roll steer.

Example 6.2

Derive the equation to show the effect of the roll steer on the steady-state turning radius when the front and rear suspensions both have the roll steer proportional to the roll angle.

Solution

As is described at the end of Section 3.3.1(2), the side-slip angles of the front and rear tires during steady-state turning with the lateral acceleration, mV^2/ρ, are described as follows:

$$\beta_f = \frac{mV^2 l_r}{2lK_f} \frac{1}{\rho}, \quad \beta_r = \frac{mV^2 l_f}{2lK_r} \frac{1}{\rho}$$

From Eqn (6.1), the roll angle in steady turning is derived:

$$\phi = \frac{\ddot{y} W_s h_s}{K_\phi} = \frac{m_S V^2 h_S}{K_\phi} \frac{1}{\rho} \quad \text{where } K_\phi = K_{\phi f} + K_{\phi r} - W_s h_s.$$

The roll steer angles of the front and rear wheels proportional to the roll angle are described as follows:

$$\alpha_f = \frac{\partial \alpha_f}{\partial \phi} \phi = \frac{\partial \alpha_f}{\partial \phi} \frac{m_S V^2 h_S}{K_\phi} \frac{1}{\rho} \tag{E6.1}$$

$$\alpha_r = \frac{\partial \alpha_r}{\partial \phi} \phi = \frac{\partial \alpha_r}{\partial \phi} \frac{m_S V^2 h_S}{K_\phi} \frac{1}{\rho} \tag{E6.2}$$

Putting the previous β_f, β_r, α_f, and α_r into Eqn (6.6) gives the following:

$$\rho = \left[1 + \left\{ -\frac{m(l_f K_f - l_r K_r)}{2l^2 K_f K_r} - \frac{m_S h_S}{l K_\phi} \left(\frac{\partial \alpha_f}{\partial \phi} - \frac{\partial \alpha_r}{\partial \phi} \right) \right\} v^2 \right] \frac{l}{\delta} \tag{E6.3}$$

6.3.4 SUSPENSION LATERAL STIFFNESS AND ITS EFFECT

Section 6.1 described how the vehicle body and the wheels are connected elastically by the suspension. This suspension system gives the vehicle wheels a displacement relative to the vehicle body mainly in the vertical direction; however, the vehicle body and the wheels are not completely connected rigidly in the lateral direction. This section will look at the effect of this suspension system lateral stiffness on the vehicle tire characteristics.

Figure 6.16 shows the connection of the vehicle body with the wheel in the horizontal plane. The lateral force does not usually act through the lateral stiffness center of the suspension system. Therefore, it produces an angular displacement of the wheel in the horizontal plane. This is called compliance steer and influences the vehicle steer characteristic.

Compliance steer is generated by the lateral tire force, which depends on the lateral acceleration. Consequently, the suspension system compliance steer is due to the lateral acceleration during cornering. If α_f and α_r in Figure 6.15 are considered as the compliance steer, the effect of the compliance steer on the vehicle steer characteristic can be studied in exactly the same way as the roll steer effect.

The compliance steer can be assumed to be proportional to the lateral force acting on the tire within a relatively small lateral acceleration range. If the cornering stiffness is K and the lateral force is proportional to the side-slip angle, then the following results:

$$\alpha = cF = cK(\beta - \alpha)$$

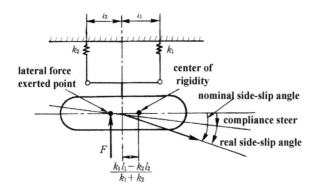

FIGURE 6.16

Suspension plane model.

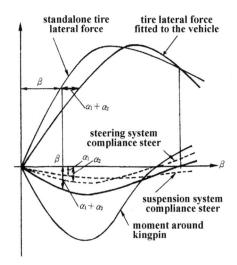

FIGURE 6.17

Tire cornering characteristics when considering compliance steer.

Here, c is a compliance coefficient of the suspension system.
Eliminating α in the previous equation gives the following:

$$F = \frac{K}{1 + cK}\beta = eK\beta$$

whereby $e = 1/(1 + cK)$. In other words, the compliance steer has the effect of changing the tire equivalent cornering stiffness from K to eK. This effect is similar to the effect of the steering system stiffness as described in Section 5.3.1.

The previous concept can be extended beyond the region where the compliance steer is proportional to the lateral force, to the nonlinear region where the lateral force is not proportional to the side-slip angle. Figure 6.17 shows how the tire cornering characteristics when considering both the suspension system lateral stiffness and the steering system stiffness equivalently vary from the original tire cornering characteristics by the compliance steer.

A tire with real side-slip angle, β, suspension system compliance steer, α_1, and steering system compliance steer, α_2, has a nominal side-slip angle of $\beta + \alpha_1 + \alpha_2$. This lateral force at a side-slip angle equal to β is equivalently regarded as the lateral force when the side-slip angle is equal to $\beta + \alpha_1 + \alpha_2$. In this manner, the equivalent tire cornering characteristics with consideration of the compliance steer are obtained, and their effect on the vehicle steer characteristics can be investigated.

6.4 EQUATIONS OF MOTION INCLUSIVE OF ROLL

Until here, the vehicle roll mechanism has only been dealt with geometrically or from a static point of view, where the roll is produced by a constant lateral acceleration. Next, this knowledge will be used to derive the vehicle equations of motion that include rolling motion. This will enable us to investigate the vehicle motion further. These equations of motion are based on those proposed by L. Segel using the fixed roll axis concept [2].

The choice of the coordinate system for dealing with the motion is not necessarily standardized, and it can vary. Here, the coordinate system that the author thinks is most easily understood is used.

6.4.1 COORDINATE SYSTEM AND DYNAMIC MODEL

A coordinate system with the X-Y plane parallel to the ground is fixed in absolute space, as shown in Figure 6.18. Point P in Figure 6.18 is where a vertical line through the C.G. crosses the roll axis. This is the origin of a coordinate system x-y-z that is fixed to the vehicle body, i.e., the sprung mass. The vehicle longitudinal direction, parallel to the ground, is taken as the x-axis (the forward direction is positive), the lateral direction perpendicular to this is taken as the y-axis (the left-hand side when the vehicle is facing to the front as positive), and the vertical direction as the z-axis (up direction as positive).

A coordinate system of x'-y'-z' is fixed to the unsprung mass with the same origin P. The vehicle longitudinal direction is the x'-axis, the lateral direction perpendicular to this is the y'-axis, and the vertical direction is the z'-axis.

The vehicle roll axis does not always coincide with the x-axis, as mentioned in Section 6.2.1. However, for simplicity, the vehicle body is assumed to roll around the x-axis. Furthermore, the roll angle is assumed to be small, and the vehicle is assumed to yaw around the z-axis. The unsprung mass roll motion is neglected, and the unsprung mass is assumed to produce the same yawing motion as the vehicle body around the z'-axis (therefore, the x-axis and the x'-axis always coincide). All the angular displacements and angular accelerations are taken as positive in the direction shown in Figure 6.18.

Figure 6.19 shows the equivalent mechanical model for the sprung mass and the unsprung mass. The mass of the vehicle body, or the sprung mass, is assumed to be distributed

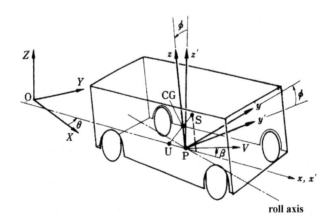

FIGURE 6.18

Coordinate axis.

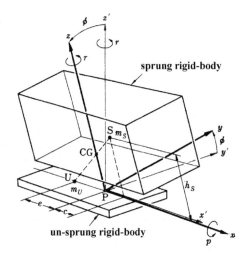

FIGURE 6.19

Equivalent dynamic model.

symmetrically to the x-z plane, and the center of gravity is taken as point S in the x-z plane. The height of the unsprung mass in the z' direction is neglected, and its mass is assumed to be distributed in the x'-y' plane with the center of gravity at point U on the x'-axis.

Here, m_S is the mass of the sprung mass, m_U is the mass of the unsprung mass, and m is the mass of the whole vehicle. h_s is the distance between the sprung mass center of gravity and the x-axis, c and e are the distances of the sprung mass and the unsprung mass center of gravity to the z-axis, respectively, r and p are the vehicle's yaw angular velocity and roll angular velocity, respectively, and ϕ is the vehicle roll angle.

6.4.2 INERTIAS

Rigid body motion can be divided into both the translational motion of the center of gravity and the rotational motion around the center of gravity. Here, these motions will be considered for the sprung mass and the unsprung mass, which are both rigid bodies.

6.4.2.1 Translational motion

The translational motion of a rigid body is the motion of the point where the entire body mass is concentrated at the center of gravity.

The coordinate system X-Y-Z is fixed in absolute space, and the coordinate system x-y-z is fixed on the moving rigid body, as shown in Figure 6.20. The position vector of the center of gravity, C, relative to the x-y-z coordinates is taken as ρ. If points P and C are described by position vectors r and R, relative to the X-Y-Z frame, then the following results:

$$r = R + \rho$$

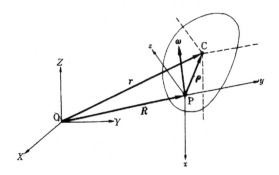

FIGURE 6.20

General motion description of rigid body.

The x-y-z coordinate system has a translational velocity, $\dot{R}$, relative to the X-Y-Z coordinates, and it also has a rotational motion with an angular velocity of ω:

$$\dot{r} = \dot{R} + \dot{\rho}_r + \omega \times \rho$$

Here, $\dot{\rho}_r$ is the relative velocity of the point C to the point P, and if point C is fixed on the x-y-z coordinate, then $\dot{\rho}_r = 0$:

$$\dot{r} = \dot{R} + \omega \times \rho \tag{6.7}$$

The position vectors of the sprung and unsprung mass centers of gravity, S and U, in the X-Y-Z coordinate system are r_S and r_U. The position vectors of point S and point U relative to the x-y-z system are ρ_S and ρ_U, and the angular velocities of the x-y-z and x'-y'-z' coordinates are ω_S and ω_U. The unit vectors in the x-y-z direction are i, j, and k, and the unit vectors in the x'-y'-z' direction are i', j', and k'. These are shown in Figure 6.21.

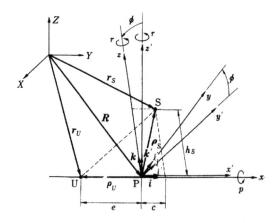

FIGURE 6.21

Motion of points S and P.

Points S and U are fixed on the x-y-z and x'-y'-z' coordinates, respectively, and the following equations are formed based on Eqn (6.7):

$$\dot{r}_S = \dot{R} + \omega_S \times \rho_S \tag{6.8}$$

$$\dot{r}_U = \dot{R} + \omega_U \times \rho_U \tag{6.9}$$

Assuming that point P has a velocity component of u in the x direction and v in the y or y' direction when ϕ is small, then the following results:

$$\dot{R} = ui + vj = ui' + vj' \tag{6.10}$$

The x-y-z coordinates that move with the sprung mass have a roll velocity of p around the x-axis and a yaw velocity of r around the z-axis:

$$\omega_S = pi + rk \tag{6.11}$$

The x'-y'-z' coordinates that move together with the unsprung mass have a yaw velocity of r around the z'-axis:

$$\omega_U = rk' \tag{6.12}$$

and, ρ_S and ρ_U can be written as follows:

$$\rho_S = ci + h_s k \tag{6.13}$$

$$\rho_U = -ei' \tag{6.14}$$

Substituting Eqns (6.10), (6.11), and (6.13) into Eqn (6.8), and Eqns (6.10), (6.12), and (6.14) into Eqn (6.9) gives $\dot{r}_S$ and $\dot{r}_U$ as follows:

$$\dot{r}_S = ui + (v - h_s p + cr)j \tag{6.15}$$

$$\dot{r}_U = ui' + (v - er)j' \tag{6.16}$$

Differentiating Eqns (6.15) and (6.16) gives the acceleration vector of point S and point U:

$$\ddot{r}_S = (\dot{u} - vr + h_s pr - cr^2)i + (\dot{v} + ur - h_s \dot{p} + c\dot{r})j + (vp + h_s p^2 + cpr)k \tag{6.17}$$

$$\ddot{r}_U = (\dot{u} - vr + er^2)i' + (\dot{v} + ur - e\dot{r})j' \tag{6.18}$$

From Eqns (6.17) and (6.18), the lateral acceleration values, in other words, the acceleration in the y and y' direction for the sprung mass and the unsprung mass, α_S and α_U, are as follows:

$$\alpha_S = \dot{v} + ur - h_s \dot{p} + c\dot{r}, \quad \alpha_U = \dot{v} + ur - e\dot{r}$$

If the side-slip angle of point P is β, $|\beta| \ll 1$, and the velocity magnitude of point P, V, is always constant, then $u \approx V$, $v \approx V\beta$, and

$$\alpha_S = V\dot{\beta} + Vr - h_s \dot{p} + c\dot{r}, \quad \alpha_U = V\dot{\beta} + Vr - e\dot{r}$$

The masses for the sprung mass and the unsprung mass are m_S and m_U, respectively. The inertia forces, Y_S and Y_U, of the sprung mass and the unsprung mass are as follows:

$$Y_S = m_S a_S = m_S V(\dot{\beta} + r) - m_S h_S \dot{p} + m_S c \dot{r}$$

$$Y_U = m_U a_U = m_U V(\dot{\beta} + r) - m_U e \dot{r}$$

Consequently, the total vehicle inertia force ΣY is the following:

$$\Sigma Y = Y_S + Y_U = (m_S + m_U)V(\dot{\beta} + r) - m_S h_S \dot{p} + (m_S c - m_U e)\dot{r}$$

Here, $m_S + m_U = m$, and the C.G. is the entire vehicle's center of gravity; therefore, $m_S c - m_U e = 0$. Thus, the following is obtained:

$$\Sigma Y = mV(\dot{\beta} + r) - m_S h_S \dot{p} \tag{6.19}$$

Example 6.3

Differentiate Eqns (6.15) and (6.16) and derive Eqns (6.17) and (6.18).

Solution

From Eqns (6.15) and (6.16),

$$\ddot{r}_S = \dot{u}\dot{i} + u\dot{i} + (\dot{v} - h_S\dot{p} + c\dot{r})\dot{j} + (v - h_S p + cr)\dot{j} \tag{E6.4}$$

$$\ddot{r}_U = \dot{u}\dot{i}' + u\dot{i}' + (\dot{v} - e\dot{r})\dot{j}' + (v - er)\dot{j}' \tag{E6.5}$$

the following is known:

$$\dot{i} = \omega_S \times i, \quad \dot{j} = \omega_S \times j$$

$$\dot{i}' = \omega_U \times i', \quad \dot{j}' = \omega_U \times j'$$

Putting Eqns (6.11) and (6.12) into the previous gives the following:

$$\dot{i} = (pi + rk) \times i = rj$$

$$\dot{j} = (pi + rk) \times j = pk - ri$$

$$\dot{i}' = rk' \times i' = rj'$$

$$\dot{j}' = rk' \times j' = -ri'$$

Using the previous, Eqns (E6.4) and (E6.5) can be rewritten as follows:

$$\ddot{r}_S = (\dot{u} - vr + h_S pr - cr^2)i + (\dot{v} + ur - h_S\dot{p} + c\dot{r})j + (vp + h_S p^2 + cpr)k \tag{6.17}$$

$$\ddot{r}_U = (\dot{u} - vr + er^2)i' + (\dot{v} + ur - e\dot{r})j' \tag{6.18}$$

6.4.2.2 Rotational motion

Generally, the moment of momentum, H_C, around center of gravity, C, of a rigid body could be written as follows:

$$H_C = I \times \omega$$
$$= \left(I_{xx}\omega_x - I_{xy}\omega_y - I_{xz}\omega_z\right)i + \left(-I_{yx}\omega_x + I_{yy}\omega_y - I_{yz}\omega_z\right)j + \left(-I_{zx}\omega_x - I_{zy}\omega_y + I_{zz}\omega_z\right)k$$

whereby I is the inertia tensor of the rigid body around point C. The elements I_{xx}, I_{xy}... are the inertia moments, or inertia products, around the axis passing through point C parallel to the x, y, z axis. The components of the angular velocity, ω, in the x, y, and z direction are ω_x, ω_y, and ω_z.

The moment of momentum of the sprung mass center of gravity, H_S, with angular velocity, ω_S, given by Eqn (6.11) is as follows:

$$H_S = I_S \times \omega_S = \left(I_{xxS}p - I_{xzS}r\right)i + \left(-I_{zzS}p + I_{zzS}r\right)k \tag{6.20}$$

Here, I_S is the inertia tensor of the sprung mass around point S. Considering that the sprung mass is symmetric in the x-z plane, $I_{yxS} = I_{yzS} = 0$.

Similarly, with the angular velocity, ω_U, given by Eqn (6.12), the moment of momentum, H_U, around the unsprung mass center of gravity, U, is as follows:

$$H_U = I_{zzU}rk' \tag{6.21}$$

Here, I_{zzU} is the moment of inertia around the axis passing through point U and parallel to the z' axis. If the unsprung mass is symmetric to the x' axis, the height is neglected, and the mass is assumed to be distributed in the x'-y' plane; all the inertia products are zero.

The time differentiation of the moment of momentum is the moment. Differentiating Eqns (6.20) and (6.21) with respect to time gives the following:

$$\dot{H}_S = \left(I_{xxS}\dot{p} - I_{xzS}\dot{r}\right)i + \left[I_{xzS}p^2 + (I_{xxS} - I_{zzS})pr - I_{xzS}r^2\right]j + \left(-I_{zzS}\dot{p} + I_{zzS}\dot{r}\right)k \tag{6.22}$$

$$\dot{H}_U = I_{zzU}\dot{r}k' \tag{6.23}$$

From these equations, the yaw and rolling moment, N_S and L_S, of the sprung mass around the axis parallel to the x- and z-axis and passing through point S are as follows:

$$N_S = -I_{zzS}\dot{p} + I_{zzS}\dot{r}$$
$$L_S = I_{xxS}\dot{p} - I_{xzS}\dot{r}$$

The yawing moment, N_U, around the axis passing through point U parallel to the z'-axis for the unsprung mass is shown:

$$N_U = I_{zzU}\dot{r}$$

From the previous, the total vehicle yawing moment, ΣN, around the z- or z'-axis and the total rolling moment, ΣL, around the x- or x'-axis are given in Eqns (6.24) and (6.25), where Y_S and Y_U act at the sprung and unsprung mass centers of gravity, S and U, respectively:

$$\Sigma N = N_S + N_U + cY_S - eY_U$$
$$= \left(I_{zzS} + I_{zzU} + m_Sc^2 + m_Ue^2\right)\dot{r} - (I_{zzS} + m_Sh_Sc)\dot{p} + (m_Sc - m_Ue)V\left(\dot{\beta} + r\right) \tag{6.24}$$
$$= I_z\dot{r} - I_{zx}\dot{p}$$

$$\Sigma L = L_S - h_s Y_S$$
$$= \left(I_{xxS} + m_s h_s^2\right)\dot{p} - (I_{xzS} + m_s h_s c)\dot{r} + m_s h_s V(\dot{\beta} + r) \qquad (6.25)$$
$$= I_x \dot{p} - I_{xz}\dot{r} - m_s h_s V(\dot{\beta} + r)$$

whereby $|\phi| \ll 1$.

$$I_z = I_{zzS} + I_{zzU} + m_s c^2 + m_U e^2$$
$$I_{zx} = I_{xz} = I_{xzS} + m_s h_s c$$
$$I_x = I_{xxS} + m_s h_s^2$$

I_z is the total yaw moment of inertia around the vertical axis passing through the vehicle center of gravity, and I_x is the rolling moment of inertia of the sprung mass around the x-axis.

6.4.3 EXTERNAL FORCE

The external forces that act on the vehicle are the lateral tire forces. As described in Section 3.2, lateral forces, which are proportional to the tire side-slip angles, act at the front and rear tires of a moving vehicle. Where roll is considered, camber thrusts also act at the tires. When the vehicle equations of motion include roll motion, these forces must be included in the external forces acting on the entire vehicle.

The front and rear roll steers are α_f and α_r. If they are assumed to be proportional to the roll angle, then the following results:

$$\alpha_f = \frac{\partial \alpha_f}{\partial \phi}\phi, \quad \alpha_r = \frac{\partial \alpha_r}{\partial \phi}\phi$$

where $\partial \alpha_f/\partial \phi$ and $\partial \alpha_r/\partial \phi$ are the front- and rear-wheel roll steering angles per unit roll angle. These are positive when the roll angle is positive, in other words, positive when the steering angle is in the anticlockwise direction.

The actual front-wheel steering angle is now changed by α_f from the steering angle δ. Similarly, a change of α_r is generated at the rear wheel. Using the same assumptions as in Section 3.2 and Eqns (3.6) and (3.7), the front- and rear-wheel tire side-slip angles are the following:

$$\beta_f = \beta + \frac{l_f}{V}r - \delta - \alpha_f = \beta + \frac{l_f}{V}r - \delta - \frac{\partial \alpha_f}{\partial \phi}\phi$$

$$\beta_r = \beta - \frac{l_r}{V}r - \alpha_r = \beta - \frac{l_r}{V}r - \frac{\partial \alpha_r}{\partial \phi}\phi$$

Therefore, the lateral forces, $2Y_f$ and $2Y_r$, acting at the front and rear wheels are as follows:

$$2Y_f = -2K_f \beta_f = 2K_f \left(\delta + \frac{\partial \alpha_f}{\partial \phi}\phi - \beta - \frac{l_f}{V}r\right)$$

$$2Y_r = -2K_r \beta_r = 2K_r \left(\frac{\partial \alpha_r}{\partial \phi}\phi - \beta + \frac{l_r}{V}r\right)$$

where the lateral force changes due to load transfer are neglected.

Assuming that the camber angle produced by the vehicle body roll is proportional to the roll angle, the camber thrust, $2Y_{cf}$ and $2Y_{cr}$, acting at the front and rear wheels, respectively, is as follows:

$$2Y_{cf} = -2K_{cf}\frac{\partial\phi_f}{\partial\phi}\phi, \quad 2Y_{cr} = -2K_{cr}\frac{\partial\phi_r}{\partial\phi}\phi$$

where K_{cf} and K_{cr} are the front- and rear-wheel tire camber thrust coefficients. $\partial\phi_f/\partial\phi$ and $\partial\phi_r/\partial\phi$ are the front- and rear-wheel camber angles per unit roll angle. These are positive if the camber angle is in the same direction as the roll. It is also assumed that a camber thrust of the same magnitude and in the same direction is produced at the left and right wheels.

The total external forces acting on the entire vehicle in the lateral direction, in other words the y-direction, are shown in Figure 6.22.

$$\Sigma F_y = 2Y_f + 2Y_r + 2Y_{cf} + 2Y_{cr}$$

$$= 2K_f\left(\delta + \frac{\partial\alpha_f}{\partial\phi}\phi - \beta - \frac{l_f}{V}r\right) + 2K_r\left(\frac{\partial\alpha_r}{\partial\phi}\phi - \beta + \frac{l_r}{V}r\right) - 2K_{cf}\frac{\partial\phi_f}{\partial\phi}\phi - 2K_{cr}\frac{\partial\phi_r}{\partial\phi}\phi$$

$$(6.26)$$

The total yaw moment around the z-axis produced by the external forces acting on the entire vehicle, ΣM_z, is shown next:

$$\Sigma M_z = 2l_f Y_f - 2l_r Y_r + 2l_f Y_{cf} - 2l_r Y_{cr}$$

$$= 2l_f K_f\left(\delta + \frac{\partial\alpha_f}{\partial\phi}\phi - \beta - \frac{l_f}{V}r\right) - 2l_r K_r\left(\frac{\partial\alpha_r}{\partial\phi}\phi - \beta + \frac{l_r}{V}r\right) - 2l_f K_{cf}\frac{\partial\phi_f}{\partial\phi}\phi + 2l_r K_{cr}\frac{\partial\phi_r}{\partial\phi}\phi$$

$$(6.27)$$

When the vehicle body rolls, it will be subjected to the reaction force by the suspension system spring and shock absorber. These reaction forces produce the rolling moment to the vehicle body around the roll axis, in other words, the x-axis. From Section 6.2.2, the rolling moment that comes from the spring is $-K_\phi\phi$, where $K_\phi = K_{\phi f} + K_{\phi r}$. Assuming that the

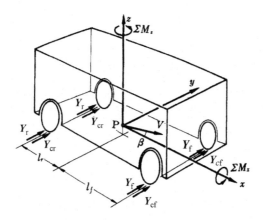

FIGURE 6.22

External forces exerted on vehicle tires.

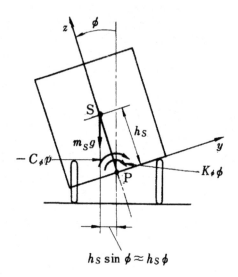

$$hs \sin \phi \approx hs\,\phi$$

FIGURE 6.23

Roll moments exerted on vehicle body.

reaction force produced by the shock absorber is proportional to the roll angular velocity, the rolling moment produced by this reaction force is written as $-C_\phi p$. Here, C_ϕ is the equivalent damping coefficient of the rolling motion, which is the sum of the rolling moment per unit roll angular velocity at the front and rear shock absorbers.

A rolling moment around the x-axis by the weight $m_S g$ also acts on the vehicle, as shown in Figure 6.23. If roll angle, ϕ, is small, this can be approximated by $m_S g h_s \phi$. The total rolling moment around the x-axis by the external forces acting on the entire vehicle, ΣM_x, is expressed as follows:

$$\Sigma M_x = \left(-K_\phi + m_S g h_s\right)\phi - C_\phi p \qquad (6.28)$$

6.4.4 EQUATIONS OF MOTION

The external forces and inertia forces of the vehicle motion inclusive of vehicle body roll can now be used in the equilibrium equations to derive the equations of motion:

$$\Sigma Y - \Sigma F_y = 0 \quad \text{(equilibrium of lateral forces)}$$
$$\Sigma N - \Sigma M_z = 0 \quad \text{(equilibrium of yawing moment)}$$
$$\Sigma L - \Sigma M_x = 0 \quad \text{(equilibrium of rolling moment)}$$

Substituting Eqns (6.19) and (6.24)–(6.28) into the previous equations derives the vehicle equations of motion inclusive of roll motion as follows. Here, $d\phi/dt = p$, and I and I_ϕ are used in replacement of I_z and I_{xx}. I is the entire vehicle yaw moment of inertia, and I_ϕ is the roll moment of inertia around the roll axis when the roll axis is assumed to coincide with the x-axis.

$$mV\left(\frac{d\beta}{dt}+r\right)-m_sh_s\frac{d^2\phi}{dt^2}=2K_f\left(\delta+\frac{\partial\alpha_f}{\partial\phi}\phi-\beta-\frac{l_f}{V}r\right)+2K_r\left(\frac{\partial\alpha_r}{\partial\phi}\phi-\beta+\frac{l_r}{V}r\right)$$
$$-2\left(K_{cf}\frac{\partial\phi_f}{\partial\phi}\phi+K_{cr}\frac{\partial\phi_r}{\partial\phi}\phi\right) \qquad (6.29)$$

$$I\frac{dr}{dt}-I_{xz}\frac{d^2\phi}{dt^2}=2K_f\left(\delta+\frac{\partial\alpha_f}{\partial\phi}\phi-\beta-\frac{l_f}{V}r\right)l_f-2K_r\left(\frac{\partial\alpha_r}{\partial\phi}\phi-\beta+\frac{l_r}{V}r\right)l_r$$
$$-2\left(l_fK_{cf}\frac{\partial\phi_f}{\partial\phi}\phi-l_rK_{cr}\frac{\partial\phi_r}{\partial\phi}\phi\right) \qquad (6.30)$$

$$I_\phi\frac{d^2\phi}{dt^2}-I_{xz}\frac{dr}{dt}-m_sh_sV\left(\frac{d\beta}{dt}+r\right)=\left(-K_\phi+m_sgh_s\right)\phi-C_\phi\frac{d\phi}{dt} \qquad (6.31)$$

Rearranging the previous equations gives the following:

$$mV\frac{d\beta}{dt}+2(K_f+K_r)\beta+\left[mV+\frac{2(l_fK_f-l_rK_r)}{V}\right]r-m_sh_s\frac{d^2\phi}{dt^2}-2Y_\phi\phi=2K_f\delta \qquad (6.29)'$$

$$2(l_fK_f-l_rK_r)\beta+I\frac{dr}{dt}+\frac{2\left(l_f^2K_f+l_r^2K_r\right)}{V}r-I_{xz}\frac{d^2\phi}{dt^2}-2N_\phi\phi=2l_fK_f\delta \qquad (6.30)'$$

$$-m_sh_sV\frac{d\beta}{dt}-I_{xz}\frac{dr}{dt}-m_sh_sVr+I_\phi\frac{d^2\phi}{dt^2}+C_\phi\frac{d\phi}{dt}+\left(K_\phi-m_sgh_s\right)\phi=0 \qquad (6.31)'$$

where

$$Y_\phi=\left(\frac{\partial\alpha_f}{\partial\phi}K_f+\frac{\partial\alpha_r}{\partial\phi}K_r\right)-\left(\frac{\partial\phi_f}{\partial\phi}K_{cf}+\frac{\partial\phi_r}{\partial\phi}K_{cr}\right) \qquad (6.32)$$

$$N_\phi=\left(\frac{\partial\alpha_f}{\partial\phi}l_fK_f-\frac{\partial\alpha_r}{\partial\phi}l_rK_r\right)-\left(\frac{\partial\phi_f}{\partial\phi}l_fK_{cf}-\frac{\partial\phi_r}{\partial\phi}l_rK_{cr}\right) \qquad (6.33)$$

These are the equations of motion of the vehicle with body roll. Here, the change of rolling resistance due to the load transfer between the left and right wheels is neglected as its effect on the vehicle yaw motion is assumed to be small.

Applying a Laplace transformation to Eqns (6.29)′–(6.31)′, the vehicle characteristic equation is obtained as follows:

$$\begin{vmatrix} mVs+2(K_f+K_r) & mV+\dfrac{2(l_fK_f-l_rK_r)}{V} & -m_sh_ss^2-2Y_\phi \\[3mm] 2(l_fK_f-l_rK_r) & Is+\dfrac{2\left(l_f^2K_f+l_r^2K_r\right)}{V} & -I_{xz}s^2-2N_\phi \\[3mm] -m_sh_sVs & -I_{xz}s-m_sh_sV & I_\phi s^2+C_\phi s+\left(K_\phi-m_sgh_s\right) \end{vmatrix}=0 \qquad (6.34)$$

The previous characteristic equation is a little complicated in this form, thus, assuming the following,

$$K_f \approx K_r \approx K, \quad l_f \approx l_r \approx \frac{l}{2}, \quad I \approx ml_f l_r \approx m\left(\frac{l}{2}\right)^2$$

and neglecting the inertia products,

$$K_\phi - m_s g h_s \approx K_\phi$$

$$I_{xz} \approx 0$$

the characteristic equation becomes the following:

$$\begin{vmatrix} mVs + 4K & mV & -m_s h_s s^2 - 2Y_\phi \\ 0 & m\left(\frac{l}{2}\right)^2 s + \dfrac{Kl^2}{V} & -2N_\phi \\ -m_s h_s Vs & -m_s h_s V & I_\phi s^2 + C_\phi s + K_\phi \end{vmatrix} = 0 \qquad (6.34)'$$

Expanding this and rearranging gives the following:

$$A_4 s^4 + A_3 s^3 + A_2 s^2 + A_1 s + A_0 = 0$$

where

$$A_0 = \frac{16K^2 K_\phi}{mV} - \frac{32 m_s h_s K N_\phi}{m l^2} V$$

$$A_1 = 8K\left(K_\phi - \frac{m_s h_s Y_\phi}{m}\right) + \frac{16K^2 C_\phi}{mV}$$

$$A_2 = \left(mK_\phi - 2m_s h_s Y_\phi\right)V + 8KC_\phi + \frac{16K^2 I_\phi}{mV}$$

$$A_3 = mC_\phi V + 4K\left(2I_\phi - \frac{m_s^2 h_s^2}{m}\right)$$

$$A_4 = m\left(I_\phi - \frac{m_s^2 h_s^2}{m}\right)V$$

$$(6.35)$$

6.5 EFFECT OF STEADY-STATE BODY ROLL ON VEHICLE DYNAMICS

In the last section, the vehicle equations of motion with vehicle body roll were derived. The equation of motion is still too complicated for analysis of the vehicle motion characteristics. In order to analyze the basic effects of vehicle body roll on vehicle motion, a two-degree-of-freedom equation of motion (vehicle lateral side-slip and yaw motion) that includes an equivalent roll effect will be derived. This is done by considering vehicle body roll due to constant lateral acceleration.

The steady-state vehicle body roll angle, from Eqn (6.31), is derived by assuming the following:

$$\frac{d\beta}{dt} = \frac{dr}{dt} = \frac{d^2\phi}{dt^2} = \frac{d\phi}{dt} = 0,$$

and, the result is as follows:

$$\phi = \frac{m_s h_s V}{K_\phi - m_s g h_s} r \tag{6.36}$$

Rewriting the right-hand side of Eqns (6.29) and (6.30) with a steady-state roll angle gives the following:

$$mV\left(\frac{d\beta}{dt} + r\right) = 2K_f\left\{\delta - \beta - \frac{l_f}{V}r + \left(\frac{\partial \alpha_f}{\partial \phi} - \frac{K_{cf}}{K_f}\frac{\partial \phi_f}{\partial \phi}\right)\phi\right\}$$
$$+ 2K_r\left\{-\beta + \frac{l_r}{V}r + \left(\frac{\partial \alpha_r}{\partial \phi} - \frac{K_{cr}}{K_r}\frac{\partial \phi_r}{\partial \phi}\right)\phi\right\}$$

$$I\frac{dr}{dt} = 2K_f\left\{\delta - \beta - \frac{l_f}{V}r + \left(\frac{\partial \alpha_f}{\partial \phi} - \frac{K_{cf}}{K_f}\frac{\partial \phi_f}{\partial \phi}\right)\phi\right\}l_f - 2K_r\left\{-\beta + \frac{l_r}{V}r + \left(\frac{\partial \alpha_r}{\partial \phi} - \frac{K_{cr}}{K_r}\frac{\partial \phi_r}{\partial \phi}\right)\phi\right\}l_r$$

ϕ in these equations is proportional to r, as given by Eqn (6.36), and the previous equations are equivalent to the following equations:

$$mV\left(\frac{d\beta}{dt} + r\right) = 2K_f\left(\delta - \beta - \frac{l_f'}{V}r\right) + 2K_r\left(-\beta + \frac{l_r'}{V}r\right) \tag{6.37}$$

$$I\frac{dr}{dt} = 2K_f\left(\delta - \beta - \frac{l_f'}{V}r\right)l_f + 2K_r\left(-\beta + \frac{l_r'}{V}r\right)l_r \tag{6.38}$$

where

$$l_f' = l_f\left(1 + B_f V^2\right) \tag{6.39}$$

$$l_r' = l_r\left(1 + B_r V^2\right) \tag{6.40}$$

$$B_f = \frac{-m_s h_s\left(\frac{\partial \alpha_f}{\partial \phi} - \frac{K_{cf}}{K_f}\frac{\partial \phi_f}{\partial \phi}\right)}{l_f\left(K_\phi - m_s g h_s\right)}, \quad B_r = \frac{m_s h_s\left(\frac{\partial \alpha_r}{\partial \phi} - \frac{K_{cr}}{K_r}\frac{\partial \phi_r}{\partial \phi}\right)}{l_r\left(K_\phi - m_s g h_s\right)} \tag{6.41}$$

When Eqns (6.37) and (6.38) are compared with Eqns (3.10) and (3.11), it is easily understood that the equivalent vehicle lateral side-slip motion and yawing motion with body roll is found by replacing the front and rear wheel distances from the vehicle center of gravity l_f and l_r with equivalent l_f' and l_r'.

Rewriting Eqns (6.37) and (6.38), the equivalent vehicle two-degree-of-freedom equations of motion with roll can be expressed as follows:

$$mV\frac{d\beta}{dt} + 2(K_f + K_r)\beta + \left[mV + \frac{2(l_f'K_f - l_r'K_r)}{V}\right]r = 2K_f\delta \tag{6.37}'$$

$$2(l_fK_f - l_rK_r)\beta + I\frac{dr}{dt} + \frac{2(l_f'l_fK_f + l_r'l_rK_r)}{V}r = 2l_fK_f\delta \tag{6.38}'$$

In steady-state cornering, $d\beta/dr = dr/dt = 0$ can be substituted into Eqns (6.37)' and (6.38)'. The yaw velocity, r, in response to a constant steering angle is shown next:

$$r = \frac{1}{1 - \frac{ml_fK_f - l_rK_r}{2l(l_f'+l_r')K_fK_r}V^2} \frac{V}{(l_f' + l_r')}\delta \tag{6.42}$$

from Eqns (6.39) and (6.40),

$$l_f' + l_r' = l(1 + BV^2) \tag{6.43}$$

where

$$B = \frac{l_fB_f + l_rB_r}{l} = \frac{msh_s}{l(K_\phi - msgh_s)}\left[\frac{\partial\alpha_r}{\partial\phi} - \frac{\partial\alpha_f}{\partial\phi} + \frac{K_{cf}}{K_f}\frac{\partial\phi_f}{\partial\phi} - \frac{K_{cr}}{K_r}\frac{\partial\phi_r}{\partial\phi}\right] \tag{6.44}$$

Substituting Eqns (3.43), (6.43), and (6.44) into Eqn (6.42) gives the following:

$$r = \frac{1}{1 + \frac{AV^2}{1+BV^2}} \frac{1}{1 + BV^2} \frac{V}{l}\delta = \frac{1}{1 + A'V^2} \frac{V}{l}\delta \tag{6.45}$$

Here, A' is the equivalent stability factor when roll is being considered.

$$A' = A + B$$
$$= -\frac{m(l_fK_f - l_rK_r)}{2l^2K_fK_r} + \frac{msh_s}{l(K_\phi - msgh_s)}\left[\frac{\partial\alpha_r}{\partial\phi} - \frac{\partial\alpha_f}{\partial\phi} + \frac{K_{cf}}{K_f}\frac{\partial\phi_f}{\partial\phi} - \frac{K_{cr}}{K_r}\frac{\partial\phi_r}{\partial\phi}\right] \tag{6.46}$$

From the previous, if $B > 0$, namely if the following is true:

$$\left[\frac{\partial\alpha_r}{\partial\phi} - \frac{\partial\alpha_f}{\partial\phi} + \frac{K_{cf}}{K_f}\frac{\partial\phi_f}{\partial\phi} - \frac{K_{cr}}{K_r}\frac{\partial\phi_r}{\partial\phi}\right] > 0$$

then the vehicle body roll changes the vehicle steer characteristics to US. More precisely, positive roll steer at the rear wheel and positive camber change at the front wheel, or negative roll steer at the front wheel and negative camber change at the rear wheel have the effect of changing the vehicle steer characteristics toward US.

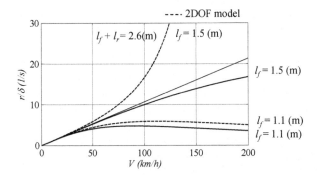

FIGURE 6.24

Effect of roll steer on steady-state cornering.

Figure 6.24 is an example of how the relation between yaw velocity, r, and the traveling speed, V, changes with the distance of the front axle from the vehicle center of gravity. The results are for steady-state cornering for a normal passenger car with rear-wheel roll steer. The rear-wheel roll steer is set so it always changes the vehicle steer characteristic to either NS or US, regardless of the distance of the front axle from the vehicle center of gravity.

In the same figure, the relationship between the yaw velocity and the traveling speed for a two-degree-of-freedom model without vehicle roll is also shown. The figure shows that with the introduction of rear-wheel roll steer, not only is the vehicle steer characteristic changed to US, but the change in the vehicle steer characteristic due to the distance between front axle and the center of gravity also becomes less prominent [3].

Applying Laplace transforms to Eqns (6.37)′ and (6.38)′, the vehicle characteristic equation is obtained, which has a general form of the following:

$$s^2 + 2D's + P'^2 = 0 \tag{6.47}$$

where

$$2D' = \frac{2m\left(l_f' l_f K_f + l_r' l_r K_r\right) + 2I(K_f + K_r)}{mIV}$$

$$= 2D + \frac{2\left(l_f^2 K_f B_f + l_r^2 K_r B_r\right)V}{I} \tag{6.48}$$

$$P'^2 = \frac{4K_f K_r l\left(l_f' + l_r'\right)}{mIV^2} - \frac{2(l_f K_f - l_r K_r)}{I}$$

$$= P^2 + \frac{4K_f K_r l^2}{mI}B \tag{6.49}$$

$2D$ and P^2 are the coefficients of the characteristic equation given by Eqns (3.56) and (3.57) when the roll is not being considered. Consequently, taking $I = mk^2$ and by Eqn (3.67), the vehicle natural frequency ω_n' when the roll is being considered is as follows:

$$
\begin{aligned}
\omega_n' = P' &= \sqrt{P + \frac{4K_f K_r l^2}{mI} B} \\
&= \sqrt{\frac{4K_f K_r l^2}{m^2 k^2} \left(\frac{1 + AV^2}{V^2}\right) + \frac{4K_f K_r l^2}{m^2 k^2} B} \\
&= \frac{2\sqrt{K_f K_r} l}{mk} \frac{\sqrt{1 + A'V^2}}{V}
\end{aligned}
\tag{6.50}
$$

Assuming $l_f \approx l_r$ and $K_f \approx K_r$, the damping ratio, ζ', when the roll is being considered is obtained:

$$
\zeta' = \frac{D'}{P'} = \left[\frac{1 + k^2/l_f l_r}{2\sqrt{k^2/l_f l_r}} + \frac{1}{2\sqrt{k^2/l_f l_r}} BV^2\right] \frac{1}{\sqrt{1 + A'V^2}}
\tag{6.51}
$$

6.6 EFFECTS OF BODY ROLL ON VEHICLE STABILITY

In the last section, only the effects of steady roll angle on the vehicle dynamics were considered for simple understanding. Therefore, the vehicle equations of motion still remain in a form of two-degree-of-freedom of side slip and yaw, even when a roll is considered, and the analytical descriptions of such parameters as stability factor, natural frequency, damping ratio, etc., are available.

Next, the transient effects of body roll on vehicle dynamics, especially on response stability of the vehicle to steering input, will be examined. For a basic understanding, the characteristic equation of vehicle motion with roll is regarded to be given again by Eqn (6.35) under the same premise for simplification taken in Section 6.4.4. As it is possible to understand from Eqns (6.32) and (6.33) that the effects of roll steer and camber change are identical, here only roll steer is considered with no camber change.

Generally, the roll steer is used for the adjustment of the vehicle steer characteristics. It is obvious from Eqn (6.46) that the roll steer has no effects on the steer characteristics if both front and rear axles have the same roll steer. Therefore, the same amount of roll steer at the front and rear axles with counter direction of each other are taken into consideration, as shown by Eqn (6.52):

$$
\frac{\partial \alpha_f}{\partial \phi} = -\frac{\partial \alpha_r}{\partial \phi} = \frac{\partial \alpha}{\partial \phi}
\tag{6.52}
$$

Putting Eqn (6.52) into Eqns (6.32) and (6.33), the following are obtained as the camber change is zero:

$$
Y_\phi = 0
\tag{6.53}
$$

$$N_\phi = \frac{\partial \alpha}{\partial \phi} l_f K_f + \frac{\partial \alpha}{\partial \phi} l_r K_r = \frac{\partial \alpha}{\partial \phi} lK \qquad (6.54)$$

As is discussed in Section 6.5, if $\partial \alpha / \partial \phi$ is positive, the vehicle becomes OS, and then negative, and then US. In addition to the previous, using the assumption $C_\phi = 0$, the characteristic equation becomes the following:

$$A_4 s^4 + A_3 s^3 + A_2 s^2 + A_1 s + A_0 = 0$$

where

$$A_0 = \frac{16K^2 K_\phi}{mV} - \frac{32 m_s h_s K^2}{ml} \frac{\partial \alpha}{\partial \phi} V$$

$$A_1 = 8KK_\phi$$

$$A_2 = mK_\phi V + \frac{16K^2 I_\phi}{mV} \qquad\qquad (6.35)'$$

$$A_3 = 4K\left(2I_\phi - \frac{m_s^2 h_s^2}{m}\right)$$

$$A_4 = m\left(I_\phi - \frac{m_s^2 h_s^2}{m}\right) V$$

As A_1, A_2, A_3, and A_4 are always positive, this is one of the stability conditions:

$$A_0 = \frac{16K^2 K_\phi}{mV}\left(1 - \frac{2 m_s h_s}{l K_\phi} \frac{\partial \alpha}{\partial \phi} V^2\right) > 0 \qquad (6.55)$$

In addition, the following condition must be satisfied for the vehicle stability:

$$A_1 A_2 A_3 - A_0 A_3^2 - A_1^2 A_4 > 0$$

Using Eqn (6.35)', the previous condition is rewritten as follows:

$$\left[1 + \frac{16K^2\left(2I_\phi - \frac{m_s^2 h_s^2}{m}\right)^2}{m m_s h_s l K_\phi^2}\frac{\partial \alpha}{\partial \phi}\right] V^2 + \frac{8K^2\left(2I_\phi - \frac{m_s^2 h_s^2}{m}\right)}{m^2 K_\phi} > 0 \qquad (6.56)$$

Now at first, from the stability condition, Eqn (6.55), if $\partial \alpha / \partial \phi$ is positive, which means the vehicle is OS, then the vehicle has a critical speed, V_{C1}, described as follows:

$$V_{C1} = \sqrt{\frac{l K_\phi}{2 m_s h_s \frac{\partial \alpha}{\partial \phi}}} \qquad (6.57)$$

The previous critical speed has the same implication as is discussed in Section 3.3.2 that if the vehicle is oversteer, there is the critical speed for stability limit, which has nothing to do with whether the vehicle motion has a roll mode or not. The vehicle is non-oscillatory unstable at the vehicle speed greater than that described by Eqn (6.57).

The next stability condition is Eqn (6.56). Because the independent term to the vehicle speed in Eqn (6.56) is always positive, the following condition must be satisfied for the vehicle stability at any speed:

$$1 + \frac{16K^2 \left(2I_\phi - \frac{m_s^2 h_s^2}{m}\right)^2}{mm_s h_s l K_\phi^2} \frac{\partial \alpha}{\partial \phi} \geq 0 \tag{6.58}$$

This is rewritten as follows:

$$\frac{\partial \alpha}{\partial \phi} \geq -\frac{mm_s h_s l K_\phi^2}{16K^2 \left(2I_\phi - \frac{m_s^2 h_s^2}{m}\right)^2} \tag{6.59}$$

and, the critical value of the roll steer is defined as follows:

$$\left(\frac{\partial \alpha}{\partial \phi}\right)_C = -\frac{mm_s h_s l K_\phi^2}{16K^2 \left(2I_\phi - \frac{m_s^2 h_s^2}{m}\right)^2} \tag{6.60}$$

As $(\partial \alpha / \partial \phi)_C$ is negative, the stability condition described by Eqn (6.56) is rewritten as the following:

$$\left[1 - \frac{\partial \alpha / \partial \phi}{(\partial \alpha / \partial \phi)_C}\right] V^2 + \frac{8K^2 \left(2I_\phi - \frac{m_s^2 h_s^2}{m}\right)}{m^2 K_\phi} > 0 \tag{6.56'}$$

This is always satisfied at any vehicle speed when $\partial \alpha / \partial \phi$ is positive, which means the roll steer makes the vehicle OS. On the other hand, if $\partial \alpha / \partial \phi$ is negative, which means the roll steer makes the vehicle US and $\partial \alpha / \partial \phi < (\partial \alpha / \partial \phi)_C$, then there is a critical vehicle speed due to Eqns (6.56) or (6.56)', described as follows:

$$V_{C2} = \sqrt{\frac{-\frac{8K^2}{m^2 K_\phi}\left(2I_\phi - \frac{m_s^2 h_s^2}{m}\right)}{1 - \frac{\partial \alpha / \partial \phi}{(\partial \alpha / \partial \phi)_C}}} \tag{6.61}$$

The vehicle is oscillatory unstable at the vehicle speed greater than that described by Eqn (6.61).

It has been shown in Chapter 3 that the understeer vehicle shows the oscillatory response to steering input at high vehicle speed; however, there is no condition to make the vehicle unstable at any speed so far as the vehicle motion is considered by the two-degree-of-freedom

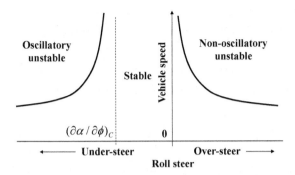

FIGURE 6.25

Image of vehicle stability limit due to roll steer.

equation of motion, side slip, and yaw. Things are different when the roll motion is taken into consideration. There is the critical vehicle speed that makes the vehicle oscillatory unstable when the vehicle is adjusted to become understeer by some roll steer.

The effects of the body roll, especially roll steer, on vehicle stability obtained in the previous are summarized as follows:

When $\partial\alpha/\partial\phi > 0$ (OS), non-oscillatory unstable at $V \geq V_{C1}$.
When $(\partial\alpha/\partial\phi)_C \leq \partial\alpha/\partial\phi \leq 0$ (NS, US), always stable at any vehicle speed.
When $\partial\alpha/\partial\phi < (\partial\alpha/\partial\phi)_C$ (US), oscillatory unstable at $V \geq V_{C2}$.

A schematic image of the stable/unstable region on the $V - \partial\alpha/\partial\phi$ plane is shown in Figure 6.25.

It is important to note that the previous analysis is based on the premise of the suspension damper equal to zero, however, there is a possibility of the existence of a practical vehicle speed for oscillatory unstable under some practically reasonable values of suspension damper and roll steer [4].

PROBLEMS

6.1 Sometimes, the roll angle of the vehicle subjected to the 0.5 g steady lateral acceleration is defined as the roll rate. Calculate the roll rate of the vehicle with the mass of the vehicle body $m_S = 1400$ kg, the body C.G. height from the roll axis $h_s = 0.52$ m, the front roll stiffness $K_{\phi f} = 65.0$ kNm/rad, and the rear roll stiffness $K_{\phi r} = 35.0$ kNm/rad.

6.2 Calculate the lateral load transfer at the front and rear suspensions, respectively, for the vehicle in Problem 6.1. Use the following vehicle parameters in addition to those in 6.1: the position of the front wheels from the C.G. $l_f = 1.1$ m, the position of the rear wheels $l_r = 1.6$ m, the front tread $d_f = 1.5$ m, the rear tread $d_r = 1.5$ m, the front roll center height from the ground $h_f = 0.05$ m, and the rear roll center height $h_r = 0.2$ m.

6.3 Calculate what percent of the cornering stiffness is equivalently reduced by the suspension compliance steer if the cornering stiffness of the original tire is 60 kN/rad and the compliance steer to a unit lateral force is 0.00185 rad/kN.

6.4 Investigate the relative value of the camber change rate and the roll steer rate that give us almost the same effects on the steer characteristics of the vehicle.

6.5 Find the roll steer rate at the rear suspension that is needed to make the OS vehicle in Figure 6.24 NS. Use the following vehicle parameters, $m = 1500$ kg, $m_S = 1400$ kg, $l_f = 1.5$ m, $l_r = 1.1$ m, $K_f = 55$ kN/rad, $K_r = 62$ kN/rad, $h_s = 0.52$ m, and $K_\phi = 100$ kNm/rad. Neglect all the camber changes and the roll steer in the front suspension.

REFERENCES

[1] Eberan R. ROLL ANGLES—the calculation of wheel loads and angular movement on curves. Automob Eng October, 1951; 379–84.

[2] Segel L. Theoretical prediction and experimental substantiation of the response of the automobile to steering control. Proc IMechE (AD) 1956–1957; 310–31.

[3] Ellis JR. Vehicle dynamics. London: London Business Book Ltd; 1969.

[4] Ishio J, Abe M. Effect of body roll on vehicle dynamics. Trans Jpn Soc Automot Eng September 2013;44(5).

VEHICLE MOTION WITH TRACTION AND BRAKING

7.1 PREFACE

Until now, the discussion of vehicle motion has been limited to the case of constant forward speed only, without consideration of the traction and braking of the vehicle. However, the ground vehicle that is the subject of study in this book accelerates and decelerates in the longitudinal direction more frequently than other modes of transportation.

Therefore, in this chapter, the basic characteristics of vehicle motion will be reconsidered to include traction and braking. In particular, it is shown that the analysis of the vehicle motion with traction and braking becomes possible if the change of vehicle speed due to traction and braking is small when the vehicle is traveling at high speed.

7.2 EQUATIONS OF MOTION INCLUSIVE OF LONGITUDINAL MOTION

From Section 3.2.1, the acceleration vector of the vehicle center of gravity P in the horizontal plane motion is given by Eqn (3.3):

$$\ddot{R} = (\dot{u} - vr)i + (\dot{v} + ur)j \tag{3.3}$$

Although v is small in comparison to u, as described previously, u is not necessarily always constant. Therefore, the use of β in the description of the vehicle motion is not always convenient. The vehicle longitudinal and lateral acceleration are better expressed by $\dot{u} - vr$ and $\dot{v} + ur$, respectively.

Using the same assumptions as in Section 3.2.1, the vehicle equations of motion in basic plane motion inclusive of the motion in the longitudinal direction can be expressed by the following equations:

$$m\left(\frac{du}{dt} - vr\right) = 2X_f + 2X_r \tag{7.1}$$

$$m\left(\frac{dv}{dt} + ur\right) = 2Y_f + 2Y_r \tag{7.2}$$

$$I\frac{dr}{dt} = 2(l_f Y_f - l_r Y_r) \tag{3.5}'$$

Here, X_f and X_r are the longitudinal forces acting on the tires. These are mainly dependent on the tire longitudinal slip ratio as discussed in Section 2.4, and they can be considered to be

187

independent of the vehicle plane motion. Y_f and Y_r are dependent on the tire side-slip angle, and when the side-slip angle is small, these forces are given by Eqns (3.8) and (3.9). The vehicle side-slip angle at this instance is $\beta \approx v/u$ when $u \gg v$. Furthermore, the lateral forces acting on the tires, Y_f and Y_r, depend on the longitudinal forces, X_f and X_r, as described in Sections 2.3 or 2.4.

It is understood that the study of the basic characteristics of vehicle motion inclusive of longitudinal motion through direct solving and analysis of Eqns (7.1), (7.2), and (3.5)′ by their original forms is impossible.

7.3 VEHICLE QUASI-STEADY-STATE CORNERING [1]

As mentioned in the previous section, the analytical study of the basic characteristics of vehicle motion with longitudinal motion based on the equations of motion is difficult, and some modification is needed.

When a vehicle undergoes steady-state cornering and constant traction/braking, the vehicle speed changes, and a steady-state condition is not fulfilled. If a very short period of time is considered, where the changes of the speed due to traction/braking can be neglected, e.g., when the vehicle is traveling at high speed, the assumption of steady-state cornering at constant longitudinal and lateral acceleration is possible. This is called quasi-steady-state cornering.

This kind of cornering condition does exist in reality. When the vehicle corners at a constant speed on a curved road with a slope while subjected to traction/braking is an example. In this case, the vehicle is capable of steady-state cornering with traction/braking.

7.3.1 EXPANSION OF THE STABILITY FACTOR FOR TRACTION/BRAKING

7.3.1.1 Tire side-slip angle during cornering

The cornering lateral acceleration of a vehicle with weight W, when accompanied by traction/braking, is defined as $\ddot{y}$, and it is the same as in previous chapters. The equilibrium equations are as follows:

$$W\ddot{y} = (K_{f1} + K_{f2})\beta_f + (K_{r1} + K_{r2})\beta_r \tag{7.3}$$

$$l_f(K_{f1} + K_{f2})\beta_f - l_r(K_{r1} + K_{r2})\beta_r = 0 \tag{7.4}$$

Here, β_f and β_r are the front and rear tire side-slip angles, K_{f1} and K_{f2} are the front left and right tire cornering stiffness, and K_{r1} and K_{r2} are the rear left and right cornering stiffness. In the following discussion, the subscript 1 will be used for the left wheel and subscript 2 for the right wheel to discriminate the characteristic difference between left and right tires.

The tire cornering stiffness depends on the tire load, and the effect of the load transfer on the cornering stiffness can be considered up to the first-order term if its effect is small. The relationship between the lateral force and the traction/braking force can be approximated using Eqn (2.40). Also, if the traction/braking force is small compared to the tire load, it can be modeled by a simple parabolic function. With these simplifications in mind, the cornering stiffness at a small side-slip angle could be written as follows, in this case for the front wheel:

$$K_{f1} \approx \left\{ K_{f0} + \frac{\partial K_f}{\partial W}\left(-\Delta W_f - \frac{\Delta W}{2}\right) \right\}\left\{ 1 - \frac{1}{2}\left(\frac{2X_f}{\mu W_f}\right)^2 \right\}$$

where ΔW_f is the load transfer across the front axle during cornering, ΔW is the load transfer between the front and rear due to traction/braking, W_f is the front axle vertical load, and μ is the friction coefficient between the tire and the road surface.

For small longitudinal and lateral accelerations,

$$\frac{\partial K_f}{\partial W} \frac{\Delta W_f}{K_{f0}}, \quad \frac{\partial K_f}{\partial W} \frac{\Delta W}{K_{f0}}, \quad \left(\frac{2X_f}{\mu W_f}\right)^2$$

These are considered as small values of the same order, giving the following:

$$K_{f1} = K_{f0} \left\{ 1 - \frac{\partial K_f}{\partial W} \frac{\Delta W_f}{K_{f0}} - \frac{\partial K_f}{\partial W} \frac{\Delta W}{2K_{f0}} - \frac{1}{2} \left(\frac{2X_f}{\mu W_f}\right)^2 \right\} \tag{7.5}$$

$$K_{f2} = K_{f0} \left\{ 1 + \frac{\partial K_f}{\partial W} \frac{\Delta W_f}{K_{f0}} - \frac{\partial K_f}{\partial W} \frac{\Delta W}{2K_{f0}} - \frac{1}{2} \left(\frac{2X_f}{\mu W_f}\right)^2 \right\} \tag{7.6}$$

$$K_{r1} = K_{r0} \left\{ 1 - \frac{\partial K_r}{\partial W} \frac{\Delta W_r}{K_{r0}} + \frac{\partial K_r}{\partial W} \frac{\Delta W}{2K_{r0}} - \frac{1}{2} \left(\frac{2X_r}{\mu W_r}\right)^2 \right\} \tag{7.7}$$

$$K_{r2} = K_{r0} \left\{ 1 + \frac{\partial K_r}{\partial W} \frac{\Delta W_r}{K_{r0}} + \frac{\partial K_r}{\partial W} \frac{\Delta W}{2K_{r0}} - \frac{1}{2} \left(\frac{2X_r}{\mu W_r}\right)^2 \right\} \tag{7.8}$$

where ΔW_r is the load transfer across the rear axle, and W_r is the rear axle vertical load.

From the previous, the equivalent cornering stiffness of the front and rear axles are obtained by summing up the left and right wheel cornering stiffness, respectively:

$$2K_f^* = K_{f1} + K_{f2} = 2K_{f0} \left\{ 1 - \frac{\partial K_f}{\partial W} \frac{\Delta W}{2K_{f0}} - \frac{1}{2} \left(\frac{2X_f}{\mu W_f}\right)^2 \right\} \tag{7.9}$$

$$2K_r^* = K_{r1} + K_{r2} = 2K_{r0} \left\{ 1 + \frac{\partial K_r}{\partial W} \frac{\Delta W}{2K_{r0}} - \frac{1}{2} \left(\frac{2X_r}{\mu W_r}\right)^2 \right\} \tag{7.10}$$

Obtaining β_f and β_r from Eqns (7.3) and (7.4) and using Eqns (7.9) and (7.10), by the same assumption of small values as previously, gives the following:

$$\beta_f \approx \frac{l_r W}{2l K_{f0}} \left\{ 1 + \frac{\partial K_f}{\partial W} \frac{\Delta W}{2K_{f0}} + \frac{1}{2} \left(\frac{2X_f}{\mu W_f}\right)^2 \right\} \ddot{y} \tag{7.11}$$

$$\beta_r \approx \frac{l_f W}{2l K_{r0}} \left\{ 1 - \frac{\partial K_r}{\partial W} \frac{\Delta W}{2K_{r0}} + \frac{1}{2} \left(\frac{2X_r}{\mu W_r}\right)^2 \right\} \ddot{y} \tag{7.12}$$

Here, $W_f = l_r W/l, W_r = l_f W/l, \Delta W = hW\ddot{x}/l, X_f = \alpha_c W\ddot{x}/2, X_r = (1 - \alpha_c)W\ddot{x}/2$. Here, α_c is the traction/braking force distribution ratio between the front and rear, $\ddot{x}$ is the longitudinal acceleration by traction/braking (in gravitational units), and h is the height of the center of gravity from the ground. Substituting these into Eqns (7.11) and (7.12) gives the following:

$$\beta_f \approx \frac{l_r W}{2lK_{f0}}\left\{1 + \frac{hW}{2lK_{f0}}\frac{\partial K_f}{\partial W}\ddot{x} + \frac{1}{2}\left(\frac{\alpha_c l}{\mu l_r}\right)^2 \ddot{x}^2\right\}\ddot{y} \tag{7.13}$$

$$\beta_r \approx \frac{l_f W}{2lK_{r0}}\left\{1 - \frac{hW}{2lK_{r0}}\frac{\partial K_r}{\partial W}\ddot{x} + \frac{1}{2}\left(\frac{(1 - \alpha_c)l}{\mu l_f}\right)^2 \ddot{x}^2\right\}\ddot{y} \tag{7.14}$$

Moreover, expressing the equivalent cornering stiffness in Eqns (7.9) and (7.10) by $\ddot{x}$, gives the following:

$$2K_f^* \approx 2K_{f0}\left[1 + \frac{hW}{2lK_{f0}}\frac{\partial K_f}{\partial W}\ddot{x} + \frac{1}{2}\left(\frac{\alpha_c l}{\mu l_r}\right)^2 \ddot{x}^2\right] \tag{7.9$'$}$$

$$2K_r^* \approx 2K_{r0}\left[1 - \frac{hW}{2lK_{r0}}\frac{\partial K_r}{\partial W}\ddot{x} + \frac{1}{2}\left(\frac{(1 - \alpha_c)l}{\mu l_f}\right)^2 \ddot{x}^2\right] \tag{7.10$'$}$$

7.3.1.2 Toe angle change and compliance steer during cornering

Sections 5.3 and 6.3 studied the effect of the tire toe angle change and compliance steer on vehicle plane motion. These effects are now included for cornering with traction/braking.

If the vehicle has longitudinal and lateral accelerations of $\ddot{x}$ and $\ddot{y}$, the vehicle will pitch and roll. This will result in changes in the suspension stroke. Assuming the pitch and the roll axes are on the ground, the toe angle change at the front wheel due to the pitching motion is $(\partial\alpha_f/\partial z)\cdot(l_f hW\ddot{x}/K_\theta)$. Similarly, the toe angle change due to roll is $(\partial\alpha_f/\partial z)\cdot(d_f hW\ddot{y}/2K_\phi)$. Here, $\partial\alpha_f/\partial z$ is the tire toe angle change per unit suspension stroke, d_f is the tread, K_θ is the pitch stiffness, and K_ϕ is the roll stiffness of the vehicle.

Furthermore, the torque, T_s, exerted on the steering system by the tire lateral force, $2Y_f = l_r W\ddot{y}$, and the longitudinal force, $X_f = \alpha_c W\ddot{x}$, acting on the front wheels is as follows:

$$T_s = 2\xi Y_f + \frac{Y_f}{K_y}2X_f = \left(\xi + \frac{\alpha_c W}{2K_y}\ddot{x}\right)\frac{l_r W}{l}\ddot{y} \tag{7.15}$$

This is shown in Figure 7.1. K_y is the tire lateral stiffness, and ξ is the sum of a pneumatic trail and a castor trail.

From the previous, the sum of the toe angle change and the compliance steer for each tire can be written as follows:

$$\alpha_{f1} = \frac{\partial\alpha_f}{\partial z}\frac{l_f hW}{K_\theta}\ddot{x} + \frac{\partial\alpha_f}{\partial z}\frac{d_f hW}{2K_\phi}\ddot{y} - \frac{\partial\alpha_f}{\partial X}\frac{\alpha_c W}{2}\ddot{x} - \frac{\partial\alpha_f}{\partial T}\left(\xi + \frac{\alpha_c W}{2K_y}\ddot{x}\right)\frac{l_r W}{l}\ddot{y} \tag{7.16}$$

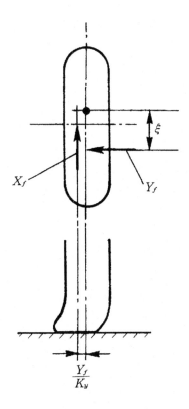

FIGURE 7.1

Steering torque by lateral and longitudinal forces.

$$\alpha_{f2} = -\frac{\partial \alpha_f}{\partial z}\frac{l_f h W}{K_\theta}\ddot{x} + \frac{\partial \alpha_f}{\partial z}\frac{d_f h W}{2K_\phi}\ddot{y} + \frac{\partial \alpha_f}{\partial X}\frac{\alpha_c W}{2}\ddot{x} - \frac{\partial \alpha_f}{\partial T}\left(\xi + \frac{\alpha_c W}{2K_y}\ddot{x}\right)\frac{l_r W}{l}\ddot{y} \quad (7.17)$$

$$\alpha_{r1} = -\frac{\partial \alpha_r}{\partial z}\frac{l_r h W}{K_\theta}\ddot{x} + \frac{\partial \alpha_r}{\partial z}\frac{d_r h W}{2K_\phi}\ddot{y} - \frac{\partial \alpha_r}{\partial X}\frac{(1-\alpha_c)W}{2}\ddot{x} \quad (7.18)$$

$$\alpha_{r2} = \frac{\partial \alpha_r}{\partial z}\frac{l_r h W}{K_\theta}\ddot{x} + \frac{\partial \alpha_r}{\partial z}\frac{d_r h W}{2K_\phi}\ddot{y} + \frac{\partial \alpha_r}{\partial X}\frac{(1-\alpha_c)W}{2}\ddot{x} \quad (7.19)$$

Here, $\partial \alpha_f / \partial X$ and $\partial \alpha_r / \partial X$ are the compliance steer per unit longitudinal force at the front and rear suspension system, $\partial \alpha_f / \partial T$ is the steering system compliance steer per unit steering torque, and α_c is the traction/braking force distribution ratio between the front and rear.

The cornering force due to these toe angle changes at the front wheel is obtained using Eqns (7.5), (7.6), (7.16), and (7.17) as follows:

$$
\begin{aligned}
K_{f1}\alpha_{f1} + K_{f2}\alpha_{f2} \\
= 2K_{f0}\left\{1 - \frac{\partial K_f}{\partial W}\frac{\Delta W}{2K_{f0}} - \frac{1}{2}\left(\frac{2X_f}{\mu W_f}\right)^2\right\}\left\{\frac{\partial \alpha_f}{\partial z}\frac{d_f hW}{2K_\phi}\ddot{y} - \frac{\partial \alpha_f}{\partial T}\left(\xi + \frac{\alpha_c W}{2K_y}\ddot{x}\right)\frac{l_r W}{l}\ddot{y}\right\} \\
- 2K_{f0}\left(\frac{\partial \alpha_f}{\partial z}\frac{l_f hW}{K_\theta}\ddot{x} - \frac{\partial \alpha_f}{\partial X}\frac{\alpha_c W}{2}\ddot{x}\right)\frac{\partial K_f}{\partial W}\frac{\Delta W_f}{K_{f0}} \quad (7.20)\\
\approx 2K_f^*\left\{\frac{\partial \alpha_f}{\partial z}\frac{d_f hW}{2K_\phi}\ddot{y} - \frac{\partial \alpha_f}{\partial T}\left(\xi + \frac{\alpha_c W}{2K_y}\ddot{x}\right)\frac{l_r W}{l}\ddot{y}\right. \\
\left. - \left(\frac{\partial \alpha_f}{\partial z}\frac{l_f hW}{K_\theta}\ddot{x} - \frac{\partial \alpha_f}{\partial X}\frac{\alpha_c W}{2}\ddot{x}\right)\frac{\partial K_f}{\partial W}\frac{\Delta W_f}{K_{f0}}\right\} = 2K_f^*\alpha_f
\end{aligned}
$$

Similarly, for the rear wheel,

$$
\begin{aligned}
K_{r1}\alpha_{r1} + K_{r2}\alpha_{r2} \approx 2K_r^*\left\{\frac{\partial \alpha_r}{\partial z}\frac{d_r hW}{2K_\phi}\ddot{y} + \left(\frac{\partial \alpha_r}{\partial z}\frac{l_r hW}{K_\theta}\ddot{x} + \frac{\partial \alpha_r}{\partial X}\frac{1 - \alpha_c}{2}W\ddot{x}\right)\frac{\partial K_r}{\partial W}\frac{\Delta W_r}{K_{r0}}\right\} \quad (7.21)\\
= 2K_r^*\alpha_r
\end{aligned}
$$

For small longitudinal and lateral accelerations, small values of second order and higher are neglected, as in the previous section.

The previous process of introducing Eqns (7.20) and (7.21) is accordingly to replace the toe angle change and compliance steer at the left and right wheels, $(\alpha_{f1}, \alpha_{f2})$ and $(\alpha_{r1}, \alpha_{r2})$, by a single equivalent steering angle, α_f and α_r, respectively. As the left and right steering due to lateral acceleration are in the same direction, it is possible to consider that a total lateral force produced by the left and right steering acts at the front and rear, respectively. In contrast, as the steering due to longitudinal acceleration is in the opposite direction, the force difference due to left and right wheel load transfer should only be considered to act on the front and the rear. This is based on using the tire cornering stiffness given by Eqns (7.9) and (7.10).

From Eqns (6.4) and (6.5), the following is obtained:

$$
\Delta W_f = \frac{hWK_{\phi f}}{d_f K_\phi}\ddot{y}, \quad \Delta W_r = \frac{hWK_{\phi r}}{d_r K_\phi}\ddot{y}
$$

The equivalent front and rear wheel steering angles α_f and α_r, with both roll centers on the ground, become the following forms:

$$
\alpha_f = (a_f + b_f\ddot{x})\ddot{y} \quad (7.22)
$$

$$
\alpha_r = (a_r + b_r\ddot{x})\ddot{y} \quad (7.23)
$$

Here,

$$
a_f = \frac{\partial \alpha_f}{\partial z}\frac{d_f hW}{2K_\phi} - \frac{\partial \alpha_f}{\partial T}\xi\frac{l_r W}{l} \quad (7.24)
$$

$$b_f = \left(\frac{\partial \alpha_f}{\partial X} \frac{\alpha_c W}{2} - \frac{\partial \alpha_f}{\partial z} \frac{l_f h W}{K_\theta} \right) \frac{\partial K_f}{\partial W} \frac{hWK_{\phi f}}{d_f K_\phi K_{f0}} - \frac{\partial \alpha_f}{\partial T} \frac{\alpha_c W}{2K_y} \frac{l_r W}{l} \tag{7.25}$$

$$a_r = \frac{\partial \alpha_r}{\partial z} \frac{d_r h W}{2K_\phi} \tag{7.26}$$

$$b_r = \left(\frac{\partial \alpha_r}{\partial X} \frac{1 - \alpha_c}{2} W + \frac{\partial \alpha_r}{\partial z} \frac{l_r h W}{K_\theta} \right) \frac{\partial K_r}{\partial W} \frac{hWK_{\phi r}}{d_r K_\phi K_{r0}} \tag{7.27}$$

7.3.1.3 Stability factor extension

Once the front and rear tire side-slip angle and steering angle are given, the relation between front steering angle, δ, and cornering radius, ρ, is as follows:

$$\delta = \frac{l}{\rho} + \beta_f - \beta_r + \alpha_f - \alpha_r \tag{6.6'}$$

Substituting Eqns (7.13), (7.14), (7.22), and (7.23) into Eqn (6.6)' gives the following:

$$\delta = \frac{l}{\rho} + \left(A_0 + A_1 \ddot{x} + A_2 \ddot{x}^2 \right) g l \ddot{y} \tag{7.28}$$

Furthermore, $g\ddot{y} = V^2/\rho$:

$$\rho = \frac{l}{\delta} \left\{ 1 + \left(A_0 + A_1 \ddot{x} + A_2 \ddot{x}^2 \right) V^2 \right\} \tag{7.29}$$

where

$$A_0 = \frac{W}{2l^2 g} \left(\frac{l_r K_{r0} - l_f K_{f0}}{K_{f0} K_{r0}} \right) + \frac{a_r - a_f}{gl} \tag{7.30}$$

$$A_1 = \frac{hW^2}{4l^3 g} \left(\frac{l_r}{K_{f0}^2} \frac{\partial K_f}{\partial W} + \frac{l_f}{K_{r0}^2} \frac{\partial K_r}{\partial W} \right) + \frac{b_r - b_f}{gl} \tag{7.31}$$

$$A_2 = \frac{W}{4\mu^2 g} \left\{ \frac{\alpha_c^2}{l_r K_{f0}} - \frac{(1 - \alpha_c^2)}{l_f K_{r0}} \right\} \tag{7.32}$$

Equation (7.28) gives the front steering angle, δ, necessary for cornering with a radius of ρ at a certain lateral acceleration with traction/braking. While Eqn (7.29) expresses the relation between the cornering radius and the vehicle speed, the following could be defined as the stability factor extended for cornering with traction/braking:

$$A^* = A_0 + A_1 \ddot{x} + A_2 \ddot{x}^2 \tag{7.33}$$

The first term, A_0, corresponds to the stability factor for steady-state cornering as defined by Eqn (3.43). The second term, A_1, is due to load transfer, toe angle change, and compliance steer caused by traction/braking. The third term, A_2, is due to tire cornering characteristics changed by traction/braking force and is proportional to the longitudinal acceleration squared, $\ddot{x}^2$.

7.3.2 EFFECT OF TRACTION AND BRAKING ON CORNERING

Transforming Eqn (7.28) and assuming $\ddot{x} = \ddot{x}_0$, $\delta = \delta_0$, and $\rho_0 = l/\delta_0$ gives the following:

$$\frac{\rho}{\rho_0} = \frac{1}{1 - \left(A_0 + A_1\ddot{x}_0 + A_2\ddot{x}_0^2\right)\rho_0 g\ddot{y}} \tag{7.34}$$

This is the relationship between cornering radius and lateral acceleration during cornering with a constant longitudinal acceleration and a constant steering angle. Similarly, when $V = V_0$ from Eqn (7.29), the following is obtained:

$$\rho = \rho_0\left\{1 + \left(A_0 + A_1\ddot{x} + A_2\ddot{x}^2\right)V_0^2\right\} \tag{7.29)'}$$

This shows how the cornering radius changes in response to longitudinal acceleration when the vehicle undergoes steady-state cornering with a constant steering angle and traction/braking. It assumes the traveling speed is high and the speed change due to traction/braking could be neglected. Here, from Eqn (7.29)', we have the following:

$$\left(\frac{\partial\rho}{\partial\ddot{x}}\right)_{\ddot{x}=0} = \rho_0 V_0^2 A_1 \tag{7.35}$$

A_1 expressed in Eqn (7.31) is defined as the sensitivity coefficient to longitudinal acceleration during cornering. The primary effect of traction/braking on cornering at small $\ddot{x}$ can be evaluated by A_1.

Figure 7.2 shows the effect of traction/braking by power-off on the cornering radius. Different types of vehicle traction modes are modeled using Eqn (7.29), neglecting the suspension toe angle change and compliance steer due to longitudinal force. In any case, when $A_1 > 0$ and is near $\ddot{x} = 0$, the cornering radius increases during acceleration and decreases during deceleration.

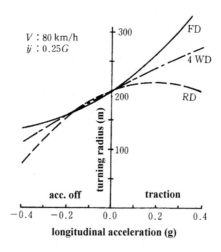

FIGURE 7.2

Effects of longitudinal acceleration on steady cornering.

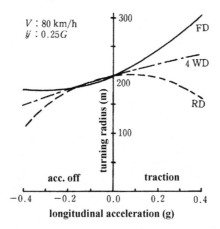

FIGURE 7.3

Turning radius to longitudinal acceleration with toe change and compliance steer.

This is prominent in front-wheel drive vehicles. For rear-wheel drive vehicles, the cornering radius decreases as $\ddot{x}$ gets larger.

Figure 7.3 shows the relationship between the cornering radius and longitudinal acceleration. Here, the toe angle change and compliance steer are set so that A_1 is as small as possible. In a practical sense, the condition of $A_1 \leq 0$ does not exist, but near $\ddot{x} = 0$, the effect of longitudinal acceleration becomes negligible. In particular, the effect of traction/braking by power-off for four-wheel drive vehicles is extremely small.

7.4 VEHICLE TRANSIENT STEERING RESPONSE [2]

The previous section extended the definition of stability factor for cornering with traction/braking. Through this, the effect of acceleration or deceleration on cornering has been studied. This extension relied on the assumption that only small accelerations take place. With this assumption, the vehicle equations of motion can be extended to include the effects of traction/braking. This is achieved by assuming that the distance of the front and rear axle from the vehicle center of gravity effectively changes in response to longitudinal acceleration. It is also shown that the analytical study of vehicle transient response to steering input with traction/braking is possible by using these equations of motion.

7.4.1 EQUATIONS OF MOTION

First, the equivalent tire cornering stiffness of the vehicle with traction/braking is given by Eqns $(7.9)'$ and $(7.10)'$ as in Section 7.3.1. Similarly, the toe angle change of a vehicle with traction/braking is expressed as follows:

$$\alpha_f = \left(p_f + q_f \ddot{x}\right)\ddot{y} \tag{7.36}$$

$$\alpha_r = \left(p_r + q_r \ddot{x}\right)\ddot{y} \tag{7.37}$$

Here, the toe angle change due to vehicle roll and pitch and the compliance steer due to the longitudinal force acting at the tires are taken into consideration, giving the following:

$$p_f = \frac{\partial \alpha_f}{\partial z} \frac{d_f h W}{2K_\phi} \tag{7.38}$$

$$q_f = \left(\frac{\partial \alpha_f}{\partial X} \frac{\alpha_c W}{2} - \frac{\partial \alpha_f}{\partial z} \frac{l_f h W}{K_\theta} \right) \frac{\partial K_f}{\partial W} \frac{h W K_{\phi f}}{d_f K_\phi K_{f0}} \tag{7.39}$$

$$p_r = \frac{\partial \alpha_r}{\partial z} \frac{d_r h W}{2K_\phi} \tag{7.40}$$

$$q_r = \left(\frac{\partial \alpha_r}{\partial X} \frac{1 - \alpha_c}{2} W + \frac{\partial \alpha_r}{\partial z} \frac{l_r h W}{K_\theta} \right) \frac{\partial K_r}{\partial W} \frac{h W K_{\phi r}}{d_r K_\phi K_{r0}} \tag{7.41}$$

Additionally, the torque acting on the steering system produced by the cornering force and the longitudinal force at the front wheels is similar to Eqn (7.15):

$$T = 2\left(\xi + \frac{\alpha_c W}{2K_y} \ddot{x} \right) Y_f$$

The front wheel cornering stiffness that accounts for the compliance steer due to the steering torque is expressed in the same manner as Section 5.3.1:

$$2eK_f^* = \frac{2K_f^*}{1 + 2\frac{\partial \alpha_f}{\partial T} \left(\xi + \frac{\alpha_c W}{2K_y} \ddot{x} \right) K_f^*} \tag{7.42}$$

Assuming the short period of time where longitudinal acceleration $\ddot{x}$ is constant and the change of the vehicle speed due to this acceleration can be neglected, the vehicle equations of motion for the plane motion in this period becomes as follows:

$$mV\left(\frac{d\beta}{dt} + r \right) = 2eK_f^*\left(\delta - \beta - \frac{l_f}{V}r + \alpha_f \right) + 2K_r^*\left(-\beta + \frac{l_r}{V}r + \alpha_r \right) \tag{7.43}$$

$$I\frac{dr}{dt} = 2l_f eK_f^*\left(\delta - \beta - \frac{l_f}{V}r + \alpha_f \right) - 2l_r K_r^*\left(-\beta + \frac{l_r}{V}r + \alpha_r \right) \tag{7.44}$$

Here, the $\ddot{y}$ term in α_f and α_r in Eqns (7.36) and (7.37) changes in response to the steering input and could be expressed as $V(d\beta/dt + r)/g$. Consequently, the equations of motion (Eqns (7.43) and (7.44)) become too complicated, and the analysis becomes too difficult. For simplicity, the toe angles, α_f and α_r in Eqns (7.36) and (7.37) are assumed to be dependent on the steady-state lateral acceleration only, $\ddot{y} = Vr/g$:

$$\alpha_f = \left(p_f + q_f \ddot{x} \right) Vr/g \tag{7.36}'$$

$$\alpha_r = \left(p_r + q_r \ddot{x} \right) Vr/g \tag{7.37}'$$

Substituting these into Eqns (7.43) and (7.44) and rearranging gives the following:

$$mV\left(\frac{d\beta}{dt}+r\right)=2eK_f^*\left(\delta-\beta-\frac{l_f^*}{V}r\right)+2K_r^*\left(-\beta+\frac{l_r^*}{V}r\right) \tag{7.43}'$$

$$I\frac{dr}{dt}=2l_feK_f^*\left(\delta-\beta-\frac{l_f^*}{V}r\right)-2l_rK_r^*\left(-\beta+\frac{l_r^*}{V}r\right) \tag{7.44}'$$

whereby

$$l_f^*=l_f\left(1-\frac{p_f+q_f\ddot{x}}{l_fg}V^2\right) \tag{7.45}$$

$$l_r^*=l_r\left(1+\frac{p_r+q_r\ddot{x}}{l_rg}V^2\right) \tag{7.46}$$

Rearranging Eqns (7.43)′ and (7.44)′ gives the following:

$$mV\frac{d\beta}{dt}+2\left(eK_f^*+K_r^*\right)\beta+\left\{mV+\frac{2\left(l_f^*eK_f^*-l_r^*K_r^*\right)}{V}\right\}r=2eK_f^*\delta \tag{7.47}$$

$$2\left(eK_f^*+K_r^*\right)\beta+I\frac{dr}{dt}+\frac{2\left(l_fl_f^*eK_f^*+l_rl_r^*K_r^*\right)}{V}r=2l_feK_f^*\delta \tag{7.48}$$

These equations of motion give us the steering response of the vehicle with traction/braking.

7.4.2 TRANSIENT RESPONSE TO STEERING INPUT

Here, the following equations are obtained as the yaw rate response to steering input by using Eqns (7.47) and (7.48):

$$\frac{r(s)}{\delta(s)}=\frac{G_\delta^r(0)(1+T_rs)}{1+\frac{2\zeta s}{\omega_n}+\frac{s^2}{\omega_n^2}} \tag{7.49}$$

where

$$\omega_n^2=\frac{4eK_f^*K_r^*l\left(l_f^*+l_r^*\right)}{mIV^2}-\frac{2\left(l_feK_f^*-l_rK_r^*\right)}{I} \tag{7.50}$$

$$2\zeta\omega_n=\frac{2m\left(l_f^*l_feK_f^*+l_r^*l_rK_r^*\right)+2I\left(eK_f^*+K_r^*\right)}{mIV} \tag{7.51}$$

$$T_r=\frac{ml_fV}{2IK_r^*} \tag{7.52}$$

$$G_\delta^r(0)=\frac{V}{\left(l_f^*+l_r^*\right)}\frac{1}{1+A^*V^2} \tag{7.53}$$

and

$$A^* = -\frac{m\left(l_f e K_f^* - l_r K_r^*\right)}{2 e K_f^* K_r^* l\left(l_f^* + l_r^*\right)} \tag{7.54}$$

By dealing with the equations under the premise of the small values as in Section 7.3, A^* is the same as the stability factor extended for the vehicle motion with traction/braking, as derived in Section 7.3. Furthermore, using Eqns (7.50–7.53), the response parameters to steering input for the vehicle with traction/braking can be obtained. This can be used to evaluate the effects of acceleration or deceleration on the transient steering response.

Figure 7.4 is the relationship of the natural frequency ω_n to the longitudinal acceleration obtained from Eqn (7.50). Figure 7.5 shows the relationship between the damping ratio, ζ, and the longitudinal acceleration. The natural frequency increases with traction and decreases with braking. In contrast, the damping ratio decreases with traction and increases with braking. In the case of rear-wheel drive, the natural frequency decreases with a large longitudinal acceleration.

The time to the first peak, t_p, for the yaw rate response to step steering input is given by Eqn (3.86). Figures 7.6 and 7.7 show how the time to the first peak, t_p, changes with longitudinal acceleration, where Eqns (7.50–7.52) are used to obtain t_p. The figures show that t_p becomes smaller with traction and larger with braking. In the case of rear-wheel drive, t_p becomes larger during traction. Furthermore, from Figure 7.7, it is possible to estimate how the effect of traction/braking on t_p can be reduced by using toe angle changes and compliance steer.

Figure 7.8 shows the yaw rate response to step steering input for the vehicle with traction/braking obtained by using Eqns (7.47) and (7.48). In the same figure, the vehicle response obtained through a numerical solution of the equations of motion (Eqns (7.1), (7.2), and (3.5)′) together with the equations of motions for vehicle roll and pitch is also plotted. The two plots agree well with each other. This proves the validity of the simplified method of dealing with the vehicle response to steering input for the vehicle with traction and braking as carried out under the assumption of quasi-steady state in this chapter.

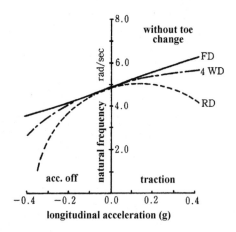

FIGURE 7.4

Effects of longitudinal acceleration on ω_n.

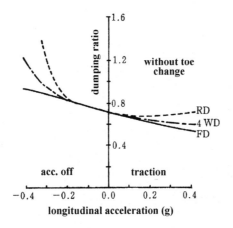

FIGURE 7.5

Effects of longitudinal acceleration on ζ.

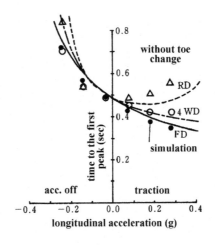

FIGURE 7.6

Effects of longitudinal acceleration on yaw rate response without toe change.

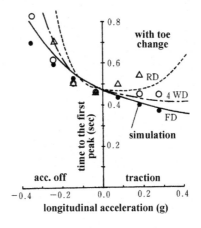

FIGURE 7.7

Effects of longitudinal acceleration on yaw rate response with toe change.

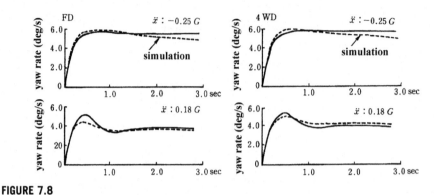

FIGURE 7.8

Effects of longitudinal acceleration on yaw rate response to step steering input, $V = 100$ km/h.

REFERENCES

[1] Abe M. On a vehicle cornering characteristics in acceleration and in braking—Part 1 theoretical analysis and extended stability factor. Trans JSAE 1988;37:134–40 [in Japanese].
[2] Abe M. On a vehicle cornering characteristics in acceleration and in braking—Part 2 transient response to steering input. Trans JSAE 1988;39:78–83 [in Japanese].

VEHICLE ACTIVE MOTION CONTROLS

8

8.1 PREFACE

The previous chapters have discussed the motion of a vehicle with only its front wheels being steered in response to the steering wheel operation. In recent years, it has been proved that vehicle dynamic characteristics can be greatly improved by active rear-wheel steering in addition to the front-wheel steering. Recently, attentions have been focused on controlling the vehicle using a direct yaw moment. This is produced by controlling the transverse split of longitudinal forces across the axles. It is called direct yaw-moment control (DYC).

This chapter will look at vehicle motion when rear-wheel steering is added to a conventional vehicle with front-wheel steering. This will also involve studying the basic vehicle motion control concept of active front-wheel steering or active front- and rear-wheel steering in response to the steering wheel. In this case, it is assumed that the front wheels have no mechanical connection to the steering wheel. The fundamental concepts of vehicle motion control using the direct yaw moment will also be considered.

8.2 VEHICLE MOTION WITH ADDITIONAL REAR-WHEEL STEERING

Assuming the front- and rear-wheel steering angles are δ_f and δ_r, then Eqns (3.8) and (3.9), which express the front- and rear-wheel lateral forces, become the following:

$$Y_f = -K_f \beta_f = -K_f \left(\beta + \frac{l_f}{V} r - \delta_f \right) \tag{8.1}$$

$$Y_r = -K_r \beta_r = -K_r \left(\beta - \frac{l_r}{V} r - \delta_r \right) \tag{8.2}$$

Substituting these into Eqns (3.4)′ and (3.5)′, the vehicle equations of motion in response to front- and rear-wheel steering are now as follows:

$$mV \frac{d\beta}{dt} + 2(K_f + K_r)\beta + \left\{ mV + \frac{2}{V}(l_f K_f - l_r K_r) \right\} r = 2K_f \delta_f + 2K_r \delta_r \tag{8.3}$$

$$2(l_f K_f - l_r K_r)\beta + I \frac{dr}{dt} + \frac{2\left(l_f^2 K_f + l_r^2 K_r \right)}{V} r = 2l_f K_f \delta_f - 2l_r K_r \delta_r \tag{8.4}$$

8.2.1 REAR-WHEEL STEERING PROPORTIONAL TO FRONT-WHEEL STEERING

There are many possible methods for simultaneous steering of both the front and rear wheels. The simplest method is considered first, where the rear-wheel steering is proportional to the front-wheel steering.

The front- and rear-wheel steering angles could be written as follows:

$$\delta_f = \frac{\delta}{n} \tag{8.5}$$

$$\delta_r = k\delta_f = \frac{k}{n}\delta \tag{8.6}$$

Here, in this chapter, δ is not the front-wheel steering angle but is used to represent the steering wheel angle, and n is the gear ratio of the front steering system.

Substituting Eqns (8.5) and (8.6) into Eqns (8.3) and (8.4) gives the response of the vehicle to steering wheel angle when the rear-wheel steering is proportional to the front-wheel steering:

$$\frac{\beta(s)}{\delta(s)} = \frac{1}{n} \frac{\begin{vmatrix} 2(K_f + kK_r) & mV + \frac{2}{V}(l_f K_f - l_r K_r) \\ 2(l_f K_f - kl_r K_r) & Is + \frac{2\left(l_f^2 K_f + l_r^2 K_r\right)}{V} \end{vmatrix}}{\begin{vmatrix} mVs + 2(K_f + K_r) & mV + \frac{2}{V}(l_f K_f - l_r K_r) \\ 2(l_f K_f - l_r K_r) & Is + \frac{2}{V}\left(l_f^2 K_f + l_r^2 K_r\right) \end{vmatrix}} \tag{8.7}$$

$$\frac{r(s)}{\delta(s)} = \frac{1}{n} \frac{\begin{vmatrix} mVs + 2(K_f + K_r) & 2(K_f + kK_r) \\ 2(l_f K_f - l_r K_r) & 2(l_f K_f - kl_r K_r) \end{vmatrix}}{\begin{vmatrix} mVs + 2(K_f + K_r) & mV + \frac{2}{V}(l_f K_f - l_r K_r) \\ 2(l_f K_f - l_r K_r) & Is + \frac{2\left(l_f^2 K_f + l_r^2 K_r\right)}{V} \end{vmatrix}} \tag{8.8}$$

Here, using Eqns (8.7) and (8.8), the transfer functions of the lateral acceleration, $V(\dot{\beta} + r)$, and yaw rate responses to steering input can be expressed in a convenient form that shows how they differ from the usual transfer function with front-wheel steering only:

$$\frac{\ddot{y}(s)}{\delta(s)} = \frac{1-k}{n} G_{\delta}^{\ddot{y}}(0) \frac{1 + (1+\lambda_1)T_{y1}s + (1+\lambda_2)T_{y2}s^2}{1 + \frac{2\zeta s}{\omega_n} + \frac{s^2}{\omega_n^2}} \tag{8.9}$$

$$\frac{r(s)}{\delta(s)} = \frac{1-k}{n} G_{\delta}^{r}(0) \frac{1 + (1+\lambda_r)T_r s}{1 + \frac{2\zeta s}{\omega_n} + \frac{s^2}{\omega_n^2}} \tag{8.10}$$

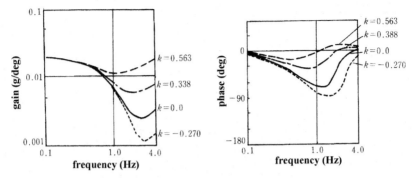

FIGURE 8.1

Effects of front wheel proportional rear-wheel steering on lateral acceleration response.

whereby

$$\lambda_1 = \frac{l}{l_r}\frac{k}{1-k}, \quad \lambda_2 = \frac{K_f + K_r}{K_r}\frac{k}{1-k}$$

$$\lambda_r = \frac{l_f K_f - l_r K_r}{l_f K_f}\frac{k}{1-k}$$

ω_n, ζ, T_{y1}, T_{y2}, T_r, $G_\delta^{\ddot{y}}(0)$, and $G_\delta^r(0)$ are the same symbols used in Section 3.4.2, for the case of a vehicle with front-wheel steering only.

Based on Eqns (8.9) and (8.10), when the rear wheel is steered proportionally to the front wheel, the overall gain is multiplied by $(1 - k)$. The coefficients of the s and s^2 terms in the numerator of the transfer function are increased by a ratio of λ_1 and λ_2. If $0 < k < 1$, λ_1 and λ_2 are positive values. This means the phase lag of the lateral acceleration to steering input is reduced when the rear wheel is steered in the same direction as the front wheel, but with a smaller angle than the front steering angle. If the vehicle steer characteristic is near NS, $\lambda_r \approx 0$, and the effect of rear-wheel steering on yaw rate response is small.

Figure 8.1 shows the influence of rear-wheel steering on the vehicle lateral acceleration response for a neutralsteer vehicle, using Eqn (8.9) [1]. The lateral acceleration response is improved by steering the rear wheel in the same direction as the front wheel.

Note, in particular, ω_n and ζ are unchanged compared to the case of the front-wheel steering vehicle by the additional rear-wheel steering proportional to front-wheel steering.

Using Eqn (8.7), the steady-state value for side-slip angle, β_s, to a constant steering angle is as follows:

$$\beta_s = \frac{1}{n}\begin{vmatrix} 2(K_f + kK_r) & mV + \dfrac{2}{V}(l_f K_f - l_r K_r) \\[2mm] 2(l_f K_f - kl_r K_r) & \dfrac{2\left(l_f^2 K_f + l_r^2 K_r\right)}{V} \\[4mm] 2(K_f + K_r) & mV + \dfrac{2}{V}(l_f K_f - l_r K_r) \\[2mm] 2(l_f K_f - l_r K_r) & \dfrac{2\left(l_f^2 K_f + l_r^2 K_r\right)}{V} \end{vmatrix}\delta \qquad (8.11)$$

From this, a value of k is sought that gives zero steady-state side slip, $\beta s = 0$. In other words, the numerator in Eqn (8.11) is zero, and the following is true:

$$k = k_0 = -\frac{l_r\left(1 - \frac{ml_f}{2l l_r K_r}V^2\right)}{l_f\left(1 + \frac{ml_r}{2l l_f K_f}V^2\right)} \tag{8.12}$$

If the proportional constant of the rear- to front-wheel steering is set as in Eqn (8.12), the side-slip angle during steady-state cornering will be zero. This means the vehicle's traveling direction is the same as its heading direction.

This proportional constant changes with the traveling speed, as shown by Figure 8.2. When k_0 is negative, the rear-wheel and front-wheel steering directions are opposite of each other. They are in the same direction if k_0 is positive. If the rear-wheel is steered in the opposite direction to the front at low speed and then in the same direction at high speed, the steady-state side-slip angle will be always zero.

8.2.2 REAR-WHEEL STEERING PROPORTIONAL TO FRONT-WHEEL STEERING FORCE

The next case is rear-wheel steering proportional to the front-wheel steering force. Assuming that the steering force of the front wheel is equal to the steering moment by the external force acting at the front wheel, then the following is obtained:

$$M = -2\xi K_f\left(\beta + \frac{l_f}{V}r - \delta_f\right) \tag{8.13}$$

And, the front and rear-wheel steering angle can be written as follows:

$$\delta_f = \frac{\delta}{n} \tag{8.5}$$

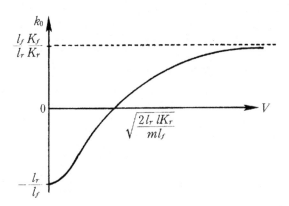

FIGURE 8.2

Rear to front steering ratio for zero side slip.

$$\delta_r = kM = -2k\xi K_f \left(\beta + \frac{l_f}{V}r - \delta_f \right) \tag{8.14}$$

where, ξ has same meaning as in Chapter 5.

Substituting these into Eqns (8.3) and (8.4) and expressing the side-slip angle and yaw rate responses in the form of a transfer function gives the following:

$$\frac{\beta(s)}{\delta(s)} = \frac{1}{n} \frac{\begin{vmatrix} 2(K_f + 2k\xi K_f K_r) & mV + \frac{2}{V}(l_f K_f - l_r K_r) + \frac{4k\xi l_f K_f K_r}{V} \\ 2(l_f K_f - 2k\xi l_r K_f K_r) & Is + \frac{2\left(l_f^2 K_f + l_r^2 K_r\right)}{V} - \frac{4k\xi l_f l_r K_f K_r}{V} \end{vmatrix}}{\begin{vmatrix} mVs + 2(K_f + K_r) + 4k\xi K_f K_r & mV + \frac{2}{V}(l_f K_f - l_r K_r) + \frac{4k\xi l_f K_f K_r}{V} \\ 2(l_f K_f - l_r K_r) - 4k\xi l_r K_f K_r & Is + \frac{2\left(l_f^2 K_f + l_r^2 K_r\right)}{V} - \frac{4k\xi l_f l_r K_f K_r}{V} \end{vmatrix}} \tag{8.15}$$

$$\frac{r(s)}{\delta(s)} = \frac{1}{n} \frac{\begin{vmatrix} mVs + 2(K_f + K_r) + 4k\xi K_f K_r & 2(K_f + 2k\xi K_f K_r) \\ 2(l_f K_f - l_r K_r) - 4k\xi l_r K_f K_r & 2(l_f K_f - 2k\xi l_r K_f K_r) \end{vmatrix}}{\begin{vmatrix} mVs + 2(K_f + K_r) + 4k\xi K_f K_r & mV + \frac{2}{V}(l_f K_f - l_r K_r) + \frac{4k\xi l_f K_f K_r}{V} \\ 2(l_f K_f - l_r K_r) - 4k\xi l_r K_f K_r & Is + \frac{2\left(l_f^2 K_f + l_r^2 K_r\right)}{V} - \frac{4k\xi l_f l_r K_f K_r}{V} \end{vmatrix}} \tag{8.16}$$

The transfer function of the lateral acceleration and yaw rate responses to steering input can be derived in the same form as in Section 8.2.1:

$$\frac{\ddot{y}(s)}{\delta(s)} = \frac{1}{n} G_\delta^{\ddot{y}}(0)^* \frac{1 + T_{y1}s + (1 + \lambda_2)T_{y2}s^2}{1 + \frac{2\zeta^* s}{\omega_n^*} + \frac{s^2}{\omega_n^{*2}}} \tag{8.17}$$

$$\frac{r(s)}{\delta(s)} = \frac{1}{n} G_\delta^r(0)^* \frac{1 + (1 + \lambda_r)T_r s}{1 + \frac{2\zeta^* s}{\omega_n^*} + \frac{s^2}{\omega_n^{*2}}} \tag{8.18}$$

Here, taking the following:

$$A^* = A + \frac{k\xi l_r m}{l^2} \tag{8.19}$$

as the stability factor,

$$G_\delta^{\ddot{y}}(0)^* = VG_\delta^r(0)^* = \frac{1}{1 + A^*V^2} \frac{V^2}{l} \tag{8.20}$$

$$\omega_n^* = \frac{2l}{V} \sqrt{\frac{K_f K_r}{mI}(1 + A^*V^2)} = \sqrt{\omega_n^2 + \frac{4k\xi l_r K_f K_r}{I}} \tag{8.21}$$

$$\zeta^* = \frac{m\left(l_f^2 K_f + l_r^2 K_r\right) + I(K_f + K_r) + 2k\xi K_f K_r(I - ml_f l_r)}{2l\sqrt{mIK_f K_r(1 + A^* V^2)}} \tag{8.22}$$

and A is the stability factor defined in Chapter 3 for a front-wheel steering vehicle. Additionally, the following are defined:

$$\lambda_2 = 2k\xi K_r, \quad \lambda_r = -\frac{2k\xi l_r K_r}{l_f}$$

and the time constants, T_{y1}, T_{y2}, and T_r, are the same as the ones used in Section 3.4.2 for the case of the vehicle with front-wheel steering only.

When the rear wheel is steered proportionally to the front-wheel steering force, both the numerator and the denominator of transfer functions (8.17) and (8.18) are changed. This gives different natural frequency and damping ratio to the vehicle with only front-wheel steering.

If $k > 0$, the stability factor becomes larger, the natural frequency increases, and the coefficient of s^2 in the numerator of lateral acceleration gets bigger, which improves the lateral acceleration response. However, the coefficient of s in the numerator of yaw rate decreases, and this deteriorates the yaw rate response. Figure 8.3 shows the influence of the proportional constant k on the vehicle lateral acceleration and yaw rate frequency responses to steering input by using Eqns (8.17) and (8.18) [2].

8.2.3 REAR-WHEEL STEERING PROPORTIONAL TO YAW RATE

The next case is rear-wheel steering proportional to the vehicle yaw rate. The front- and rear-wheel steering angles are written as shown next:

$$\delta_f = \frac{\delta}{n} \tag{8.5}$$

$$\delta_r = kr \tag{8.23}$$

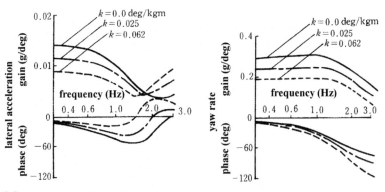

FIGURE 8.3

Effects of front steering force proportional rear steering on frequency responses.

Substituting these into Eqns (8.3) and (8.4) and expressing the side-slip angle and yaw rate response to steering input in the form of transfer functions gives the following:

$$\frac{\beta(s)}{\delta(s)} = \frac{1}{n} \frac{\begin{vmatrix} 2K_f & mV + \frac{2}{V}(l_f K_f - l_r K_r) - 2kK_r \\ 2l_f K_f & Is + \frac{2}{V}(l_f^2 K_f + l_r^2 K_r) + 2kl_r K_r \end{vmatrix}}{\begin{vmatrix} mVs + 2(K_f + K_r) & mV + \frac{2}{V}(l_f K_f - l_r K_r) - 2kK_r \\ 2(l_f K_f - l_r K_r) & Is + \frac{2}{V}(l_f^2 K_f + l_r^2 K_r) + 2kl_r K_r \end{vmatrix}} \tag{8.24}$$

$$\frac{r(s)}{\delta(s)} = \frac{1}{n} \frac{\begin{vmatrix} mVs + 2(K_f + K_r) & 2K_f \\ 2(l_f K_f - l_r K_r) & 2l_f K_f \end{vmatrix}}{\begin{vmatrix} mVs + 2(K_f + K_r) & mV + \frac{2}{V}(l_f K_f - l_r K_r) - 2kK_r \\ 2(l_f K_f - l_r K_r) & Is + \frac{2}{V}(l_f^2 K_f + l_r^2 K_r) + 2kl_r K_r \end{vmatrix}} \tag{8.25}$$

Equations (8.24) and (8.25) can be used to derive the transfer functions of the lateral acceleration and yaw rate responses to a steering input. In these forms, they can be easily compared with the usual transfer functions for front-wheel steering only.

$$\frac{\ddot{y}(s)}{\delta(s)} = \frac{1}{n} G_\delta^{\ddot{y}}(0) * \frac{1 + (1 + \lambda_1)T_{y1}s + T_{y2}s^2}{1 + \frac{2\zeta^* s}{\omega_n^*} + \frac{s^2}{\omega_n^{*2}}} \tag{8.26}$$

$$\frac{r(s)}{\delta(s)} = \frac{1}{n} G_\delta^r(0) * \frac{1 + T_r s}{1 + \frac{2\zeta^* s}{\omega_n^*} + \frac{s^2}{\omega_n^{*2}}} \tag{8.27}$$

Here, taking the following:

$$A^* = A + \frac{k}{l}\frac{1}{V} \tag{8.28}$$

as the stability factor for this case, and the following:

$$G_\delta^{\ddot{y}}(0)^* = VG_\delta^r(0)^* = \frac{1}{1 + A^*V^2}\frac{V^2}{l} \tag{8.29}$$

$$\omega_n^* = \frac{2l}{V}\sqrt{\frac{K_f K_r}{mI}(1 + A^*V^2)} = \sqrt{\omega_n^2 + \frac{4klK_f K_r}{mIV}} \tag{8.30}$$

$$\zeta^* = \frac{m\left(l_f^2 K_f + l_r^2 K_r\right) + I(K_f + K_r) + kml_r K_r V}{2l\sqrt{mIK_f K_r(1 + A^*V^2)}} \tag{8.31}$$

where,

$$\lambda_1 = \frac{kV}{l_r}$$

In this manner, the rear-wheel steering proportional to the yaw rate changes the vehicle stability, as in Eqn (8.28), and the natural frequency and damping ratio as well. If $k > 0$, the natural frequency increases, and the coefficient of s in the numerator of lateral acceleration becomes larger too. Here the vehicle response could be considered to be improved.

8.3 REAR-WHEEL STEERING CONTROL FOR ZERO SIDE-SLIP ANGLE

Section 8.2.1 shows that a vehicle with zero side-slip angle at the center of gravity, during a steady-state cornering, could be realized by selecting the appropriate proportional constant for the rear- to front-wheel steering. Further consideration is taken into the control logic of the rear-wheel steering such that the side-slip angle of the center of gravity is always zero in response to any steering input.

8.3.1 REAR-WHEEL STEERING IN RESPONSE TO FRONT-WHEEL STEERING ANGLE

Consider that the relationship between the rear and front wheels is expressed by a general transfer function, rather than a simple proportional constant. The rear-wheel steering could be written as follows:

$$\delta_r(s) = k(s)\delta_f(s) = \frac{k(s)}{n}\delta(s) \tag{8.6$'$}$$

Here, $k(s)$ is the transfer function of the rear-wheel steering to front-wheel steering.

The side-slip angle response to steering wheel steering input is the same as obtained by Eqn (8.7):

$$\frac{\beta(s)}{\delta(s)} = \frac{1}{n} \frac{\begin{vmatrix} 2\{K_f + k(s)K_r\} & mV + \dfrac{2}{V}(l_fK_f - l_rK_r) \\ 2\{l_fK_f - k(s)l_rK_r\} & Is + \dfrac{2\left(l_f^2K_f + l_r^2K_r\right)}{V} \end{vmatrix}}{\begin{vmatrix} mVs + 2(K_f + K_r) & mV + \dfrac{2}{V}(l_fK_f - l_rK_r) \\ 2(l_fK_f - l_rK_r) & Is + \dfrac{2}{V}\left(l_f^2K_f + l_r^2K_r\right) \end{vmatrix}} \tag{8.7$'$}$$

Consequently, if $k(s)$ is chosen so the numerator of Eqn (8.7)$'$ is zero, then the transfer function of side-slip angle response to steering wheel steering input will become zero:

$$\begin{vmatrix} 2\{K_f + k(s)K_r\} & mV + \dfrac{2}{V}(l_fK_f - l_rK_r) \\ 2\{l_fK_f - k(s)l_rK_r\} & Is + \dfrac{2\left(l_f^2K_f + l_r^2K_r\right)}{V} \end{vmatrix} = 0 \tag{8.32}$$

Expanding Eqn (8.32) gives $k(s)$ as follows:

$$k(s) = -\frac{l_r - \frac{ml_f}{2lK_r}V^2 + \frac{IV}{2lK_r}s}{l_f + \frac{ml_r}{2lK_f}V^2 + \frac{IV}{2lK_f}s} = \frac{k_0}{1 + T_e s} - \frac{K_f}{K_r}\frac{T_e s}{1 + T_e s} \tag{8.33}$$

Here, k_0 is given by Eqn (8.12), and the following is obtained:

$$T_e = \frac{IV}{2ll_f K_f + ml_r V^2} \tag{8.34}$$

In other words, if the rear wheel is steered in response to the front wheel, as described by the transfer function given by Eqn (8.33), the side-slip angle of the center of gravity is always equal to zero. This ensures the vehicle traveling direction and heading direction are always the same. The yaw rate response to steering input in this case becomes the following:

$$\frac{r(s)}{\delta(s)} = \frac{1}{n}\frac{1}{1 + \frac{ml_r}{2ll_f K_f}V^2}\frac{V}{l_f}\frac{1}{1 + T_e s} \tag{8.35}$$

8.3.2 REAR-WHEEL STEERING PROPORTIONAL TO FRONT-WHEEL STEERING AND YAW RATE

The center of gravity side-slip angle can be made equal to zero by steering the rear-wheel an amount equal to the sum of the front-wheel steering and yaw rate proportional terms:

$$\delta_r = k_\delta \delta_f + k_r r = \frac{k_\delta}{n}\delta + k_r r \tag{8.36}$$

Obtaining the side-slip angle response to steering input gives the following:

$$\frac{\beta(s)}{\delta(s)} = \frac{1}{n}\frac{\begin{vmatrix} 2(K_f + k_\delta K_r) & mV + \frac{2}{V}(l_f K_f - l_r K_r) - 2k_r K_r \\ 2(l_f K_f - k_\delta l_r K_r) & Is + \frac{2}{V}\left(l_f^2 K_f + l_r^2 K_r\right) + 2k_r l_r K_r \end{vmatrix}}{\begin{vmatrix} mVs + 2(K_f + K_r) & mV + \frac{2}{V}(l_f K_f - l_r K_r) - 2k_r K_r \\ 2(l_f K_f - l_r K_r) & Is + \frac{2}{V}\left(l_f^2 K_f + l_r^2 K_r\right) + 2k_r l_r K_r \end{vmatrix}} \tag{8.37}$$

If the following is fulfilled, the side-slip angle will be zero:

$$\begin{vmatrix} 2(K_f + k_\delta K_r) & mV + \frac{2}{V}(l_f K_f - l_r K_r) - 2k_r K_r \\ 2(l_f K_f - k_\delta l_r K_r) & Is + \frac{2}{V}\left(l_f^2 K_f + l_r^2 K_r\right) + 2k_r l_r K_r \end{vmatrix} = 0 \tag{8.38}$$

Expanding Eqn (8.38) gives a first-order equation in s. Letting the coefficient of s and the constant term equal zero gives k_δ and k_r:

$$k_\delta = -\frac{K_f}{K_r} \tag{8.39}$$

$$k_r = \frac{mV^2 + 2(l_f K_f - l_r K_r)}{2K_r V} \tag{8.40}$$

By choosing these proportional constants, $\beta(s)/\delta(s)$ is always zero, and the vehicle side-slip angle of the center of gravity is always zero. The yaw rate response to steering input is the same as the previous case and is expressed by Eqn (8.35).

8.4 MODEL FOLLOWING CONTROL BY REAR-WHEEL STEERING

There is, in general, a concept of model following control in control engineering. It is possible to realize the intentional vehicle response to the steering wheel input by using the model following rear-wheel steering control. One of the easy ways to set an intentional model of vehicle motion is to use a yaw rate model response. The yaw rate model following rear-wheel steering will be studied in this section. Also, it will be shown that a lateral acceleration model following is impossible by rear-wheel steering control.

8.4.1 FEED-FORWARD YAW RATE MODEL FOLLOWING CONTROL

Applying the Laplace transform to the vehicle equations of motion for the steering wheel angle and rear-wheel steering angle, Eqns (8.3) and (8.4), and solving for the yaw rate response to the steering angles gives the following:

$$r(s) = \frac{\frac{1}{n}G_\delta^r(0)(1 + T_r s)\delta(s) + G_{\delta_r}^r(0)(1 + T_{rr}s)\delta_r(s)}{1 + \frac{2\zeta}{\omega_n}s + \frac{1}{\omega_n^2}s^2} \tag{8.41}$$

$$= \frac{\omega_n^2 H_r(s)}{s^2 + 2\zeta\omega_n s + \omega_n^2}$$

where, $G_\delta^r(0)$ and T_r are already defined in Chapter 3 and

$$H_r(s) = \frac{1}{n}G_\delta^r(0)(1 + T_r s)\delta(s) + G_{\delta_r}^r(0)(1 + T_{rr}s)\delta_r(s) \tag{8.42}$$

$$G_{\delta_r}^r(0) = -\frac{1}{1 + AV^2}\frac{V}{l} = -G_\delta^r(0)$$

$$T_{rr} = \frac{ml_r V}{2lK_f}$$

The first-order lag yaw rate model is used to determine the intentional yaw rate:

$$r_m(s) = \frac{1}{n} \frac{G_e}{1 + T_e s} \delta(s) \tag{8.43}$$

If this response is identical with the response described by Eqn (8.41), the following equation is obtained:

$$\frac{\frac{1}{n}G_\delta^r(0)(1 + T_r s)\delta(s) - G_\delta^r(0)(1 + T_{rr} s)\delta_r(s)}{1 + \frac{2\zeta}{\omega_n}s + \frac{1}{\omega_n^2}s^2} = \frac{1}{n}\frac{G_e}{1 + T_e s}\delta(s)$$

It is possible to obtain $\delta_r(s)$ from the previous equation as follows:

$$\delta_r(s) = \frac{1}{n}\left\{ -\frac{G_e\left(1 + \frac{2\zeta}{\omega_n}s + \frac{1}{\omega_n^2}s^2\right)}{G_\delta^r(0)(1 + T_e s)(1 + T_{rr} s)} + \frac{1 + T_r s}{1 + T_{rr} s} \right\}\delta(s) \tag{8.44}$$

This is the rear-wheel steering control law for the feed-forward yaw rate model following control.

8.4.2 FEED-FORWARD AND YAW RATE FEED-BACK MODEL FOLLOWING CONTROL

The same yaw rate model response is used as in Eqn (8.43). This gives the following expression:

$$\left(s + \frac{1}{T_e}\right)r_m(s) = \frac{1}{n}\frac{G_e}{T_e}\delta(s) \tag{8.43'}$$

Here, the error, e, between the model yaw rate, r_m, and the vehicle yaw rate with the control, r, is defined as follows:

$$e(s) = \left(s + \frac{1}{T_e}\right)(r(s) - r_m(s))$$

To make the error converge to zero in the first-order lag response, the following equation must be satisfied:

$$\left(s + \frac{1}{T_g}\right)e(s) = 0$$

From the two previous equations, the following expression can be obtained:

$$\left(s + \frac{1}{T_g}\right)\left(s + \frac{1}{T_e}\right)(r(s) - r_m(s)) = 0$$

Taking Eqn (8.41) into consideration, $(s + 1/T_g)(s + 1/T_e)r(s)$ is divided into two parts, which are the part described by the steering wheel angle, δ, and the rear-wheel steering angle, δ_r, and the remnant term as follows:

$$\left(s + \frac{1}{T_g}\right)\left(s + \frac{1}{T_e}\right)r(s) = \left\{s^2 + 2\zeta\omega_n s + \omega_n^2 + \left(\frac{1}{T_g} + \frac{1}{T_e} - 2\zeta\omega_n\right)s + \frac{1}{T_g T_e} - \omega_n^2\right\}r(s)$$

$$= \omega_n^2 H_r(s) + (c_1 s + c_0)r(s)$$

where,

$$c_1 = \frac{1}{T_g} + \frac{1}{T_e} - 2\zeta\omega_n, \quad c_0 = \frac{1}{T_g T_e} - \omega_n^2$$

From Eqn (8.43)' we can get the following:

$$\left(s + \frac{1}{T_g}\right)\left(s + \frac{1}{T_e}\right)r_m(s) = \left(s + \frac{1}{T_g}\right)\frac{1}{n}\frac{G_e}{T_e}\delta(s)$$

Consequently,

$$\left(s + \frac{1}{T_g}\right)\left(s + \frac{1}{T_e}\right)(r(s) - r_m(s)) = \omega_n^2 H_r(s) + (c_1 s + c_0)r(s) - \frac{1}{n}\frac{G_e}{T_e}\left(s + \frac{1}{T_g}\right)\delta(s) = 0$$

Putting Eqn (8.42) into the previous equation results in the following:

$$\frac{1}{n}G_\delta^r(0)(1 + T_r s)\delta(s) - G_\delta^r(0)(1 + T_{rr}s)\delta_r(s) + \frac{1}{\omega_n^2}(c_1 s + c_0)r(s) - \frac{1}{n}\frac{G_e}{\omega_n^2 T_e}\left(s + \frac{1}{T_g}\right)\delta(s) = 0$$

The rear-wheel steering angle needed to follow the model yaw rate response is obtained from the previous equation as follows:

$$\delta_r(s) = \frac{1}{n}\left\{\frac{1 + T_r s}{1 + T_{rr}s} - \frac{G_e}{\omega_n^2 T_g T_e G_\delta^r(0)}\frac{1 + T_g s}{1 + T_{rr}s}\right\}\delta(s) + \frac{c_0}{\omega_n^2 G_\delta^r(0)}\frac{1 + \frac{c_1}{c_0}s}{1 + T_{rr}s}r(s) \qquad (8.45)$$

This is a rear-wheel steering control law for the feed-forward and yaw rate feed-back model following control.

8.4.3 FEED-FORWARD LATERAL ACCELERATION MODEL FOLLOWING CONTROL

The lateral acceleration response to steering wheel angle and rear-wheel steering angle inputs is obtained using Eqns (8.3) and (8.4) as follows:

$$\ddot{y}(s) = \frac{\frac{1}{n}G_\delta^{\ddot{y}}(0)\left(1 + \frac{l_r}{V}s + \frac{I}{2lK_r}s^2\right)\delta(s) + G_{\delta_r}^{\ddot{y}}(0)\left(1 - \frac{l_r}{V}s - \frac{I}{2lK_f}s^2\right)\delta_r(s)}{1 + \frac{2\zeta}{\omega_n}s + \frac{1}{\omega_n^2}s^2} \qquad (8.46)$$

On the other hand, a model response of lateral acceleration is introduced as follows:

$$\ddot{y}_m(s) = \frac{1}{n}\frac{G_y}{1 + T_y s}\delta(s) \qquad (8.47)$$

If this is identical with Eqn (8.46), as follows:

$$\frac{\frac{1}{n}G_\delta^{\ddot{y}}(0)\left(1 + \frac{l_r}{V}s + \frac{I}{2lK_r}s^2\right)\delta(s) + G_{\delta_r}^{\ddot{y}}(0)\left(1 - \frac{l_r}{V}s - \frac{I}{2lK_f}s^2\right)\delta_r(s)}{1 + \frac{2\zeta}{\omega_n}s + \frac{1}{\omega_n^2}s^2} = \frac{1}{n}\frac{G_y}{1 + T_y s}\delta(s)$$

then, we can formally obtain the following:

$$\delta_r(s) = \frac{G_y\left(1 + \frac{2\zeta}{\omega_n}s + \frac{1}{\omega_n^2}s^2\right) - G_{\delta_r}^{\ddot{y}}(0)(1 + T_y s)\left(1 + \frac{l_f}{V}s + \frac{I}{2lK_r}s^2\right)}{nG_{\delta_r}^{\ddot{y}}(0)(1 + T_y s)\left(1 - \frac{l_r}{V}s - \frac{I}{2lK_r}s^2\right)}\delta(s) \tag{8.48}$$

This is a rear-wheel steering control law of the feed-forward lateral acceleration model following control; however, there is a positive pole in the transfer function of rear-wheel steering to steering wheel input, and the control system is unstable and practically impossible. This is due to the fact that there is a positive zero in the numerator of the lateral acceleration transfer function to rear-wheel steering input, and the things are same as for the feed-forward and lateral acceleration feed-back model following control of rear-wheel steering.

8.5 MODEL FOLLOWING FRONT-WHEEL ACTIVE STEERING CONTROL

Front wheels of normal vehicles are mechanically connected to the steering wheel and steered proportionally to the steering wheel angle. In this chapter, regarding the front wheels as being disconnected from the steering wheel, a model following front-wheel active steering control will be discussed.

8.5.1 FEED-FORWARD YAW RATE MODEL FOLLOWING CONTROL

A yaw rate response to the front-wheel steering angle, δ_f, is obtained from Eqns (8.3) and (8.4) as follows:

$$r(s) = \frac{\omega_n^2 G_\delta^r(0)(1 + T_r s)}{s^2 + 2\zeta\omega_n s + \omega_n^2}\delta_f(s) \tag{8.49}$$

Here, again, we introduce the yaw rate model response described by Eqn (8.43). Making this identical to the yaw rate described by Eqn (8.49), we have the following:

$$\frac{\omega_n^2 G_\delta^r(0)(1 + T_r s)}{s^2 + 2\zeta\omega_n s + \omega_n^2}\delta_f(s) = \frac{1}{n}G_e\frac{1}{1 + T_e s}\delta(s)$$

and

$$\delta_f(s) = \frac{1}{n}\frac{G_e(s^2 + 2\zeta\omega_n s + \omega_n^2)}{\omega_n^2 G_\delta^r(0)(1 + T_r s)(1 + T_e s)}\delta(s) \tag{8.50}$$

This is a control law of the feed-forward yaw rate model following front-wheel active steering.

8.5.2 FEED-FORWARD AND YAW RATE FEED-BACK MODEL FOLLOWING CONTROL

Giving the yaw rate model response as in Eqn (8.43), the same introduction process as in Section 8.4.2 brings us the following:

$$\left(s + \frac{1}{T_g}\right)\left(s + \frac{1}{T_e}\right)r(s) = \omega_n^2 G_\delta^r(0)(1 + T_r s)\delta_f(s) + \left\{\left(\frac{1}{T_g} + \frac{1}{T_e} - 2\zeta\omega_n\right)s + \frac{1}{T_g T_e} - \omega_n^2\right\}r(s)$$

$$\left(s + \frac{1}{T_g}\right)\left(s + \frac{1}{T_e}\right)r_m(s) = \left(s + \frac{1}{T_g}\right)\frac{1}{n}\frac{G_e}{T_e}\delta(s)$$

Accordingly, the condition of the error between $r(s)$ and $r_m(s)$ converging to zero is expressed as follows:

$$\left(s + \frac{1}{T_g}\right)\left(s + \frac{1}{T_e}\right)(r(s) - r_m(s)) = \omega_n^2 G_\delta^r(0)(1 + T_r s)\delta_f(s) + (c_1 s + c_0)r(s)$$

$$- \frac{1}{n}\frac{G_e}{T_e}\left(s + \frac{1}{T_g}\right)\delta(s) = 0$$

Thus, the control law for the feed-forward and yaw rate feed-back model following front-wheel active steering is the following:

$$\delta_f(s) = \frac{1}{n}\frac{G_e}{\omega_n^2 T_g T_e G_\delta^r(0)}\frac{1 + T_g s}{1 + T_r s}\delta(s) - \frac{c_0}{\omega_n^2 G_\delta^r(0)}\frac{1 + \frac{c_1}{c_0}s}{1 + T_r s}r(s) \qquad (8.51)$$

8.5.3 FEED-FORWARD LATERAL ACCELERATION MODEL FOLLOWING CONTROL

A lateral acceleration response to the front-wheel steering input is described by the following:

$$\ddot{y}(s) = G_\delta^{\ddot{y}}(0)\frac{1 + \frac{l_r}{V}s + \frac{I}{2lK_r}s^2}{1 + \frac{2\zeta}{\omega_n}s + \frac{1}{\omega_n^2}s^2}\delta_f(s)$$

Let the lateral acceleration model response be expressed by Eqn (8.47) again and be equal to the previous equation, then we can obtain the following:

$$G_\delta^{\ddot{y}}(0)\frac{1 + \frac{l_r}{V}s + \frac{I}{2lK_r}s^2}{1 + \frac{2\zeta}{\omega_n}s + \frac{1}{\omega_n^2}s^2}\delta_f(s) = \frac{1}{n}\frac{G_y}{1 + T_y s}\delta(s)$$

which brings us the following:

$$\delta_f(s) = \frac{1}{n}\frac{G_y}{G_\delta^{\ddot{y}}(0)}\frac{1 + \frac{2\zeta}{\omega_n}s + \frac{1}{\omega_n^2}s^2}{\left(1 + T_y s\right)\left(1 + \frac{l_r}{V}s + \frac{I}{2lK_r}s^2\right)}\delta(s) \qquad (8.52)$$

This is a control law for the feed-forward lateral acceleration model following active front-wheel steering. A control law for the feed-forward and lateral acceleration feed-back model following control will be possible to obtain by the same method as in the Section 8.5.2.

8.6 FRONT- AND REAR-WHEEL ACTIVE STEERING CONTROL

Previously, the vehicle motion control was examined for front wheels that are steered in proportion to the steering wheel angle, as normal, and rear wheels that are steered in response to some control laws or front-wheels that are only actively controlled. This section will consider

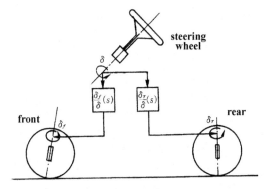

FIGURE 8.4

Front- and rear-wheel active steering.

both the front and rear wheels as being steered according to some control laws. Here, the front-wheels are regarded to be disconnected from the steering wheel. The concept of this is shown in Figure 8.4.

Applying Laplace transforms to Eqns (8.3) and (8.4) and expressing the relations among the side-slip angle, the yaw rate, the front-wheel steering angle, and the rear-wheel steering angle responses to the steering wheel angle input gives the following:

$$\{mVs + 2(K_f + K_r)\}\frac{\beta(s)}{\delta(s)} + \left\{mV + \frac{2}{V}(l_fK_f - l_rK_r)\right\}\frac{r(s)}{\delta(s)} = 2K_f\frac{\delta_f(s)}{\delta(s)} + 2K_r\frac{\delta_r(s)}{\delta(s)} \quad (8.53)$$

$$2(l_fK_f - l_rK_r)\frac{\beta(s)}{\delta(s)} + \left\{Is + \frac{2\left(l_f^2K_f + l_r^2K_r\right)}{V}\right\}\frac{r(s)}{\delta(s)} = 2l_fK_f\frac{\delta_f(s)}{\delta(s)} - 2l_rK_r\frac{\delta_r(s)}{\delta(s)} \quad (8.54)$$

Usually, the vehicle motion is studied in response to given front- and rear-wheel active steering angles. This is given by the equations of motion (8.3) and (8.4). For control purposes, the opposite is required: the front- and rear-wheel active steering angles for a desired vehicle response to steering wheel angle, $\beta(s)/\delta(s)$ and $r(s)/\delta(s)$, are required. This could be achieved by taking $\delta_f(s)/\delta(s)$ and $\delta_r(s)/\delta(s)$ as variables, rewriting Eqns (8.53) and (8.54), and solving the following first-order algebraic equations:

$$\begin{bmatrix} 2K_f & 2K_r \\ 2l_fK_f & -2l_rK_r \end{bmatrix}\begin{bmatrix} \dfrac{\delta_f(s)}{\delta(s)} \\ \dfrac{\delta_r(s)}{\delta(s)} \end{bmatrix} = \begin{bmatrix} \{mVs + 2(K_f + K_r)\}\dfrac{\beta(s)}{\delta(s)} + \left\{mV + \dfrac{2}{V}(l_fK_f - l_rK_r)\right\}\dfrac{r(s)}{\delta(s)} \\ 2(l_fK_f - l_rK_r)\dfrac{\beta(s)}{\delta(s)} + \left\{Is + \dfrac{2}{V}\left(l_f^2K_f + l_r^2K_r\right)\right\}\dfrac{r(s)}{\delta(s)} \end{bmatrix}$$

$$(8.55)$$

In other words, the front and rear wheels should be steered according to the following steering control laws to have a desired vehicle response of $\beta(s)/\delta(s)$ and $r(s)/\delta(s)$:

$$\frac{\delta_f(s)}{\delta(s)} = \frac{ml_rVs + 2lK_f}{2lK_f}\frac{\beta(s)}{\delta(s)} + \frac{Is + ml_rV + \frac{2l_flK_f}{V}}{2lK_f}\frac{r(s)}{\delta(s)} \quad (8.56)$$

$$\frac{\delta_r(s)}{\delta(s)} = \frac{ml_f Vs + 2lK_r}{2lK_r} \frac{\beta(s)}{\delta(s)} - \frac{Is - ml_f V + \frac{2l_r l K_r}{V}}{2lK_r} \frac{r(s)}{\delta(s)} \tag{8.57}$$

The previous equation shows that any arbitrary response of side-slip angle and yaw rate could be available to be used in the front- and rear-wheel active steering control laws of Eqns (8.56) and (8.57).

Assuming that the desired responses for the side-slip angle and the yaw rate are a first-order lag system with the same time constant, T, then the following is obtained:

$$\frac{\beta(s)}{\delta(s)} = \frac{G_\beta}{1+Ts} \tag{8.58}$$

$$\frac{r(s)}{\delta(s)} = \frac{G_r}{1+Ts} \tag{8.59}$$

And, if $G_\beta = 0$, the side-slip angle is always zero, and the lateral acceleration response is as follows:

$$\frac{\ddot{y}(s)}{\delta(s)} = \frac{G_r V}{1+Ts} \tag{8.60}$$

Obtaining the front- and rear-wheel active steering control laws by using Eqns (8.56) and (8.57) gives the following:

$$\frac{\delta_f(s)}{\delta(s)} = \frac{G_r}{2lK_f}\left[\frac{I}{T} + \left(ml_r V + \frac{2ll_f K_f}{V} - \frac{I}{T}\right)\frac{1}{1+Ts}\right] \tag{8.61}$$

$$\frac{\delta_r(s)}{\delta(s)} = \frac{G_r}{2lK_r}\left[-\frac{I}{T} + \left(ml_f V - \frac{2ll_r K_r}{V} + \frac{I}{T}\right)\frac{1}{1+Ts}\right] \tag{8.62}$$

These are the front- and rear-wheel active steering control laws in order for the side-slip angle to be always equal to zero and the yaw rate response to steering input to be a first-order lag system.

Moreover, from Eqns (8.58) and (8.59), the lateral acceleration response is described as the following:

$$\frac{\ddot{y}(s)}{\delta(s)} = V\left\{s\frac{\beta(s)}{\delta(s)} + \frac{r(s)}{\delta(s)}\right\} = \frac{G_r V\left(1 + \frac{G_\beta}{G_r}s\right)}{1+Ts}$$

If $G_\beta = G_r T$, then

$$\frac{\ddot{y}(s)}{\delta(s)} = G_r V \tag{8.63}$$

In other words, the lateral acceleration response is completely proportional to the steering input.

The front- and rear-wheel active steering control laws for this case are obtained by assuming $G_\beta = G_r T$ and substituting Eqns (8.58) and (8.59) into Eqns (8.56) and (8.57):

$$\frac{\delta_f(s)}{\delta(s)} = \frac{G_r}{2lK_f}\left[\frac{I}{T} + ml_r V + \left(\frac{2ll_f K_f}{V} + 2lK_f T - \frac{I}{T}\right)\frac{1}{1+Ts}\right] \tag{8.64}$$

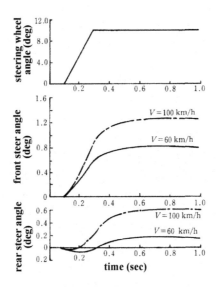

FIGURE 8.5

Front and rear steering angle.

$$\frac{\delta_r(s)}{\delta(s)} = \frac{G_r}{2lK_r}\left[-\frac{I}{T} + ml_fV + \left(-\frac{2ll_rK_r}{V} + 2lK_rT + \frac{I}{T}\right)\frac{1}{1+Ts}\right] \qquad (8.65)$$

Here, by using Eqns (8.61) and (8.62), Figure 8.5 shows exactly how the front- and rear-wheel active steering angles respond to the steering wheel angle when a ramped step input of the steering wheel angle is given. The frequency response is also shown in Figure 8.6. If the front and rear wheels are steered like this in response to the steering wheel angle, then the yaw rate responds as in Eqn (8.59). The side-slip angle is always equal to zero, and the lateral acceleration response is obtained by Eqn (8.60).

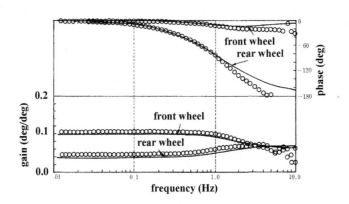

FIGURE 8.6

Front and rear steering frequency responses to steering wheel angle input.

8.7 DIRECT YAW-MOMENT CONTROL

Recently, attentions have focused on the vehicle lateral motion control called DYC. This uses the yaw moment produced by the longitudinal forces of the tires, especially for stabilizing the vehicle motion in critical conditions [3]. Of course, the critical vehicle motions mostly occur in the tire nonlinear region. However, to understand the fundamental concept of the vehicle motion control by the yaw moment, the linear equations of vehicle motion will be effectively used.

8.7.1 PASSIVE-TYPE YAW-MOMENT CONTROL

The vehicle right and left wheels can normally rotate independently. To achieve this, the traction wheels at front, rear, or both are connected to the driving shaft through the differential. If the differential is locked or viscously coupled during cornering, the longitudinal forces arise on the right and left tires in counter directions to each other, which produces the yaw moment to control the lateral motion of the vehicle. This yaw moment and its effect on the vehicle's motion will be studied in this subsection.

8.7.1.1 Effects of locked differential

There is a limited-slip differential in which the differential gears are locked until the transfer torque through the differential is within a limit. This state is called the differential lock. Figure 8.7 shows the vehicle plane motion with locked differential.

Depending on the definition by Eqn (2.46) or (2.66), the longitudinal slips of the right and left tires with the locked differential during cornering, s_L and s_R, are expressed by the following:

$$s_L = \frac{V - \frac{d}{2}r - R_0\omega}{V - \frac{d}{2}r} \approx -\frac{d}{2V}r \tag{8.66}$$

$$s_R = \frac{V + \frac{d}{2}r - R_0\omega}{V + \frac{d}{2}r} \approx \frac{d}{2V}r \tag{8.67}$$

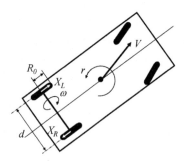

FIGURE 8.7

Vehicle turning with locked differential.

where, ω is axis rotational speed, d is wheel tread, and R_0 is the effective tire radius. Though there is a difference in the definitions of longitudinal slip under braking or traction, the previous definition is applied to both under the assumption that $V \approx R_0\omega$.

The longitudinal forces produced by the slip are proportional to the longitudinal slip when it is small. The longitudinal forces exerted on the left and right tires are expressed as follows:

$$X_L = -K_S s_L = \frac{K_S d}{2V}r$$

$$X_R = -K_S s_R = -\frac{K_S d}{2V}r$$

Here, K_S is the tire longitudinal force per unit slip as defined by Eqn (2.54). Accordingly, the yaw moment, M, produced by these forces is as follows:

$$M = -\frac{d}{2}X_L + \frac{d}{2}X_R = -\frac{K_S d^2}{2V}r \tag{8.68}$$

This is a yaw moment in the opposite direction to the yaw rate and proportional to it. This is called a yaw damping. This yaw moment is inversely proportional to the vehicle speed and decreases as the speed increases.

8.7.1.2 Effect of viscous coupling differential

There is a differential gear mechanism in which the left and right wheels are viscously connected with each other. This is called a viscous coupling differential.

Figure 8.8 shows the vehicle plane motion with the left and right wheels coupled with a viscous torque. Using the same symbols as in the previous section, the longitudinal slips of the left and right tires are as follows:

$$s_L = \frac{V - \frac{d}{2}r - R_0\left(\omega - \frac{\Delta\omega}{2}\right)}{V - \frac{d}{2}r} \approx \frac{-dr + R_0\Delta\omega}{2V} \tag{8.66}'$$

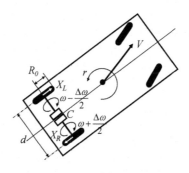

FIGURE 8.8

Vehicle turning with viscous coupling differential.

$$s_R = \frac{V + \frac{d}{2}r - R_0\left(\omega + \frac{\Delta\omega}{2}\right)}{V + \frac{d}{2}r} \approx \frac{dr - R_0\Delta\omega}{2V} \tag{8.67}'$$

where $\Delta\omega$ is rotational speed difference between left and right wheels.

The axle torque due to the longitudinal forces produced by these longitudinal slips is balanced with the viscous coupling torque. This leads to the following equations, where C is the viscous coupling torque coefficient:

$$-R_0 K_S s_L = R_0 K_S s_R = C\Delta\omega$$

Putting Eqn (8.66)' or (8.67)' into the previous equation gives the following:

$$R_0 K_S \frac{dr - R_0\Delta\omega}{2V} = C\Delta\omega$$

From this equation, $\Delta\omega$ can be obtained as follows:

$$\Delta\omega = \frac{1}{1 + \frac{2CV}{R_0^2 K_S}} \frac{dr}{R_0} \tag{8.69}$$

The longitudinal forces exerted upon the left and right tires are shown next:

$$X_L = -K_S s_L = \frac{C\Delta\omega}{R_0} = \frac{1}{1 + \frac{R_0^2 K_S}{2CV}} \frac{K_S d}{2V} r$$

$$X_R = -K_S s_R = -\frac{C\Delta\omega}{R_0} = -\frac{1}{1 + \frac{R_0^2 K_S}{2CV}} \frac{K_S d}{2V} r$$

And, the yaw moment exerted upon the vehicle is now shown:

$$M = -\frac{d}{2}X_L + \frac{d}{2}X_R = -\frac{C}{C + \frac{R_0^2 K_S}{2V}} \frac{K_S d^2}{2V} r \tag{8.68}'$$

It is understood that the yaw moment is in the opposite direction to the yaw rate and proportional to it, which is the same as the case of the locked differential. Also, it is easy to see from Eqn (8.68)' that the yaw moment can be changed by the viscous torque coefficient. The equation shows that when C is equal to zero, the yaw moment, M, is zero and increases with C until it reaches the same value as the locked differential case, where C becomes substantially very large. The yaw moment decreases with increasing vehicle speed, which is also the same as the locked differential case.

8.7.1.3 Effect of yaw rate proportional yaw moment on vehicle dynamics

The equations of vehicle motion with the additional yaw moment described by the following:

$$M = -\frac{k_r}{V} r$$

are obtained from Eqns (3.12) and (3.13). Putting this yaw moment on the right-hand side of Eqn (3.13) gives the following:

$$mV\frac{d\beta}{dt} + 2(K_f + K_r)\beta + \left\{ mV + \frac{2}{V}(l_f K_f - l_r K_r) \right\} r = 2K_f \frac{\delta}{n} \tag{8.70}$$

$$2(l_f K_f - l_r K_r)\beta + I\frac{dr}{dt} + \left\{ \frac{2\left(l_f^2 K_f + l_r^2 K_r\right) + k_r}{V} \right\} r = 2l_f K_f \frac{\delta}{n} \tag{8.71}$$

The side-slip response to the steering input is obtained in Laplace transform form as follows:

$$\frac{\beta(s)}{\delta(s)} = \frac{1}{n} \frac{\begin{vmatrix} 2K_f & mV + \frac{2}{V}(l_f K_f - l_r K_r) \\ \\ 2l_f K_f & Is + \frac{2\left(l_f^2 K_f + l_r^2 K_r\right) + k_r}{V} \end{vmatrix}}{\begin{vmatrix} mVs + 2(K_f + K_r) & mV + \frac{2}{V}(l_f K_f - l_r K_r) \\ \\ 2(l_f K_f - l_r K_r) & Is + \frac{2\left(l_f^2 K_f + l_r^2 K_r\right) + k_r}{V} \end{vmatrix}} \tag{8.72}$$

$$= \frac{1}{n} G_\delta^{*\beta}(0) \frac{1 + T_\beta^* s}{1 + \frac{2\zeta^* s}{\omega_n^*} + \frac{s^2}{\omega_n^{*2}}}$$

whereby

$$\omega_n^{*2} = \frac{4K_f K_r l^2}{mIV^2}\left(1 + AV^2 + \frac{K_f + K_r}{2l^2 K_f K_r}k_r\right)$$

$$\zeta^* = \frac{m\left(l_f^2 K_f + l_r^2 K_r\right) + I(K_f + K_r) + \frac{mk_r}{2}}{2l\sqrt{mIK_f K_r\left(1 + AV^2 + \frac{K_f + K_r}{2l^2 K_f K_r}k_r\right)}}$$

$$G_\delta^{*\beta}(0) = \frac{1 - \frac{m}{2l}\frac{l_f}{l_r K_r}V^2 + \frac{k_r}{2ll_r K_r}}{1 + AV^2 + \frac{K_f + K_r}{2l^2 K_f K_r}k_r}\frac{l_r}{l}$$

$$T_\beta^* = \frac{IV}{2ll_r K_r}\frac{1}{1 - \frac{m}{2l}\frac{l_f}{l_r K_r}V^2 + \frac{k_r}{2ll_r K_r}}$$

The yaw rate response is also obtained as follows:

$$
\frac{r(s)}{\delta(s)} = \frac{1}{n} \frac{\begin{vmatrix} mVs + 2(K_f + K_r) & 2K_f \\ 2(l_f K_f - l_r K_r) & 2l_f K_f \end{vmatrix}}{\begin{vmatrix} mVs + 2(K_f + K_r) & mV + \frac{2}{V}(l_f K_f - l_r K_r) \\ 2(l_f K_f - l_r K_r) & Is + \frac{2\left(l_f^2 K_f + l_r^2 K_r\right) + k_r}{V} \end{vmatrix}}
$$

$$
= \frac{1}{n} G_\delta^{*r}(0) \frac{1 + T_r s}{1 + \frac{2\zeta^* s}{\omega_n^*} + \frac{s^2}{{\omega_n^*}^2}}
$$

(8.73)

whereby

$$
G_\delta^{*r}(0) = \frac{1}{1 + AV^2 + \frac{K_f + K_r}{2l^2 K_f K_r} k_r} \frac{V}{l}
$$

It is clear from the previous that the vehicle motion with the yaw moment in the opposite direction and proportional to the yaw rate becomes robust to disturbances because the yaw damping is increased and yaw rate gain to steering input is reduced. In other words, the vehicle stability is improved, and the handling response is impaired by the yaw moment.

8.7.2 DYC FOR ZERO SIDE-SLIP ANGLE

Next, let us look at the active vehicle control using the yaw moment obtained from the tire longitudinal forces. It has already been shown that zero side-slip response is possible by steering the rear wheel in response to the front-wheel steering and the yaw rate. The possibility of having zero side-slip motion by DYC is examined in this section.

First, consider the yaw moment proportional to steering angle and yaw rate:

$$
M = k_\delta \delta_f + k_r r = \frac{k_\delta}{n} \delta + k_r r
$$

(8.74)

The vehicle equations of motion are obtained by putting this yaw moment into Eqn (3.13):

$$
mV \frac{d\beta}{dt} + 2(K_f + K_r)\beta + \left\{ mV + \frac{2}{V}(l_f K_f - l_r K_r) \right\} r = 2K_f \delta_f
$$

(8.75)

$$
2(l_f K_f - l_r K_r)\beta + I\frac{dr}{dt} + \left\{ \frac{2\left(l_f^2 K_f + l_r^2 K_r\right)}{V} - k_r \right\} r = (2l_f K_f + k_\delta)\delta_f
$$

(8.76)

To find the condition for the steady-state side-slip angle to be equal to zero, $d\beta/dt$ and dr/dt are set equal to zero in the previous equations. The equations are then solved for β:

$$\beta = \frac{\begin{vmatrix} 2K_f & mV + \dfrac{2}{V}(l_f K_f - l_r K_r) \\[2mm] 2l_f K_f + k_\delta & \dfrac{2\left(l_f^2 K_f + l_r^2 K_r\right)}{V} - k_r \end{vmatrix}}{\begin{vmatrix} 2(K_f + K_r) & mV + \dfrac{2}{V}(l_f K_f - l_r K_r) \\[2mm] 2(l_f K_f - l_r K_r) & \dfrac{2\left(l_f^2 K_f + l_r^2 K_r\right)}{V} - k_r \end{vmatrix}} \delta_f$$

Developing the previous equation and setting it equal to zero gives the following conditions for steady-state zero side-slip angle:

$$2K_f k_r + \left\{ mV + \frac{2}{V}(l_f K_f - l_r K_r) \right\} k_\delta = \frac{4 l l_r K_f K_r}{V}\left(1 - \frac{m l_f}{2 l l_r K_r}V^2\right) \tag{8.77}$$

For example, setting $k_r = 0$ in the previous equation yields the following:

$$k_\delta = \frac{4 l l_r K_f K_r \left(1 - \frac{m l_f}{2 l l_r K_r}V^2\right)}{mV^2 + 2(l_f K_f - l_r K_r)} \tag{8.78}$$

The yaw rate response to steering input, $r(s)/\delta(s)$, can be obtained by setting $k_r = 0$, substituting Eqn (8.78) into Eqns (8.75) and (8.76), and applying Laplace transforms:

$$\frac{r(s)}{\delta(s)} = \frac{1}{n} \frac{2K_f V}{mV^2 + 2(l_f K_f - l_r K_r)} \left\{ \frac{1 + \frac{m\left(l_f^2 K_f + l_r^2 K_r\right)V}{2l^2 K_f K_r(1+AV^2)}s}{1 + \frac{2\zeta}{\omega_n}s + \frac{1}{\omega_n^2}s^2} \right\} \tag{8.79}$$

This is the yaw rate response to the steering input for the vehicle with the steady-state zero side-slip, which is produced by DYC with the yaw moment proportional to the steering angle. By setting $k_\delta = 0$ in Eqn (8.77), we obtain the following:

$$k_r = \frac{2 l l_r K_r}{V}\left(1 - \frac{m l_f}{2 l l_r K_r}V^2\right) \tag{8.80}$$

and, the yaw rate response to steering input can also be found by putting Eqn (8.80) into Eqns (8.75) and (8.76) and applying the Laplace transforms:

$$\frac{r(s)}{\delta(s)} = \frac{1}{n} \frac{2K_f V}{mV^2 + 2(l_f K_f - l_r K_r)} \left\{ \frac{1 + \frac{m l_f V}{2 l K_r}s}{1 + d_1 s + d_2 s^2} \right\} \tag{8.81}$$

where

$$d_1 = \frac{ml_f\{mV^2 + 2(l_fK_f - l_rK_r)\} + 2I(K_f + K_r)}{2IK_r\{mV^2 + 2(l_fK_f - l_rK_r)\}}V$$

$$d_2 = \frac{mIV^2}{2IK_r\{mV^2 + 2(l_fK_f - l_rK_r)\}}$$

This is the yaw rate response to the steering input for the vehicle with the steady-state zero side slip controlled by DYC using the yaw moment proportional to the yaw rate algorithm.

By adjusting the parameters k_δ and k_r, as shown previously, the steady-state zero side slip vehicle motion and yaw rate responses described by Eqns (8.79) and (8.81) can be achieved. However, it is important to understand from these equations that if the vehicle speed is low, the denominator in the yaw rate response gain is equal to zero at the speed $V = \sqrt{-2(l_fK_f - l_rK_r)/m}$ and changes the sign. Also, if the vehicle speed is less than this, the constant term in the characteristic equation of the vehicle response becomes negative, and the vehicle tends to be unstable. Therefore, it should be understood that the zero side-slip control mentioned previously makes sense at vehicle speeds high enough for the conditions to be physically reasonable.

It seems through this analysis that zero side-slip control by DYC is practical at speeds above those where the steady-state side-slip response to the front-wheel steering is changed from positive to negative. This is expressed by the following inequality:

$$1 - \frac{ml_f}{2Il_rK_r}V^2 \leq 0$$

Here, the DYC can control the negative side-slip angle to the front steering input and stabilize the vehicle motion at high speed.

Next is to look at the yaw moment that gives a side-slip angle always equal to zero at any time during vehicle motion at high speeds:

$$M(s) = k(s)\delta_f(s) = k(s)\frac{\delta(s)}{n} \tag{8.82}$$

The Laplace transforms of the equations of vehicle motion, with this yaw moment, are written as follows:

$$mVs\beta(s) + 2(K_f + K_r)\beta(s) + \left\{mV + \frac{2}{V}(l_fK_f - l_rK_r)\right\}r(s) = 2K_f\delta_f(s) \tag{8.83}$$

$$2(l_fK_f - l_rK_r)\beta(s) + Isr(s) + \frac{2\left(l_f^2K_f + l_r^2K_r\right)}{V}r(s) = \{2l_fK_f + k(s)\}\delta_f(s) \tag{8.84}$$

Solving the previous equation for the side-slip angle and setting it equal to zero gives the following equation:

$$\begin{vmatrix} 2K_f & mV + \frac{2}{V}(l_fK_f - l_rK_r) \\ 2l_fK_f + k(s) & Is + \frac{2}{V}\left(l_f^2K_f + l_r^2K_r\right) \end{vmatrix} = 0$$

It is possible to solve this equation for $k(s)$, thereby obtaining the following yaw moment, $M(s)$, to have the side-slip response to steering input always equal to zero:

$$M(s) = k(s)\frac{\delta(s)}{n} = \frac{1}{n}\left\{ \frac{4ll_rK_fK_r\left(1 - \frac{ml_f}{2ll_rK_r}V^2\right)}{mV^2 + 2(l_fK_f - l_rK_r)} + \frac{2IK_fV}{mV^2 + 2(l_fK_f - l_rK_r)}s \right\}\delta(s) \qquad (8.85)$$

The yaw rate response to the steering input is obtained by putting $k(s)$ in the previous equation and $\beta(s) = 0$ into Eqn (8.84):

$$\frac{r(s)}{\delta(s)} = \frac{1}{n}\frac{2K_fV}{mV^2 + 2(l_fK_f - l_rK_r)} \qquad (8.86)$$

Though it is possible to formally obtain the yaw moment for zero side-slip response and the desired yaw rate response to the steering input, as in Eqns (8.85) and (8.86), attention must be given to the fact that these equations are reasonable at high vehicle speeds. Even so, it is clear that the yaw rate response gain of the vehicle with zero side-slip DYC is fairly small compared with that of the normal front-wheel steering vehicle and decreases with increasing vehicle speed significantly.

8.7.3 MODEL FOLLOWING DYC

Here, the model following control by DYC will be investigated as is done for the rear-wheel steering in Section 8.4.

The vehicle responses to the steering wheel and the yaw moment, M, produced by the longitudinal forces of the right and left tires are as follows:

$$mV\frac{d\beta}{dt} + 2(K_f + K_r)\beta + \left\{ mV + \frac{2}{V}(l_fK_f - l_rK_r) \right\}r = 2K_f\frac{\delta}{n} \qquad (8.87)$$

$$2(l_fK_f - l_rK_r)\beta + I\frac{dr}{dt} + \frac{2\left(l_f^2K_f + l_r^2K_r\right)}{V}r = 2l_fK_f\frac{\delta}{n} + M \qquad (8.88)$$

Applying Laplace transforms to these equations and solving the equations for the yaw rate response, $r(s)$, to the steering and yaw moment input gives the following equations:

$$r(s) = \frac{\frac{1}{n}G_\delta^r(0)(1 + T_rs)\delta(s) + G_M^r(0)(1 + T_Ms)M(s)}{1 + \frac{2\zeta}{\omega_n}s + \frac{1}{\omega_n^2}s^2}$$

$$= \frac{\omega_n^2H_M(s)}{s^2 + 2\zeta\omega_n + \omega_n^2} \qquad (8.89)$$

where

$$H_M(s) = \frac{1}{n}G_\delta^r(0)(1 + T_rs)\delta(s) + G_M^r(0)(1 + T_Ms)M(s) \qquad (8.90)$$

$$G_M^r(0) = \frac{(K_f + K_r)V}{2l^2 K_f K_r (1 + AV^2)}$$

$$T_M = \frac{mV}{2(K_f + K_r)}$$

8.7.3.1 Feed-forward yaw rate model following DYC

Here, the same yaw rate model response is adopted as in the rear-wheel steering:

$$r_m(s) = \frac{1}{n} \frac{G_e}{1 + T_e s} \delta(s) \tag{8.43}$$

If this response is identical with the response described by Eqn (8.89), the following equation is obtained:

$$\frac{\frac{1}{n} G_\delta^r(0)(1 + T_r s)\delta(s) + G_M^r(0)(1 + T_M s)M(s)}{1 + \frac{2\gamma}{\omega_n} s + \frac{1}{\omega_n^2} s^2} = \frac{1}{n} \frac{G_e}{1 + T_e s} \delta(s)$$

It is possible to obtain $M(s)$ from the previous equation as follows:

$$M(s) = \frac{1}{n} \left\{ \frac{G_e \left(1 + \frac{2\gamma}{\omega_n} s + \frac{1}{\omega_n^2} s^2\right)}{G_M^r(0)(1 + T_M s)(1 + T_e s)} - \frac{G_\delta^r(0)}{G_M^r(0)} \frac{1 + T_r s}{1 + T_M s} \right\} \delta(s) \tag{8.91}$$

This is the required direct yaw-moment control law for the feed-forward control to follow the first-order lag yaw rate model response.

8.7.3.2 Feed-forward and yaw rate feed-back model following DYC

The same yaw rate model response as in Section 8.7.3.1 is adopted again. Using Eqn (8.43), Eqn (8.43)′ is obtained as follows:

$$\left(s + \frac{1}{T_e}\right) r_m(s) = \frac{1}{n} \frac{G_e}{T_e} \delta(s) \tag{8.43}'$$

Again, the error, e, between the model yaw rate, r_m, and the vehicle yaw rate response to the steering and yaw moment input, r, is as follows:

$$e(s) = \left(s + \frac{1}{T_e}\right)(r(s) - r_m(s))$$

To make the error converge to zero in the first-order lag response, the following is set:

$$\left(s + \frac{1}{T_g}\right) e(s) = 0$$

From the previous two equations, the following expression can be obtained:

$$\left(s + \frac{1}{T_g}\right)\left(s + \frac{1}{T_e}\right)(r(s) - r_m(s)) = 0$$

Taking Eqn (8.89) into consideration, $(s + 1/T_g)(s + 1/T_e)r(s)$ is rewritten in the same way as in Section 8.4.2 as follows, where c_1 and c_0 are the same as defined in Section 8.4.2:

$$\left(s + \frac{1}{T_g}\right)\left(s + \frac{1}{T_e}\right)r(s) = \left\{s^2 + 2\zeta\omega_n s + \omega_n^2 + \left(\frac{1}{T_g} + \frac{1}{T_e} - 2\zeta\omega_n\right)s + \frac{1}{T_g T_e} - \omega_n^2\right\}r(s)$$

$$= \omega_n^2 H_M(s) + (c_1 s + c_0)r(s)$$

From Eqn (8.43)$'$, we can get the following:

$$\left(s + \frac{1}{T_g}\right)\left(s + \frac{1}{T_e}\right)r_m(s) = \left(s + \frac{1}{T_g}\right)\frac{1}{n}\frac{G_e}{T_e}\delta(s)$$

Accordingly, the condition of the error between $r(s)$ and $r_m(s)$ converging to zero is expressed as follows:

$$\left(s + \frac{1}{T_g}\right)\left(s + \frac{1}{T_e}\right)(r(s) - r_m(s)) = \omega_n^2 H_M(s) + (c_1 s + c_0)r(s) - \frac{1}{n}\frac{G_e}{T_e}\left(s + \frac{1}{T_g}\right)\delta(s) = 0$$

Putting Eqn (8.90) into the previous equation gives the following relationship:

$$\frac{1}{n}G_\delta^r(0)(1 + T_r s)\delta(s) + G_M^r(0)(1 + T_M s)M(s) + \frac{1}{\omega_n^2}(c_1 s + c_0)r(s) - \frac{1}{n}\frac{G_e}{\omega_n^2 T_e}\left(s + \frac{1}{T_g}\right)\delta(s) = 0$$

Thereby, the yaw moment needed for the DYC to follow the model yaw rate response is obtained from the previous equation as follows:

$$M(s) = \frac{1}{n}\left\{-\frac{G_\delta^r(0)}{G_M^r(0)}\frac{1 + T_r s}{1 + T_M s} + \frac{G_e}{\omega_n^2 T_g T_e G_M^r(0)}\frac{1 + T_g s}{1 + T_M s}\right\}\delta(s) - \frac{c_0}{\omega_n^2 G_M^r(0)}\frac{1 + \frac{c_1}{c_0}s}{1 + T_M s}r(s)$$

$$(8.92)$$

This is a direct yaw-moment control law for the feed-forward and yaw rate feed-back control to follow the first-order lag yaw rate model response to the steering wheel input.

8.7.3.3 Feed-forward lateral acceleration model following DYC

The lateral acceleration response to steering wheel angle and yaw moment inputs is obtained using Eqns (8.3) and (8.4) as follows:

$$\ddot{y}(s) = \frac{\frac{1}{n}G_\delta^{\ddot{y}}(0)\left(1 + \frac{l_r}{V}s + \frac{l}{2lK_r}s^2\right)\delta(s) + G_M^{\ddot{y}}(0)\left\{1 + \frac{l_r K_r - l_f K_f}{(K_f + K_r)V}s\right\}M(s)}{1 + \frac{2\zeta}{\omega_n}s + \frac{1}{\omega_n^2}s^2}$$

$$(8.93)$$

The same model response of lateral acceleration as the previous one is shown here:

$$\ddot{y}_m(s) = \frac{1}{n} \frac{G_y}{1 + T_y s} \delta(s)$$

If this is identical with Eqn (8.93), as shown next:

$$\frac{\frac{1}{n} G_\delta^{\ddot{y}}(0) \left(1 + \frac{l_r}{V} s + \frac{I}{2lK_r} s^2\right) \delta(s) + G_M^{\ddot{y}}(0) \left\{1 + \frac{l_r K_r - l_f K_f}{(K_f + K_r) V} s\right\} M(s)}{1 + \frac{2\gamma}{\omega_n} s + \frac{1}{\omega_n^2} s^2} = \frac{1}{n} \frac{G_y}{1 + T_y s} \delta(s)$$

then we have the following:

$$M(s) = \frac{G_y \left(1 + \frac{2\gamma}{\omega_n} s + \frac{1}{\omega_n^2} s^2\right) - G_\delta^{\ddot{y}}(0)(1 + T_y s)\left(1 + \frac{l_r}{V} s + \frac{I}{2lK_r} s^2\right)}{n G_M^{\ddot{y}}(0)(1 + T_y s)\left\{1 + \frac{l_r K_r - l_f K_f}{(K_f + K_r) V} s\right\}} \delta(s) \qquad (8.94)$$

This is a control law of the feed-forward lateral acceleration model following DYC, and the feed-forward and lateral acceleration feed-back model following control can also be introduced by the same method as the previous one. However, it is obvious that as the coefficient of s in the denominator of Eqn (8.94), $(l_r K_r - l_f K_f)/((K_f + K_r)V)$, is considerably small and close to zero, the control requires high responsiveness with significantly high gain in a high frequency range. Therefore, the lateral acceleration model following control by DYC is assumed to be not necessarily practical.

8.7.4 ADVANTAGE OF DYC [4]

Until now, various DYC control algorithms have been dealt with under a premise of linear characteristics of tire lateral force to side-slip angle. Furthermore, it is important to perceive that DYC has a distinctive feature regarding tire longitudinal forces that produce the yaw moment needed for the motion control, especially in a range of near limit of the tire lateral force where the control is often required.

One is that the longitudinal force of the tire available for the control is larger than the additional lateral force in a near limit of the tire lateral force. Figure 8.9 schematically shows the previous content by a friction circle. It is understandable that a margin of the lateral force to increase is not necessarily enough, yet the available range of the longitudinal force for the control remains more or less sufficient around the operating point. Consequently, it seems reasonable to intentionally use the tire longitudinal force rather than lateral one for the vehicle motion control as DYC in a near-limit region of vehicle motion.

Another point is that the tire longitudinal force that generates the yaw moment for the control is produced by directly applying the brake/traction torque command to the tire wheel spinning motion. A slip between spinning and transversal motions of the wheel is produced by the torque command until the longitudinal force of the tire balances with the input torque command no matter how the vehicle motion is. The spinning motion of the wheel is basically independent from the subject vehicle motion to be controlled. Therefore, the tire longitudinal force produced has no direct influence of the tire vertical load, the tire lateral force, or the vehicle motion itself so far as the vertical load and the friction coefficient between tire and road surface are large enough to produce the tire longitudinal force to balance with the command torque.

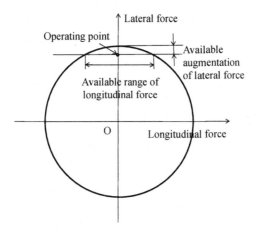

FIGURE 8.9

Available range of longitudinal force for DYC in friction circle.

On the other hand, the lateral force for the motion control is produced by steering angle command and is directly influenced by the vehicle lateral motion, the tire vertical load, and the longitudinal force of the tire. Consequently, it is assumed that the tire longitudinal force for the motion control will be more accurately produced by the torque command compared with the tire lateral force for the control by the steering command. This is a big advantage of DYC over the other active steering controls in the vehicle motion controls.

REFERENCES

[1] Furukawa Y. Improvement of vehicle handling by 4 wheel steering system. JSAE J 1986; 40(3) [in Japanese].
[2] Irie N, Shibahata Y. Improving vehicle stability and controllability through active Rea wheel steering. JSAE J 1986;40(3) [in Japanese].
[3] Abe M. Vehicle dynamics and control for improving handling and active safety – from 4WS to DYC. In: Proceedings of the IMechE Part K, vol. 213; 1999.
[4] Abe M. 32. Integrated controls, road and off-road vehicle system dynamics handbook. Taylor and Francis; October 2013.

VEHICLE MOTION WITH ALL-WHEEL CONTROL

9

9.1 PREFACE

In Chapter 8, the fundamentals of vehicles with active motion control, especially for conventional vehicles, have been discussed. Meanwhile, in the actual world, numerous research and development efforts of vehicle motion controls, especially chassis controls for high vehicle performance and active safety, have been done the past two decades. On the other hand, the issue of environmental problems accelerates the number of electric-powered vehicles (EPV) in the market, which are promising vehicles from the viewpoint of motion control as well.

Once an EPV is equipped with four in-wheel motors for each wheel, it is easy to control the four tire longitudinal forces independently for more sophisticated vehicle motion control. In addition, all of the front and rear wheels of the vehicle are able to be steered independently by some electromagnetic actuators that control the tire lateral forces of right and left, and front and rear independently for the motion control. Thus, the EPV will easily become a full drive-by-wire vehicle—all-wheel control vehicle—that has the eight independently controllable tire forces, four longitudinal forces, and four lateral ones. The all-wheel control vehicle will dramatically extend the possibility of vehicle motion control and must be an ultimate active motion control vehicle for a new era [1].

This chapter will offer a trial example in detail of this promising vehicle motion control in order to understand what sort of vehicle motion control is available for such a full drive-by-wire vehicle and what effects are possible for us to expect from the control [2,3].

9.2 TIRE FORCE DISTRIBUTION

A vehicle's plane motion is composed of three components: longitudinal, lateral, and yaw motions, as described in Section 7.2. They are, respectively, determined by a total tire force in the longitudinal direction, a total tire force in the lateral direction, and a total yaw moment produced by the tire forces, which are shown in Figure 9.1. Therefore, even though eight forces of the tires are to be determined arbitrarily, total tire forces in the longitudinal and lateral direction, respectively, and a total yaw moment must satisfy some conditions in order to have the intended vehicle response to the driver's commands of steering and acceleration/braking. They become three constraint conditions among the independently controllable eight forces, and there eventually remains a five-degree-of-freedom redundancy in determining the eight tire forces to give the vehicle response to the driver's command.

Then, a problem arises with how the forces should be distributed reasonably to each wheel depending on this redundancy. To eliminate the redundancy, some additional constraints are

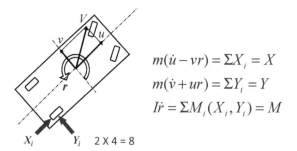

$$m(\dot{u} - vr) = \Sigma X_i = X$$
$$m(\dot{v} + ur) = \Sigma Y_i = Y$$
$$I\dot{r} = \Sigma M_i(X_i, Y_i) = M$$

FIGURE 9.1

Vehicle plane motion.

available to apply. Also, if the number of constraints is less than number of variables, some reasonable norm for definitely determining the eight tire forces is essential.

9.2.1 CONSTRAINTS TO HAVE AN INTENDED VEHICLE RESPONSE

Since the lateral and longitudinal forces of the four wheels of the electric vehicle considered here are independently controllable according to steering and acceleration/brake commands, theoretically any vehicle response is available for this vehicle by appropriately controlling the lateral and longitudinal forces to follow a given model response. The lateral force and the yaw moment required for the vehicle to follow the model response are available by applying an inverse method to the vehicle plane motion. To have the same response of the controlled vehicle as that of the ordinary vehicle, the response of the two-degree-of-freedom ordinary vehicle to steering wheel input, δ_h, is introduced for the model vehicle response with the tire force distribution control, which has been given in Section 3.4.2 as follows:

$$\frac{\beta}{\delta_h}(s) = \frac{1}{n} \frac{G_\delta^\beta(1 + T_\beta s)}{1 + \frac{2\zeta}{\omega_n}s + \frac{1}{\omega_n^2}s^2} \tag{3.77}''$$

$$\frac{r}{\delta_h}(s) = \frac{1}{n} \frac{G_\delta^r(1 + T_r s)}{1 + \frac{2\zeta}{\omega_n}s + \frac{1}{\omega_n^2}s^2} \tag{3.78}''$$

where n means the steering gear ratio.

The side-slip and yaw motions are described as follows:

$$mV(\dot{\beta} + r) = Y, \quad I\dot{r} = M$$

The following equations are obtained by putting Eqn (3.77)″ and (3.78)″ into the previous equation, which are the required total lateral force, Y, and yaw moment, M, to follow the given vehicle response:

$$Y = \frac{\omega_n^2 mV}{n} \left\{ G_\delta^\beta T_\beta + \frac{\left(G_\delta^\beta + G_\delta^r T_r - 2\zeta\omega_n G_\delta^\beta T_\beta\right)s + G_\delta^r - \omega_n^2 G_\delta^\beta T_\beta}{s^2 + 2\zeta\omega_n s + \omega_n^2} \right\} \delta_h \tag{9.1}$$

$$M = \frac{\omega_n^2 G_\delta^r I}{n} \left\{ T_r - \frac{(2\zeta\omega_n T_r - 1)s + \omega_n^2 T_r}{s^2 + 2\zeta\omega_n s + \omega_n^2} \right\} \delta_h \tag{9.2}$$

A total longitudinal force, X, required for the vehicle motion is given by the driver's command to braking or acceleration. The lateral force, the yaw moment, and the longitudinal force obtained previously must, respectively, be equal to the total force of the tire lateral forces, the total yaw moment produced by the tire forces, and the total tire longitudinal forces. They are the three described constraints for the tire force distribution controls to be satisfied by distributed tire forces, which are described as follows:

$$Y_1 + Y_2 + Y_3 + Y_4 = Y \tag{9.3}$$

$$l_f(Y_1 + Y_2) - l_r(Y_3 + Y_4) + d_f(X_2 - X_1) + d_r(X_4 - X_3) = M \tag{9.4}$$

$$X_1 + X_2 + X_3 + X_4 = X \tag{9.5}$$

where, X_i and Y_i ($i = 1 - 4$) are the longitudinal and lateral force of each tire, respectively, required at least to have the model response of the vehicle.

9.2.2 DISTRIBUTION NORM

If there is no more constraint other than Eqns (9.3)–(9.5) for ordering the vehicle motion, there still remains a five-degree-of-freedom redundancy. Therefore, in order to definitely determine the eight forces and allocate them to each tire, it is essential to introduce some reasonable norm for the distribution.

Because a horizontal force of the tire is limited by the tire's vertical load, it is reasonable to use the tires equally as much as possible depending on their vertical loads and to avoid a specific tire experiencing an extraordinary severe load condition during vehicle motion. It is expected, consequently, to extend the limit performance of the vehicle. Base on the previous view, the following norm minimizing a square sum of tire workload is introduced as one of the distribution norm. As the tire workload of each tire is defined as follows, where Z_i is a vertical load of the tire:

$$\mu_i = \frac{\sqrt{X_i^2 + Y_i^2}}{Z_i} \tag{9.6}$$

the following objective function of minimization norm is defined:

$$J = \sum_{i=1}^{4} \mu_i^2 = \sum_{i=1}^{4} \frac{X_i^2 + Y_i^2}{Z_i^2} \tag{9.7}$$

The vertical load of each tire needed in the norm function during the vehicle motion is regarded to be determined by load transfers by lateral and longitudinal accelerations and estimated, respectively, as follows:

$$Z_1 = \frac{mgl_r}{2l} - \frac{ma_x}{2l}h - k_f\frac{ma_y}{d_f}h$$

$$Z_2 = \frac{mgl_r}{2l} - \frac{ma_x}{2l}h + k_f\frac{ma_y}{d_f}h$$

$$Z_3 = \frac{mgl_f}{2l} + \frac{ma_x}{2l}h - k_r\frac{ma_y}{d_r}h$$

$$Z_4 = \frac{mgl_f}{2l} + \frac{ma_x}{2l}h + k_r\frac{ma_y}{d_r}h$$

where, k_f and k_r are the front/rear roll stiffness rates, and a_x and a_y are the longitudinal/lateral accelerations.

When the vehicle is controlled by tire forces, the tires dissipate the slip energy because the slip and the slip force arise at the tire contact surface to the ground. A product of the slip force in a slip region and the slip speed is the dissipation energy rate due to the tire slip. We already know by using a brush tire model with combined slip in Section 2.4 that the contact surface is divided into adhesion and slip regions, as well as how the slip features and the lateral and longitudinal forces in the slip region are.

Referring to Section 2.4.2, during braking for instance, it is possible to describe the ratios of the slip force to the total force in longitudinal and lateral directions, respectively, as follows:

$$\alpha_{sx} = -6\mu F_z \cos\theta\left(\frac{1}{6} - \frac{1}{2}\xi_s^2 + \frac{1}{3}\xi_s^3\right)/F_x,$$

$$\alpha_{sy} = -6\mu F_z \sin\theta\left(\frac{1}{6} - \frac{1}{2}\xi_s^2 + \frac{1}{3}\xi_s^3\right)/F_y$$

When the whole contact surface is in slip, then the following is obtained:

$$\alpha_{sx} = \alpha_{sy} = 1$$

Also, the slip speeds at the slip region in the longitudinal and lateral directions are described as follows:

$$v_{sx} = us, \quad v_{sy} = u\tan\beta$$

Then, the expected energy dissipation rate for the i-th wheel is described as follows:

$$W_{sxi} = v_{sxi}\alpha_{sxi}X_i, \quad W_{syi} = v_{syi}\alpha_{syi}Y_i, \quad (i = 1 \sim 4)$$

It is reasonable to control the vehicle motion with less slip energy consumption as much as possible. The following are defined as another one of the distribution norms minimizing the tire energy dissipation during the motion control:

$$J = \sum_{i=1}^{4}\left(W_{sxi}^2 + W_{syi}^2\right) = \sum_{i=1}^{4}\left(v_{sxi}^2\alpha_{sxi}^2X_i^2 + v_{syi}^2\alpha_{syi}^2Y_i^2\right) \tag{9.8}$$

This is supposed to be effective in reducing tire wear of the vehicle.

9.2.3 DISTRIBUTION CONTROL

As the form of both of these objective functions is a second-order algebraic equation of the tire lateral and longitudinal forces as follows:

$$J = \sum_{i=1}^{4} \left(a_i X_i^2 + b_i Y_i^2\right) \tag{9.9}$$

the problem becomes a minimization of the second-order function of eight variables with three constraints, whichever the objective function is. From Eqns (9.3)–(9.5), the following are obtained:

$$Y_4 = Y - Y_1 - Y_2 - Y_3 \tag{9.10}$$

$$X_3 = -\frac{d_f + d_r}{2d_r} X_1 + \frac{d_f - d_r}{2d_r} X_2 + \frac{1}{d_r} Y_1 + \frac{1}{d_r} Y_2 + \frac{X}{2} - \frac{M}{d_r} - \frac{l_r Y}{d_r} \tag{9.11}$$

$$X_4 = \frac{d_f - d_r}{2d_r} X_1 - \frac{d_f + d_r}{2d_r} X_2 - \frac{1}{d_r} Y_1 - \frac{1}{d_r} Y_2 + \frac{X}{2} + \frac{M}{d_r} + \frac{l_r Y}{d_r} \tag{9.12}$$

Putting the previous equations into Eqn (9.9), the objective function of the five variables, X_1, X_2, Y_1, Y_2, and Y_3 is obtained. Thus, the linear first-order algebraic equations of five variables are given by a partial differential of the objective function as the minimizing condition as follows:

$$\frac{\partial J}{\partial X_1} = 0, \quad \frac{\partial J}{\partial X_2} = 0, \quad \frac{\partial J}{\partial Y_1} = 0, \quad \frac{\partial J}{\partial Y_2} = 0, \quad \frac{\partial J}{\partial Y_3} = 0 \tag{9.12}$$

These equations are easy for us to solve analytically. By solving the equations with the constraints, the lateral and longitudinal forces of each tire (the tire forces to be distributed to each tire) are obtained at each moment during the vehicle motion to follow the model response. Figure 9.2(a) shows schematic image of sharing X, Y, and M by four tires.

Then, the traction torque and steering angle of each tire to generate the previous forces obtained are determined as control commands of the distribution control as shown in Figure 9.2(b). Given the lateral and longitudinal forces, Y_i and X_i, to be distributed to each tire, the steering angle command, δ_i, of each wheel is determined by the following equations based on the assumption of linear tire, where K_i is the cornering stiffness of each tire:

$$Y_1 = -K_1(\beta + l_f r/V - \delta_1), \quad Y_2 = -K_2(\beta + l_f r/V - \delta_2)$$
$$Y_3 = -K_3(\beta - l_r r/V - \delta_3), \quad Y_4 = -K_4(\beta - l_r r/V - \delta_4) \tag{9.13}$$

The motor torque command, T_i, is determined by the following:

$$T_i = R_i X_i \quad (i = 1 \sim 4) \tag{9.14}$$

where, R_i is a wheel radius.

The previous is the feed-forward type of an all-wheel control system with the norm minimizing the objective function, of which the outline is shown by a block diagram in Figure 9.3. The brush tire model is to be used for estimation of tire forces and their slip energy dissipation.

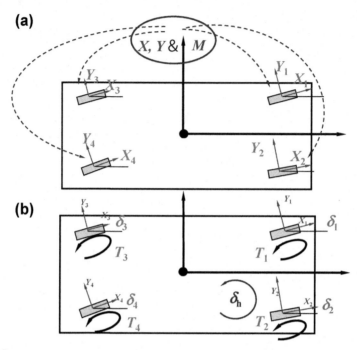

FIGURE 9.2

Tire force distribution and all-wheel control, (a) force distribution and (b) steer angle and torque distribution.

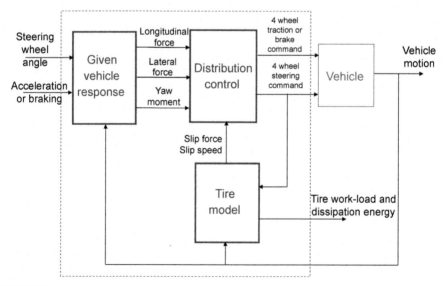

FIGURE 9.3

Feed-forward type of all-wheel control system.

Basically, the inputs of the system are the driver steering wheel angle and acceleration/braking commands, and the outputs are the steering and torque commands to the four wheels and also the workload and the energy dissipation rate of each tire.

9.3 EFFECTS OF DISTRIBUTION CONTROL

9.3.1 EFFECT OF DISTRIBUTION CONTROL MINIMIZING TIRE WORKLOAD

It is expected that the norm minimizing square sum of the tire workload contributes to avoiding a specific tire experiencing extraordinary severe load condition during controlled vehicle motion. Therefore, it makes the controlled vehicle more stable than the vehicle without control.

Figure 9.4 shows the experimental results of lane change behaviors with an experimental, small-size full drive-by-wire electric vehicle, comparing the motions of the vehicles with and without distribution control to each other. The workloads of each wheel of the vehicles during the lane change with braking are also compared in Figure 9.5. The vehicle behaviors are almost the same as shown in Figure 9.4; however, the tire workload of the vehicle without control as a whole is higher than that of the vehicle with the control. Especially, the tire workload of the inner-rear wheel during lane change reaches 1.0. As the experimental lane change is done on a dry road surface and the tire–road friction coefficient is roughly 1.0, the tire workload of 1.0 indicates almost the limit of the tire horizontal force available. Therefore, it seems that the vehicle without control is more liable to approach unstable conditions.

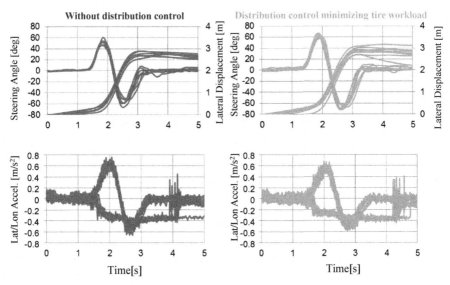

FIGURE 9.4

Vehicle lane change with braking with/without distribution control.

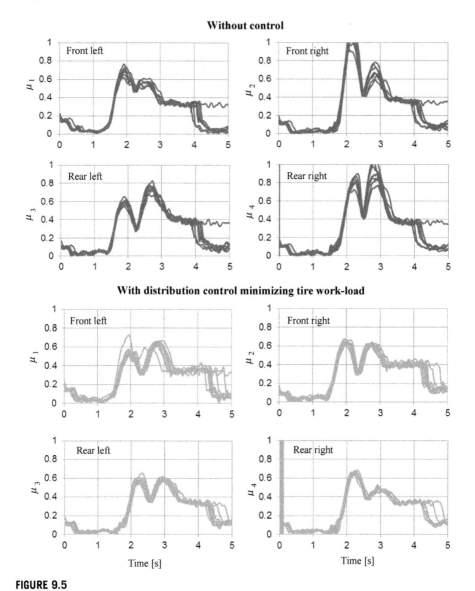

FIGURE 9.5

Workload of each tire during lane change with braking.

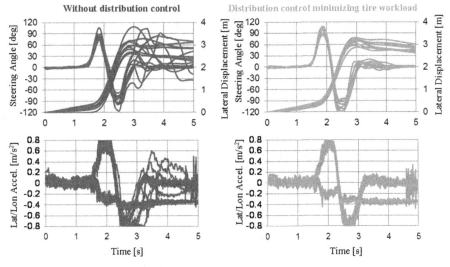

FIGURE 9.6

Severe lane change behaviors with braking.

Figures 9.6 and 9.7 are the results of more severe lane changes with braking. The lane change behaviors of the vehicle with the distribution control still remain in stable motion; on the other hand, the vehicle motions without the control are mostly unstable at the latter part of the lane change. Almost all the tire workloads of the vehicle without control also reach 1.0 and are sometimes over 1.0, whereas all the tire workloads of the vehicle with the control are still less than 1.0, even under such a severe lane change.

These results convince us that using tires equally in terms of horizontal force as much as possible is significantly reasonable in vehicle motion control.

9.3.2 EFFECT OF DISTRIBUTION CONTROL MINIMIZING TIRE SLIP ENERGY DISSIPATION

From Section 9.2.2, the energy dissipation rate of each tire during vehicle motion is described as follows:

$$W_i = v_{sxi}\alpha_{sxi}X_i + v_{syi}\alpha_{syi}Y_i, \quad (i = 1 \sim 4)$$

And, the summation of W_i is a total tire dissipation rate of the vehicle, and the time integration of W_i throughout the motion is the tire slip energy dissipation of the vehicle.

The tire energy dissipation rate and the dissipation energy of the vehicles with and without distribution controls measured during the experimental lane changes shown in Figure 9.8 are compared in Figure 9.9. The lane change behaviors of the vehicles are almost similar; even so, there is a difference in the energy dissipation rate and the dissipation energy. More specifically, the time histories of the energy dissipation rate and the integrated dissipation energy of each tire for the vehicles with and without distribution controls during the lane change are respectively

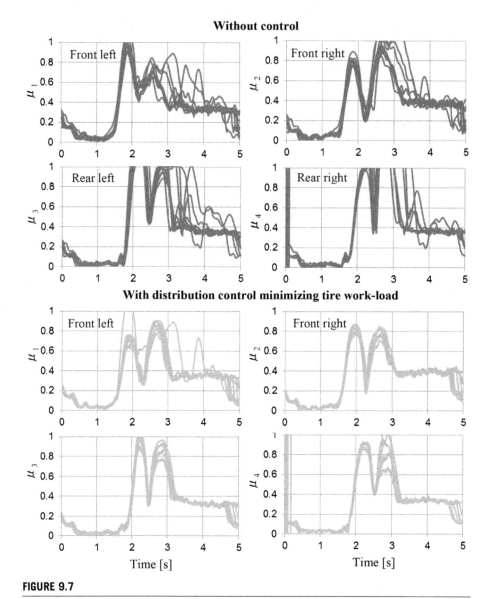

FIGURE 9.7

Workload of each tire during severe lane change with braking.

compared in Figure 9.9. Also, the total amount of dissipation energy of the four wheels as the tire energy dissipation of the vehicles with and without distribution controls are compared with each other.

The vehicle behavior during more severe lane change with higher peak lateral acceleration is shown in Figure 9.10. The energy dissipation rates and the dissipation energies for the respective vehicles are also compared in Figure 9.11 in the same way as in Figure 9.9.

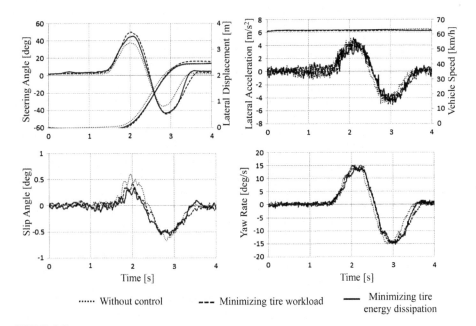

FIGURE 9.8

Lane change of the vehicles with/without distribution controls.

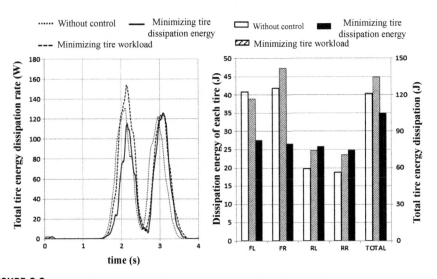

FIGURE 9.9

Tire energy dissipation of the vehicles with/without distribution controls.

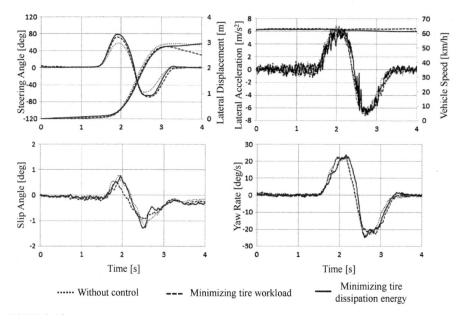

FIGURE 9.10

Severe lane change of the vehicles with/without distribution controls.

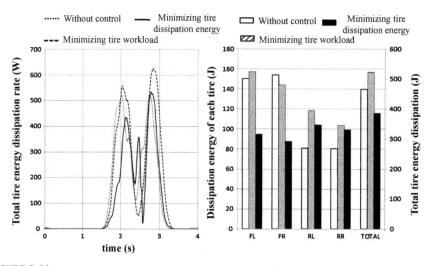

FIGURE 9.11

Tire energy dissipation of the vehicles during severe lane change.

From the previous figures, it seems that the tire energy dissipation due to tire slip during lane change increases dramatically with the increase of the peak lateral acceleration. The energy dissipation increases by around five times, with the increase of the peak lateral acceleration from 0.4 G to 0.6 G in this case.

Figures 9.9 and 9.11 show that the distribution control with the norm minimizing the square sum of energy dissipation rate is consistently effective in reducing the tire energy dissipation; on the other hand, the distribution control with the norm minimizing the tire workload has a negative effect on reducing the tire energy dissipation, though the distribution control is effective in stabilizing the vehicle motion as shown in Figure 9.6. It is recognized that the vehicle with the distribution control minimizing energy dissipation dissipates the tire energy more equally at each tire during the lane change compared with the other vehicles. Consequently, it achieves almost 15 to 20% reduction of the tire energy dissipation compared with the vehicle without the distribution control.

9.4 ROLL CONTROL INTEGRATED WITH TIRE FORCE DISTRIBUTION CONTROL

As the tire longitudinal force brings us the vertical forces exerted upon the vehicle body through some suspension mechanism, it is possible for us to control the body roll of the all-wheel control vehicle with four in-wheel motors, for instance, by controlling the longitudinal forces of the tires appropriately for improving vehicle agility [4]. However, the additional tire longitudinal forces for the roll control may cause the specific tire experiencing extraordinary severe load conditions during vehicle motions and impair the vehicle stability.

Section 9.3.1 showed that the tire force distribution control minimizing the tire workload of the all-wheel control vehicle can significantly improve the vehicle's stability. The integration of the tire force distribution control with the roll control arises in this section for higher vehicle agility and stability, avoiding possible ill effects of the roll control.

9.4.1 ADDITIONAL CONSTRAINT TO CONTROL VEHICLE BODY ROLL

The tire longitudinal forces are able to produce vertical forces that act on the vehicle body through the suspension mechanism as shown in Figure 9.12. If a roll moment produced by the forces, which is given by the left side of Eqn (9.15), satisfies the condition described by Eqn (9.15), it can control the roll motion of the vehicle body caused by the lateral acceleration during vehicle motion:

$$\frac{d_{sf}}{2}(X_2 - X_1)\tan\theta_f - \frac{d_{sr}}{2}(X_4 - X_3)\tan\theta_r = -\alpha m h_s \ddot{y} \tag{9.15}$$

where, d_{sf} and d_{sr} are the front and rear spring treads, θ_f and θ_r are the front and rear anti-dive/squat angles, h_s is the C.G. height from the roll axis, α is the roll control rate, and $\ddot{y}$ is the lateral acceleration. This becomes the fourth constraint among the eight tire forces, in addition to the three constraints described by Eqns (9.3)–(9.5) in Section 9.2.1 for the tire force distribution control minimizing the tire workload to follow the intended vehicle response.

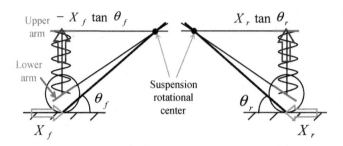

FIGURE 9.12

Vertical force produced by tire longitudinal force.

9.4.2 DISTRIBUTION CONTROL

From Eqns (9.3)–(9.5) and (9.15), the following are obtained:

$$Y_4 = Y - Y_1 - Y_2 - Y_3 \tag{9.16}$$

$$X_2 = X_1 - \frac{2d_{sr}\tan\theta_r}{d_r d_{sf}\tan\theta_f + d_f d_{sr}\tan\theta_r}\left\{l(Y_1 + Y_2) - l_r Y - M + \frac{\alpha d_r m h_s \ddot{y}}{d_{sr}\tan\theta_r}\right\} \tag{9.17}$$

$$X_3 = -X_1 - \frac{(d_f - d_r)d_{sr}\tan\theta_r}{d_r(d_r d_{sf}\tan\theta_f + d_f d_{sr}\tan\theta_r)}\left\{l(Y_1 + Y_2) - l_r Y - M + \frac{\alpha d_r m h_s \ddot{y}}{d_{sr}\tan\theta_r}\right\}$$
$$+ \frac{l}{d_r}(Y_1 + Y_2) - \frac{l_r}{d_r}Y + \frac{X}{2} - \frac{M}{d_r} \tag{9.18}$$

$$X_4 = -X_1 + \frac{(d_f + d_r)d_{sr}\tan\theta_r}{d_r(d_r d_{sf}\tan\theta_f + d_f d_{sr}\tan\theta_r)}\left\{l(Y_1 + Y_2) - l_r Y - M + \frac{\alpha d_r m h_s \ddot{y}}{d_{sr}\tan\theta_r}\right\}$$
$$- \frac{l}{d_r}(Y_1 + Y_2) + \frac{l_r}{d_r}Y + \frac{X}{2} + \frac{M}{d_r} \tag{9.19}$$

As the distribution norm for minimizing tire workload is described by Eqn (9.7), putting the previous equations into Eqn (9.7) gives us the distribution norm as a function of four variables, X_1, Y_1, Y_2, and Y_3. Thus, the partial derivative of the objective function brings us the linear first-order algebraic equations of four variables as the minimizing condition as follows:

$$\frac{\partial J}{\partial X_1} = 0, \quad \frac{\partial J}{\partial Y_1} = 0, \quad \frac{\partial J}{\partial Y_2} = 0, \quad \frac{\partial J}{\partial Y_3} = 0 \tag{9.20}$$

Solving the equations with the constraints, the lateral and longitudinal tire forces of each tire (the tire forces to be distributed to each tire) are obtained at each moment during the vehicle motion to follow the model response with the roll control. Then, the traction torque and steering angle of each tire to generate the previous forces obtained are determined as control commands of the vehicle motion control in the same way as described in Section 9.2.3.

9.4.3 **EFFECTS OF ROLL CONTROL**

Figure 9.13 shows a typical result of the lane change using a small-size experimental vehicle at a speed of 40 km/h. The body roll of the vehicle with the roll control only and the roll control integrated with tire force distribution is reduced by 50% as expected, though the plane motion of the vehicles with and without control, especially the lateral accelerations, are almost the same. The time histories of the tire workload of each tire of the vehicles with and without control during the lane change are shown in Figure 9.14. It is seen that the rear-inner tire of the vehicle with the roll control only is subjected to severe load conditions during the lane change. On the other hand, any tire of the vehicle with the integrated control seems not to experience such a severe load condition, even though the additional longitudinal forces for the roll control are exerted on the tires.

The lane change behaviors of the respective vehicles without control, with roll control only, and with the integrated control are compared in Figure 9.15. The measured time histories of 10 lane change trials for each vehicle with and without controls are superimposed on each figure. Though a stand-alone use of the roll control worsens the vehicle stability, the integration with tire force distribution control improves it. Figure 9.16 shows each tire workload of the vehicles without control, with the roll control only, and with the integrated control. The tire workload of the vehicle with the integrated control is lower than those of the other vehicles especially at the rear wheels, which is supposed to be effective in stabilizing the vehicle motion.

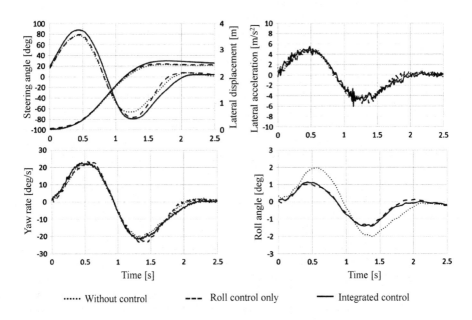

FIGURE 9.13

Lane change of the vehicles with/without roll/integrated controls.

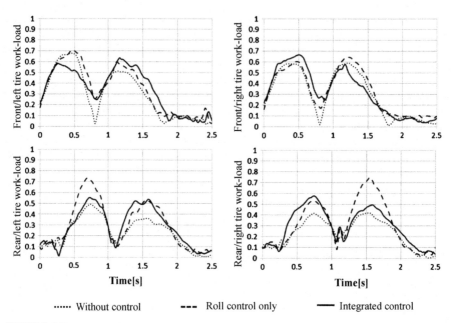

······ Without control   --- Roll control only   —— Integrated control

FIGURE 9.14

Workload of each tire during lane change.

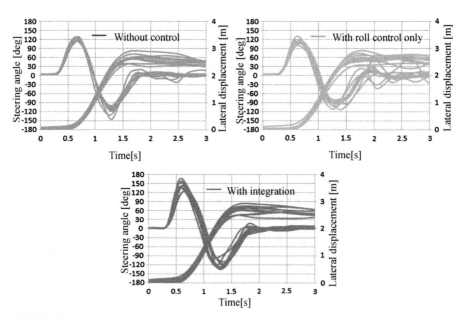

FIGURE 9.15

Effects of control integration on vehicle stability.

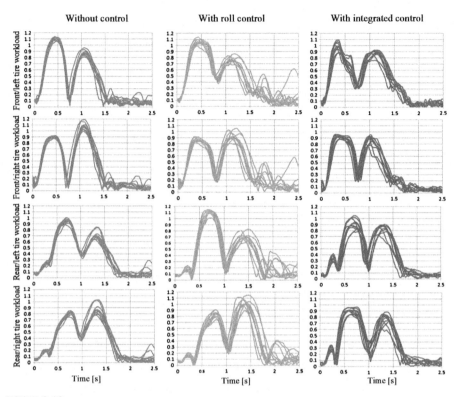

FIGURE 9.16

Effects of control integration on tire workload.

9.5 FURTHER DEVELOPMENT OF ALL-WHEEL CONTROL

Until now, some possibilities of the vehicle motion control by the tire force distribution control to all four wheels have been examined. As basically any vehicle response to steering input is available for the vehicle with the distribution control, another possibility of extending the tire force distribution control is to adjust the model response to follow for the vehicle with distribution control. The model response is reflected to the three essential constraints for the intended vehicle plane motion. If the driver prefers the vehicle with high-handling agility, a large value of stationary gain and lead time constant of yaw rate response will be recommended. If a driver likes a gentle vehicle response, a relatively large value of natural frequency with appropriate damping coefficient of the model response will be reasonable. These will contribute to the vehicle chassis design for enjoyable driving. (Chapter 12 will contribute to this discussion as well.)

Some extended motion control by the tire force distribution controls with some different reasonable norm from the discussed ones and an additional constraint to the three essential ones other than or in addition to the roll control constraint studied in Section 9.4.1 can be suggested. It will be able to bring us a vehicle with higher agility and/or active safety because four or five degree-of-freedom redundancy still remains.

It is possible to apply the control contents studied in this chapter to the vehicles that do not necessarily have eight controllable forces of all wheels but have six or even four controllable forces. This is more feasible from a practical point of view.

REFERENCES

[1] Mokhiamar O, Abe M. Simultaneous optimal distribution of lateral and longitudinal tire forces for the model following control transactions of the ASME. J Dyn Syst Meas Contr December 2004;126:753–63.
[2] Suzuki Y, Kano Y, Abe M. A study on tyre force distribution controls for full drive-by-wire electric vehicle. Veh Syst Dyn 2014;52(Suppl.):235–50.
[3] Suzuki Y, Kano Y, Abe M, Sugai T, Mogi K, Hirata J, et al. Roll control using tire longitudinal forces integrated with tire force distribution of full drive-by-wire electric vehicle. In: Proc. of AVEC14; September 2014 [Tokyo].
[4] Murata S. Innovation by in-wheel-motor drive unit. Veh Syst Dyn June 2012;50(6):807–30.

VEHICLE MOTION WITH HUMAN DRIVER

10

10.1 PREFACE

The vehicle dealt with in this book is that capable of moving freely, in a plane, by the forces generated from the motion of the vehicle body itself, similar to aircraft and ships.

The vehicle's lateral motion, yaw motion, and roll motion arise through steering action or by disturbances and external forces. The vehicle motion is controlled through the actions of the driver who observes the vehicle's motion and carries out suitable steering to achieve his intended path.

The previous chapters have studied the inherent motion characteristics of the vehicle in considerable detail, focusing on the vehicle's response to steering and disturbances. However, the previous chapters are based on the assumption that the driver is not actively steering to control the vehicle motion intentionally.

The vehicle motion response to steering and disturbances, based on a ground-fixed coordinate frame, is expressed by the transfer functions given in Eqns $(3.87)'$, $(3.88)'$, (4.11), and (4.12). These transfer functions include $1/s^2$ and $1/s$ forms of integration of the vehicle lateral displacement and yaw angle response to steering and disturbances. This shows that these responses to the steering input will increase their value over time if suitable control is not added. By this analysis, the vehicle motion relative to the coordinate axis fixed onto the absolute space could be said to be "unstable" by nature. This is why the vehicle can move freely in plane when suitable steering is given, i.e., the vehicle has freedom of motion that is unrestricted and unstable.

Until now, the unstable vehicle motion characteristics have not been treated as a major factor. However, this kind of treatment is incomplete in the discussion of vehicle dynamics. The vehicle is capable of meaningful motion that follows the driver's intention through a suitable steering action. It is important to understand the vehicle motion in the absolute space when steering is given intentionally by the driver in response to the vehicle's motion.

10.2 HUMAN CONTROL ACTION

In situations where a human controls the machine, the human controller plays an important role in deciding the motion. In such situations, examination of the moving body will always have the problem of how to describe the human as the controller.

From this point of view, pioneering researches concerning the functions of human operators have been carried out by several researchers, from which valuable results have been obtained. However, even when concerned with the human controller function only, humans are not consistent, and the same treatment may not be able to describe every condition. Moreover, the vehicle motion that is dealt with in the previous chapters could well be derived by theoretical expansion of the basic equations of motion based on Newton's Law, even if the secondary factors that govern the motion are omitted. In other words, there is a theoretical guarantee to the nature of the motion, which could be expressed by mathematical equations. On the contrary to this, with regard to humans, there is no certain guarantee that the human control action can be described completely using mathematical equations.

Nevertheless, if observed carefully, a human operator exhibits some regular behavior from the viewpoint of machine control. A model that simulates the human's regularity can be derived and made to represent the human functions as the controller. This is the basic way of thinking when dealing with the human control function.

This is similar to the concept of the black box used in control engineering. The fact that automatic control theory can now be applied easily to human control behavior allows researchers to understand the human control function and develop effective designs of machines controlled by a human operator. One popular method is treating the human control behavior as a linear continuous feedback control, and expressing it as a transfer function. Various transfer functions have then been proposed to suit different conditions.

Table 10.1 shows the proposed transfer functions of a human controller. All these transfer functions have almost the same characteristics. Here, the following transfer function is used as the example:

$$H(s) = h\left(\tau_D s + 1 + \frac{1}{\tau_I s}\right)e^{-\tau_L s} \tag{10.1}$$

The physical meaning of human control can now be studied in more detail.

First, a human operator will have a time lag to make an action (output) under a given stimulus (input). This is represented by $e^{-\tau_L s}$, and the time lag is expressed by τ_L. The control action that the human operator can do at ease and with the least workload is the kind of action that gives an output signal proportional to the input signal, in other words, a proportional action. This is represented by the proportional constant h. The human operator is also capable of the control action that predicts a change in the input. Here, the output signal is normally proportional to the input rate or the differential value of the input. In other words, this is called the derivative action. This is represented by the derivative time constant, τ_D. The human operator is also capable of making integral action where the output signal is proportional to the integrated

Table 10.1 Transfer Functions of Human Controller		
1	$K\frac{(1+Ts)}{s}e^{-\tau s}$	Tustin
2	$K\frac{(T_1 s+1)e^{-\tau s}}{(T_2 s+1)(T_3 s+1)}$	McRuer and Krendel
3	$K\left(T_1 s + 1 + \frac{1}{T_2 s}\right)e^{-\tau s}$	Ragazzini
4	$K\frac{(A_n s^n + ... + A_0)}{s^i(B_m s^m + ... + B_0)}e^{-\tau s}$	Jackson

value of the input signal. This shows that the human operator is able to relocate the original point when it is deviated away from. This is represented by the integral time constant, τ_I.

Combining all these control actions, the human transfer function $H(s)$ can be represented by Eqn (10.1).

The human operator is different from normal control systems. The distinctive characteristics of the human controller lie in the fact that the transfer function constants h, τ_D, and τ_I can be changed for appropriate control within a possible range. This characteristic is generally called the control action adaptability of the control system. Among h, τ_D, and τ_I, the human operator is able to change the constant of the proportional action, h, substantially without workload, while the other constants have some limitations. In particular, the increase of derivative action could give considerable workload to the human operator. If the control target needs intense derivative action up to a certain level, the human may not be able to carry out the control anymore.

The control action that the human controller is capable of performing without much workload and continuously for an extended time is considered to be the proportional action, with very weak derivative action or an integral one.

10.3 VEHICLE MOTION UNDER DRIVER CONTROL
10.3.1 DRIVER MODEL

Using the previous ideas for general human control behavior, the control behavior of a human driver controlling the vehicle motion through the steering angle will be considered.

Expressing the transfer function of the vehicle lateral motion to steering angle in absolute space will give something like Eqn (3.87) or (3.87)'. The equation shows that the vehicle path, the control target of the human driver, has characteristics near to the double integration of the steering angle. Generally, for a control target with transfer function of $1/s^2$, derivative action is needed, in addition to proportional action, to stabilize the system.

A driver could also sense the vehicle attitude, i.e., yaw angle, as well as the lateral displacement. The transfer function of the vehicle yaw angle response to the steering angle in the absolute space is given by Eqn (3.88) or (3.88)' and has characteristics near the single integration of the steering angle. If the driver makes the control action by sensing the yaw angle, rather than the lateral displacement, it is expected that the driver will be able to control the vehicle more easily, even without the derivative action, but through some actions equivalent to the derivative action.

Based on this argument, the driver inside the vehicle is assumed to look forward to L(m) ahead of the vehicle and estimate the deviation of the vehicle's lateral displacement relative to the target course. The estimated lateral displacement is a lateral component of the vehicle traveling distance in a time of L/V at the current attitude. The driver makes the feedback control based on this deviation. The scenario is shown in Figure 10.1. This approach now becomes the most fundamental understanding of the human driver's control model for dealing with the vehicle motion. Figure 10.2 shows the block diagram of such an approach. Here, L is called the look ahead distance, and the L(m) point is called the look ahead point.

At this time, since the vehicle has strong integration characteristics, some derivative control element is included in the transfer function of the human driver's control behavior. Consequently, the human driver's transfer function becomes the following:

$$H(s) = h(1 + \tau_D s)e^{-\tau_L s} \qquad (10.2)$$

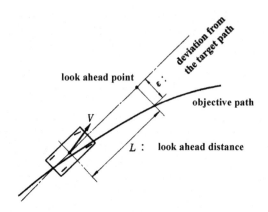

FIGURE 10.1

Course error at *L(m)* ahead point.

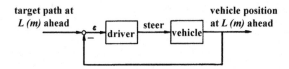

FIGURE 10.2

Closed-loop driver-vehicle system.

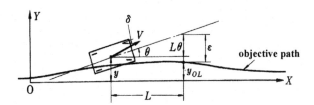

FIGURE 10.3

Vehicle motion controlled by driver with nearly straight objective path.

10.3.2 VEHICLE MOTION FOLLOWING A COURSE

To study the vehicle motion under the control, a target course is defined that consists of gentle curves and is almost a straight line, as shown in Figure 10.3. The vehicle receives the control action from the driver and moves along this target course. If the X-axis is the direction along the target course and the Y-axis is the direction perpendicular to it, the vehicle state of motion will be as shown in Figure 10.3.

Here, y is the vehicle lateral displacement, θ is the yaw angle, and y_{OL} is the lateral displacement of the target course at the look ahead point. Since $|\theta| \ll 1$, the supposed lateral deviation of the vehicle from the target course at the look ahead point is as follows:

$$\varepsilon = y + L\theta - y_{OL} \tag{10.3}$$

Furthermore, if the vehicle motion is expressed in terms of the steering angle δ, the transfer functions of y and θ to δ, $G_y(s)$, and $G_\theta(s)$ are given by Eqn (3.87)′ and (3.88)′.

$$G_y(s) = \frac{y(s)}{\delta(s)} = G_\delta^y(0) \frac{1 + T_{y1}s + T_{y2}s^2}{s^2 \left(1 + \frac{2\zeta s}{\omega_n} + \frac{s^2}{\omega_n^2}\right)} \tag{10.4}$$

$$G_\theta(s) = \frac{\theta(s)}{\delta(s)} = G_\delta^r(0) \frac{1 + T_r s}{s \left(1 + \frac{2\zeta s}{\omega_n} + \frac{s^2}{\omega_n^2}\right)} \tag{10.5}$$

The vehicle motion, y, in relation to the given target course, y_{OL}, is expressed in a block diagram form like Figure 10.4. The vehicle under driver control, even when deviating away from the target, will eventually converge to and follow the target course if the system as shown in the figure is stable. At this instance, the vehicle motion is also affected by the vehicle's inherent dynamic characteristics, expressed by $G_y(s)$ and $G_\theta(s)$, and by the human driver's control characteristics, expressed by the form of $H(s)$ and the value of L(m).

Figure 10.5 shows a computer simulation of vehicle motion based on the preceding approach. The driver's transfer function, $H(s) = h/(1 + \tau_L s)$, instead of Eqn (10.2), is used to investigate the basic motion of the vehicle under control by the human driver. The figure indicates that the vehicle motion is greatly affected by how the human controls the vehicle and the vehicle's inherent motion characteristics.

These results are for vehicle motion when acted on by a sudden external force in the lateral direction. The force acts for a short period of time at the center of gravity of the vehicle traveling in a straight line. Figure 10.5(a) is the vehicle motion when the driver's look ahead distance, L, is changed, whereas Figure 10.5(b) is the vehicle motion when the vehicle steer characteristics are changed. Here, y_{OL} is always zero.

It is shown that if the look ahead distance, L, is too short, the vehicle motion oscillates and becomes unstable. If L is big, the vehicle is stable and could return to the course. For the

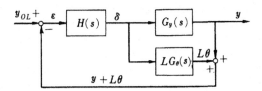

FIGURE 10.4

Block diagram of driver-vehicle system.

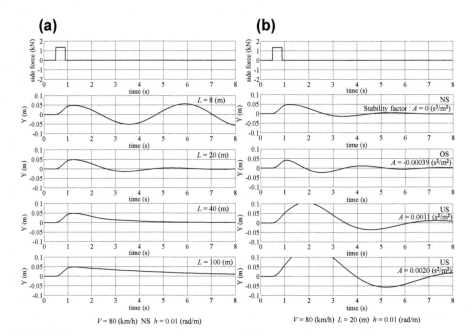

FIGURE 10.5

Vehicle motion controlled by modeled driver, (a) effect of L and (b) effect of A.

vehicle motion to be stable enough, L must increase with speed. The oscillation with poor damping occurs for the OS characteristic vehicle; whereas in the case of US characteristics, the motion is damped and stable. However, vehicles that exhibit a strong US characteristic have large lateral displacements and deteriorated damping. Furthermore, with concern to the controlled vehicle tracking the target course, it is also shown that if the look ahead distance is too large or the vehicle exhibits too much US characteristic, the target course tracking will be impaired as well.

10.3.3 **MOTION STABILITY**

Here, the stability of the vehicle motion, under the human driver's control, described in Section 10.3.2 is studied in more detail.

The transfer function of y to y_{OL} is derived from the block diagram of Figure 10.4 as follows:

$$\frac{y(s)}{y_{OL}(s)} = \frac{G_y(s)H(s)}{1 + H(s)\left[G_y(s) + LG_\theta(s)\right]} \tag{10.6}$$

Therefore, the characteristic equation for the controlled vehicle is shown next:

$$1 + H(s)\lfloor G_y(s) + LG_\theta(s)\rfloor = 0 \tag{10.7}$$

Substituting Eqns (10.2), (10.4), and (10.5) into Eqn (10.7) results in the final characteristic equation. However, the characteristic equation is again too complicated. The vehicle characteristics must be first simplified before understanding the basic characteristics of vehicle stability under a driver's control.

The yaw rate response to the front-wheel steering input is approximated from Eqn (3.78) under the assumptions of $K_f = K_r = K$, $l_f = l_r = l/2$:

$$\frac{r(s)}{\delta(s)} = \frac{V}{l}\frac{1}{1 + t_r s} \tag{10.8}$$

where

$$t_r = \frac{mV}{4K} \tag{10.9}$$

The yaw angle response to steer input is described as follows:

$$G_\theta(s) = \frac{\theta(s)}{\delta(s)} = \frac{V}{l}\frac{1}{1 + t_r s}\frac{1}{s} \tag{10.10}$$

Assuming that the side-slip angle is very small, then the lateral acceleration is expressed as follows:

$$d^2y/dt^2 = Vr = Vd\theta/dt$$

and so, the lateral displacement to steering input is described by the following:

$$G_y(s) = \frac{y(s)}{\delta(s)} = \frac{V^2}{l}\frac{1}{1 + t_r s}\frac{1}{s^2} \tag{10.11}$$

Putting Eqns (10.2), (10.10), and (10.11) into Eqn (10.7), gives the following characteristic equation:

$$s^2 + \frac{V^2}{l}\left(\frac{L}{V}s + 1\right)\frac{h(1 + \tau_D s)e^{-\tau_L s}}{1 + t_r s} = 0 \tag{10.12}$$

For further simplification, the driver is considered without derivative control action, which means $\tau_D = 0.0$, under the long-time continuous control task. Furthermore, the driver time lag is approximated as follows:

$$e^{-\tau_L s} \approx \frac{1}{1 + \tau_L s} \tag{10.13}$$

Now, the characteristic equation becomes the following:

$$s^2 + \frac{V^2}{l}\left(\frac{L}{V}s + 1\right)\frac{h}{(1 + t_r s)(1 + \tau_L s)} = 0 \tag{10.14}$$

The development of Eqn (10.13) leads to the following:

$$A_4 s^4 + A_3 s^3 + A_2 s^2 + A_1 s + A_0 = 0 \qquad (10.14)'$$

where

$$A_4 = \tau_L t_r$$

$$A_3 = \tau_L + t_r$$

$$A_2 = 1$$

$$A_1 = \frac{hL}{l} V$$

$$A_0 = \frac{h}{l} V^2 \qquad (10.15)$$

As all the coefficients A_i ($i = 1 \sim 4$) are positive, and the stability conditions of the driver-vehicle system is obtained:

$$\begin{vmatrix} A_1 & A_0 & 0 \\ A_3 & A_2 & A_1 \\ 0 & A_4 & A_3 \end{vmatrix} = A_1 A_2 A_3 - A_0 A_3^2 - A_1^2 A_4 > 0 \qquad (10.16)$$

Substituting Eqn (10.15) into the inequality (10.16) gives the following:

$$\frac{h}{l} V \left\{ (\tau_L + t_r) L - (\tau_L + t_r)^2 V - \frac{hL^2 V}{l} \tau_L t_r \right\} \geq 0 \qquad (10.16)'$$

As all, h, l, and V, are positive:

$$h \leq \frac{\tau_L + t_r}{\tau_L t_r} \frac{l}{LV} \left\{ 1 - \frac{V(\tau_L + t_r)}{L} \right\} \qquad (10.17)$$

This inequality gives us the stable region of the driver parameters, on an h-L plane, for the given values of τ_L, t_r, l, and V.

If it is possible that both τ_L and t_r are small enough to be neglected, the following characteristic equation can be used instead of Eqn (10.14)':

$$s^2 + \frac{hL}{l} Vs + \frac{h}{l} V^2 = 0 \qquad (10.14)''$$

As all the coefficients of this equation are positive, the driver-vehicle is stable at any positive h and L. This means that the driver-vehicle system becomes unstable only when there is a response delay in either the driver or the vehicle response.

The driver time lag, τ_L, cannot be completely equal to zero. Also, the mechanical system from the driver's hands to the front wheels must lead to some response delay of the front steering angle responding to the driver's input to the steering wheel. Thus, it is not reasonable to set $\tau_L = 0$. On the other hand, the vehicle response delay, t_r, as expressed by Eqn (10.9), can be regarded as zero when V is low and m is small relative to K. Under this condition, the characteristic equation changes to the following form:

$$A_3 s^3 + A_2 s^2 + A_1 s + A_0 = 0 \qquad (10.14)'''$$

where

$$A_3 = \tau_L$$

$$A_2 = 1$$

$$A_1 = \frac{hL}{l}V$$

$$A_0 = \frac{h}{l}V^2 \qquad (10.15)'$$

The stability condition of the preceding is as follows:

$$A_1A_2 - A_0A_3 \geq 0$$

Putting Eqn (10.15)' into the previous gives the following:

$$\frac{hV}{l}(L - V\tau_L) \geq 0$$

Namely, the driver-vehicle system is stable when L is greater than $V\tau_L$.

It is clear that for a stable system, there is a lower limit of the driver's look ahead distance (preview distance). This limit is due to the response delay of the driver, and the critical preview distance is equal to the distance the vehicle moves, with speed V, during the delay time, τ_L. In other words, L/V can be called the preview time of the driver, and if the preview time is greater than the delay time, τ_L, the system is stable.

The driver-vehicle system is always stable if the vehicle response delay is negligible and either (a) the look ahead distance is longer than the distance moved during his/her delay time, or (b) the preview time is larger than the delay. If either case is true, there is no limit of the gain constant, h, for assuring stability.

When both the driver and vehicle response delays are not negligible, the stability condition is described by Eqn (10.17), which gives us the stability region shown previously. Figure 10.6 shows the calculation results of the stability region on the h-L plane. It is understandable that the upper limit of the driver control gain, h, arises for the first time when the response delay of the vehicle becomes significant, especially with regard to vehicle speed. The upper limit of h rapidly decreases with the increase of the vehicle speed.

Figure 10.7 is the calculation results for $y_0 = 3.5$-m-wide lane change behavior at the vehicle speed of 20 m/s, $t_r = 0.1$ s, and $\tau_L = 0.1$ s using Eqns (10.6), (10.10), and (10.11) for the driver-vehicle system. The following equation is used as the driver transfer function, with the parameters corresponding to points A, B, and C, respectively, in Figure 10.6.

$$H(s) \approx \frac{h}{1 + \tau_L s} \qquad (10.18)$$

It is interesting to see that the driver-vehicle system with driver parameters located out of the stable region (points A and C) is oscillatory and unstable.

FIGURE 10.6

Stable region on *h-L* plane.

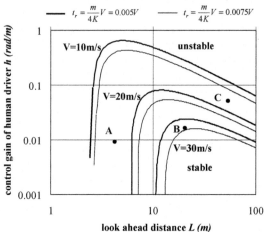

FIGURE 10.7

Simplified driver-vehicle system lane change response. (a), (b) and (c) correspond with A, B and C on the figure 10.6 respectively.

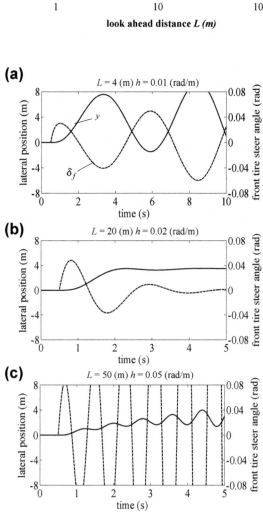

Example 10.1

Investigate how the derivative control action of the driver contributes to stabilizing the driver-vehicle system.

Solution

Adopting Eqns (10.10) and (10.11) as the vehicle response to the steering input and using $e^{-\tau_L s} = 1/(1 + \tau_L s)$, the characteristic equation of the driver-vehicle system becomes the following:

$$s^2 + \frac{V^2}{l}\left(\frac{L}{V}s + 1\right)\frac{h(1 + \tau_D s)}{(1 + t_r s)(1 + \tau_L s)} = 0 \tag{E10.1}$$

Here, if the driver controls the vehicle with the derivative time constant, τ_D, almost the same as the lag time, τ_L, then the characteristic equation becomes the following:

$$t_r s^3 + s^2 + \frac{hLV}{l}s + \frac{hV^2}{l} = 0$$

and, the stability condition is shown next:

$$\frac{hV}{l}(L - Vt_r) \geq 0 \quad \text{or} \quad L \geq Vt_r \tag{E10.2}$$

Or, if the driver's derivative time is almost equal to the response time of the vehicle, t_r, then the characteristic equation is as follows:

$$\tau_L s^3 + s^2 + \frac{hLV}{l}s + \frac{hV^2}{l} = 0$$

and, the stability condition becomes the following:

$$\frac{hV}{l}(L - V\tau_L) \geq 0 \quad \text{or} \quad L \geq V\tau_L \tag{E10.3}$$

It is understood that if the driver uses a derivative time almost equal to their lag time or the vehicle response time, the stability limit in the *h-L* plane can be widened, and the upper limit of the gain, *h*, is eliminated.

So far, the region for the proportional constant *h* and the look ahead distance *L* has been chosen in order for the vehicle motion under human control to be stable. However, while this can distinguish whether the motion is stable or not, it cannot determine the level of stability.

Yamakawa, took the human driver's transfer function as $H(s) = h(1 + \tau_D s)e^{-\tau_L s}$ and, using equations equivalent to Eqns (10.4) and (10.5), determined the roots of the characteristic equation of a vehicle under the control of a human driver [1]. Two of the results are shown in Figure 10.8 and Figure 10.9.

Figure 10.8 is the root locus when the human time lag, τ_L, and proportional constant, *h*, are changed. From the figure, the existence of time lag, τ_L, is the basic cause for the vehicle motion to become unstable. If the proportional constant, *h*, is too large, the vehicle motion will become unstable. Figure 10.9 is the root locus when the vehicle SM and the proportional constant, *h*, are changed. From the figure, if the vehicle has a US characteristic, the larger the SM, the more stable the vehicle motion will become.

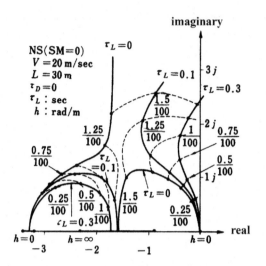

FIGURE 10.8

Root locus of driver-vehicle system for various τ_L.

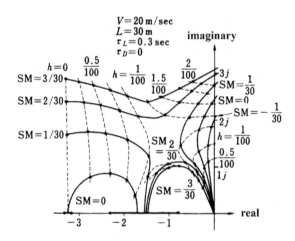

FIGURE 10.9

Root locus of driver-vehicle system for various SM.

10.4 HUMAN ADAPTATION TO VEHICLE CHARACTERISTICS AND LANE CHANGE BEHAVIOR

Until here, the characteristics of vehicle motion and human control behavior have been regarded as independent of each other. However, unlike mechanical controllers, a human's distinctive characteristic is the ability to change control behavior to produce an appropriate control that suits the control objective.

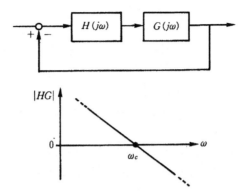

FIGURE 10.10

Crossover model.

For simplicity, the following characteristic equation can be found by putting $t_r = 0$ in Eqn (10.12):

$$s^2 + \frac{V^2}{l}\left(\frac{L}{V}s + 1\right)h(\tau_D s + 1)e^{-\tau_L s} = 0 \qquad (10.12)'$$

If the human driver changes h proportionally to the inverse vehicle speed squared, V^2, and changes the look ahead distance L proportionally to V, then the preceding equation will not change even if the vehicle speed changes. In this condition, the vehicle motion will not differ greatly with the traveling speed. The human driver is not independent of the vehicle motion characteristics and behaves as a controller that skilfully changes the control characteristics to suit the vehicle characteristics.

A way of understanding the adaptive behavior of human controllers has been proposed. McRuer has modeled human control behavior and shows it adapts fully to the machine characteristics. This is done by assuming the human controller is a continuous linear feedback controller as described in Section 10.2 [2]. As shown in Figure 10.10, the human controller adapts to the changes of a control objective, $G(j\omega)$, by adjusting human control behavior, $H(j\omega)$, such as that near the crossover frequency, ω_c, where the open-loop frequency response gain $|H(j\omega)G(j\omega)|$ is one; then, the following is given:

$$H(j\omega)G(j\omega) \approx \frac{\omega_c e^{j\omega\tau}}{j\omega} \qquad (10.19)$$

Here, τ is the time lag of the human muscles. This is called the crossover model. By applying this method, it might be possible to consider the human adaptation to the vehicle characteristic changes.

Figure 10.11 shows the simulated results of a vehicle lane change for several combinations of the driver parameters h-L and various vehicle speeds. Here, the feasibility that a human driver changes their parameters for different vehicle speeds is seen. As was predicted at the beginning in this section, it is obvious that the parameters h and L should be adapted to the change of vehicle speed in order to maintain consistent, ordinary driving behavior during the lane change.

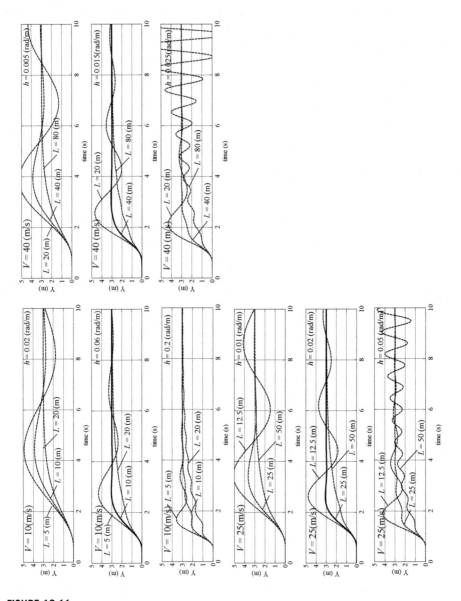

FIGURE 10.11

Effects of driver parameters on vehicle motion under various running speed.

Example 10.2

Execute the Matlab-Simulink simulation of the vehicle lane change behaviors, as in Fig. 9.11, to compare the results of the high US and OS vehicle with the normal vehicle. Investigate how the driver should adapt to the vehicles in order to keep consistent lane change behaviors under variations of the steer characteristics.

Solution

It is convenient to use Eqns (3.21) and (3.22) to simulate the vehicle motion during a lane change on a straight road. Equation (10.2) is used to find the course error, $L(m)$ ahead, during the lane change. Equation (10.18) acts as the driver transfer function. The previous equations are rewritten for the integral type of block diagram required for simulation:

$$\frac{dv}{dt} = -\frac{2(K_f + K_r)}{mV}v - \frac{2(l_f K_f - l_r K_r)}{mV}r + \frac{2(K_f + K_r)}{m}\theta + \frac{2K_f}{m}\delta \tag{E10.4}$$

$$\frac{dr}{dt} = -\frac{2(l_f K_f - l_r K_r)}{IV}\frac{dy}{dt} - \frac{2\left(l_f^2 K_f + l_r^2 K_r\right)}{IV}\frac{d\theta}{dt} + \frac{2(l_f K_f - l_r K)}{I}\theta + \frac{2l_f K_f}{I}\delta \tag{E10.5}$$

$$\frac{dy}{dt} = v \tag{E10.6}$$

$$\frac{d\theta}{dt} = r \tag{E10.7}$$

$$\frac{d\delta}{dt} = -\frac{1}{\tau_L}\delta - h(y + L\theta - y_{OL}) \tag{E10.8}$$

From these equations, the block diagram for simulation is obtained, as shown in Figure E10.2(a). The driver-vehicle system parameters for the simulation are as in Figure E10.2(b). The simulation program is shown in Figure E10.2(c), and Figure E10.2(d) is a result of the simulation.

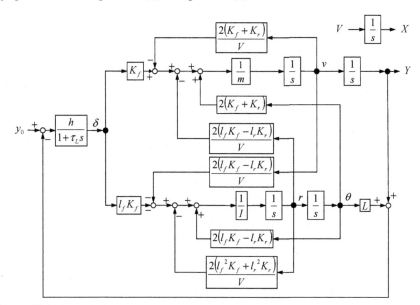

FIGURE E10.2(a)

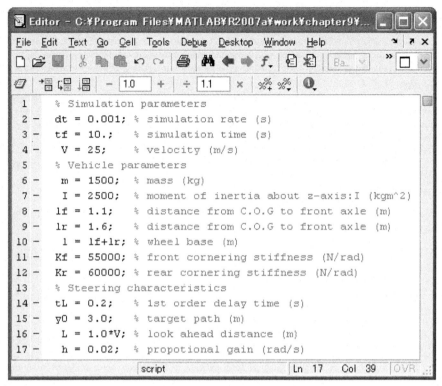

FIGURE E10.2(b)

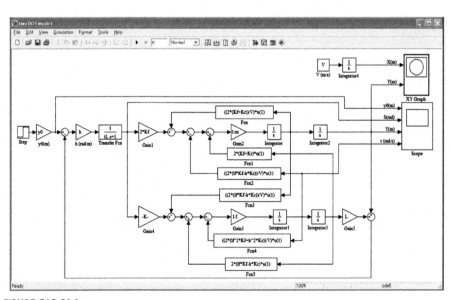

FIGURE E10.2(c)

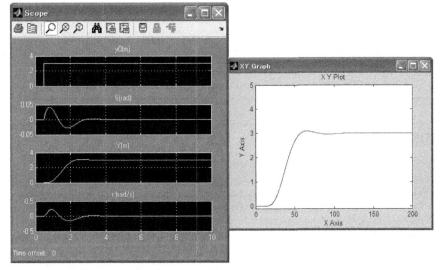

FIGURE E10.2(d)

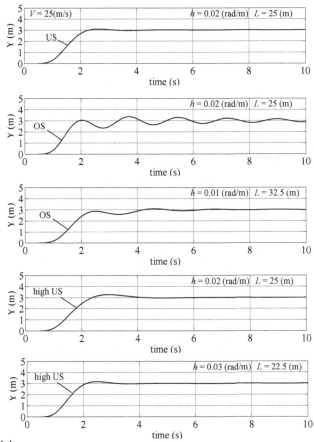

FIGURE E10.2(e)

The results are summarized in Figure E10.2(e). It is found from this figure that if the vehicle is highly US, the driver should increase the gain, h, and decrease the preview distance, L. On the other hand, if the vehicle becomes OS, then the driver should decrease the gain, h, and increase the preview distance, L, compared with the case of the normal steer characteristic vehicle.

PROBLEMS

10.1 Even though the transient characteristics are neglected, the response of the lateral displacement of the vehicle to the steering input is basically a second-order integral and is described as $y(s)/\delta(s) = V^2/(ls^2)$. So, confirm that a proportional control action is not enough and some kind of derivative control action is essential for the driver to stabilize the lateral motion of the vehicle.

10.2 The yaw rate response is approximated by Eqn (10.8) in Section 10.3.3. Try to calculate the time constant of this first-order lag response for a normal passenger vehicle at vehicle speeds 10 km/h $\sim$ 100 km/h. Find the vehicle speed range at which the response time constant becomes significant compared with the response time of the human driver.

10.3 Try to calculate the range of the parameter L for the driver-vehicle system to be stable, use Eqn (10.17) with some specific values of h, τ_L, t_r, l, and V.

10.4 Execute the Matlab-Simulink simulation of the driver-vehicle system behavior subjected to the gust wind force with the same conditions as in Example 4.3 under the fixed driver parameters. Refer to the Example 10.2 for the driver parameters to be used in the simulation.

10.5 Try to mention possible driver models to simulate steering behaviors of the drivers other than a linear continuous PD (proportional + drivative) control model.

REFERENCES

[1] Yamakawa S. An investigation of vehicle under-steer/over-steer characteristics with a driver steering control. JSAE J 1964;18(11) [In Japanese].

[2] McRuer DT. A review of Quasi-linear pilot models. IEEE Trans Hum Factors Electron 1969; HFE-8(3).

VEHICLE HANDLING QUALITY

11

11.1 PREFACE

Previously, the study has focused on the motion characteristics of the vehicle itself, the kind of motion the vehicle exhibits, and the vehicle's motions when it is controlled by a human driver. Not only is it important to characterize such motion so that it can be estimated theoretically and observed objectively, but it is also important to evaluate how easy it is for a driver to control the vehicle, in another words, a handling quality evaluation. This is important both objectively and subjectively.

For normal machines, a human operator (as a third party) observes and also theoretically estimates a machine's performance and function to evaluate the machine objectively. However, for the case of an airplane and also the vehicle, where the machine function is only fulfilled with the direct control of the human operator, the subjective rating of the pilot/driver toward the controllability is very important.

This chapter will investigate the relationship of the fundamental motion characteristics of the vehicle with the driver's subjective rating of the vehicle handling quality. Nevertheless, at this stage now, a theoretical methodology that could express the vehicle controllability analytically is not yet established. For that reason, only a methodology based on actual practical facts and experiences can be applied. Here, the vehicle's controllability is concluded by the driver's evaluation when the vehicle motion characteristics are changed.

11.2 VEHICLE CONTROLLABILITY

As mentioned, the theoretical methodology that could express the vehicle's controllability analytically is not yet established. At this moment, there is also no general methodology to deal with the subjective evaluation of the driver or any relationship between the vehicle motion characteristics and the vehicle's controllability. Various methodologies are applied, and each corresponds to and suits different cases.

In contrast, in the field of aerospace, pilot rating (PR) has been used for a long time as the method of expressing qualitatively the pilot's subjective evaluation of handling and stability. The PR has been continuously modified, and it is now relied on as the general assessment method to evaluate the airplane handling and stability in the design of easy-to-control airplanes.

In a similar way to PR, trials have been done to measure the subjective evaluation of the vehicle's controllability based on the driver's evaluation. By changing the vehicle motion characteristics and examining systematically the driver's evaluation of the vehicle's controllability when he/she actually drives the vehicle, the relation between vehicle characteristics and the controllability becomes clear to a certain extent. Because the vehicle's controllability is ultimately the subjective evaluation of the driver, this method is practical and straightforward. However, this method is also easily subjected to individual differences in drivers who make the evaluation. Hence, the objectiveness and generality of the results becomes weak. Persisting with this method will bring difficulties in deriving the theoretical relation between the vehicle motion characteristics and the vehicle's controllability. The prior estimation of the vehicle's controllability to new changes in the vehicle motion characteristics will be difficult too.

Task performance is another method to evaluate the vehicle's controllability. This is a method where a target course is set and an actual experiment is carried out, for example, to see how well the vehicle could pass through this course in a certain amount of time at a certain speed without error. This method has an advantage of objective results, but questions remain on how to set the target course and evaluate the result, and whether this corresponds fully to the driver's subjective evaluation or not. Of course, the theoretical study of the relationship between the vehicle motion characteristics and the vehicle's controllability is not necessarily easy.

There is also a method to evaluate the vehicle's controllability by measuring the biological response of the driver, for example, heart beat, metabolism, perspiration rate (skin current), and so forth, which depend on the driver's workload. By changing the vehicle motion characteristics systematically and investigating the changes in the driver's biological response, it is possible to find the vehicle characteristics that make the vehicle easier to control. Although this method gives us the objective measurement results, the results itself are easily affected by various causes, and hence, the definite relation between biological responses and vehicle's controllability is difficult to establish.

These are simple methods of looking at the driver's evaluation of controllability through experiments when the vehicle motion characteristics are being changed. Of course, by changing the vehicle motion characteristics and investigating the vehicle's controllability, it is possible to establish a general form of the relationship between vehicle motion characteristics and controllability to a certain degree. It still does not fully provide us a method for estimating the vehicle's controllability when subjected to changes in the vehicle motion characteristics.

Even though the general methodology to deal with the vehicle's controllability is not established yet, the relation between vehicle motion characteristics and controllability is becoming clearer with the different evaluation methods as described. The following sections will introduce and expand on these results.

11.3 VEHICLE MOTION CHARACTERISTICS AND CONTROLLABILITY

11.3.1 STEER CHARACTERISTICS AND CONTROLLABILITY

Chapters 3 and 4 have shown that the vehicle steer characteristic is an important factor that influences the vehicle motion characteristics. First, the relationship between vehicle steer characteristic and the vehicle's controllability will be considered.

Figure 11.1 shows the frequency distribution of the steering angle correction for a US characteristic vehicle and an OS characteristic vehicle, respectively [1]. This is based on the actual measurement of the driver's steering angle correction during high-speed traveling. The vehicle is a normal passenger car with the aerodynamic center coincident with the center of gravity that is subjected to random lateral wind. From the figure, a US characteristic vehicle with a large steering angle has a small correction frequency; whereas, an OS characteristic vehicle has a smaller steering angle, and the correction frequency is large. This is because the more US

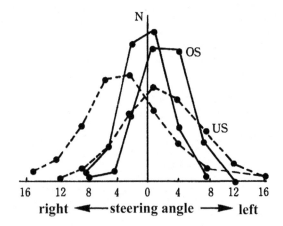

FIGURE 11.1

Steering correction distribution.

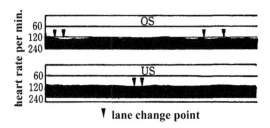

FIGURE 11.2

Driver heart rate change at $V = 140$ km/h.

the vehicle is, the smaller the gain of the vehicle motion to steering input. This is also seen in Chapter 4, particularly in Section 4.2.3, concerning the difference in vehicle motion to disturbances due to steer characteristics.

Furthermore, Figure 11.2 is the measurement data of the driver's heart beat during the same test. The heart beat of the driver when driving a US characteristic vehicle is at an average of 120; whereby in relation to that, the heart beat of the driver when driving an OS characteristic vehicle is at around 130–140. It is implied that an OS characteristic vehicle requires a larger control workload to the driver compared to a US characteristic vehicle.

From the previous, it is assumed that the US characteristic vehicle is easier for the driver to control compared to the OS characteristic vehicle.

An extent of US/OS characteristic of the vehicle is determined and shown by the stability factor, A.

$$A = -\frac{m}{2l^2} \frac{l_f K_f - l_r K_r}{K_f K_r} \tag{3.43}$$

There is an example to examine the relation between the derivative control action and the stability factor A. In this example, the driver's steering angle when driving a vehicle along

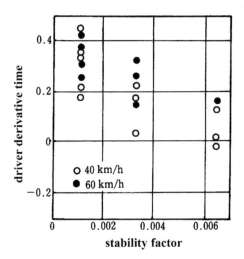

FIGURE 11.3

Effect of stability factor on derivative control action.

a given course is measured, and the equivalent derivative action that should be included in the driver's control action is calculated [2]. The result is shown in Figure 11.3. As explained in Chapter 10, derivative control action is one of the causes of the workload to the driver, and thus, the vehicle that could be controlled with less derivative control action is considered to be easier to control. The horizontal axis shows the stability factor A, which is larger than zero and expresses the extent of the US characteristic. Based on these, Figure 11.3 shows the relationship between the vehicle's US characteristic and the vehicle controllability.

The figure shows a US characteristic vehicle is easier to control. However, there is no guarantee that the driver will evaluate the vehicle's controllability based only on the amount of derivative action required in his control action. Moreover, the stability factors A or SM, etc. show the vehicle steer characteristic and are the only factors directly affecting the vehicle's steady-state characteristic. However, if the vehicle steer characteristic changes, the vehicle's inherent dynamic characteristic will change also. Besides the steer characteristic, there are other factors that determine the vehicle dynamic characteristics. Therefore, it is unsuitable for us to judge the vehicle's controllability based only on the steer characteristics and to draw the conclusion that a US characteristic vehicle is easier to control.

11.3.2 DYNAMIC CHARACTERISTICS AND CONTROLLABILITY

If a driver gives a sudden steer of a fixed angle to a vehicle originally traveling in a straight line, and then removes his/her hands (free control), a stable vehicle will reach a steady state while making some oscillatory yaw motion. One of the indices that determines the vehicle motion characteristic in such conditions is the yaw damping. The yaw damping is affected by and changes with the vehicle steer characteristics, the steering system characteristics, and the vehicle body roll, etc. Bergman has investigated the relation of the measured actual vehicle yaw damping with the subjective rating of the driver based on a 10-point rating system [3].

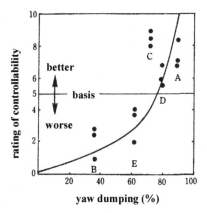

FIGURE 11.4

Controllability to yaw damping.

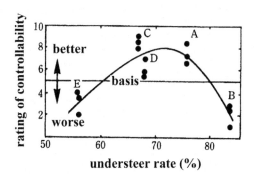

FIGURE 11.5

Controllability to understeer rate.

The result is shown in Figure 11.4. From the figure, it could be considered that a vehicle with larger yaw damping is easier to control.

Figure 11.5 is the relation of the controllability with the vehicle understeer rate, U_R, for the same vehicle. From the figure, it is clear that it is not always the vehicle with a strong US characteristic that is easier for the driver to control.

One of the common methods of studying the vehicle dynamic characteristics is to observe the vehicle response to periodic steering, as described in Section 3.4.3. In particular, the driver is sensitive to the yaw rate changes in relation to the steering angle. Some examples study the vehicle's relation between vehicle controllability and yaw rate gain for a periodic steering input.

When the parameters that govern the vehicle motion characteristics are changed, the shape of the yaw rate gain, for example the one shown in Figure 3.30 in Section 3.4.3, will change. Sugimoto uses the ratio of the peak gain value to the gain value at low frequency ($f = 0.2$ Hz),

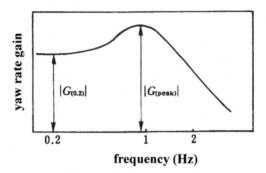

FIGURE 11.6

Frequency response gain ratio.

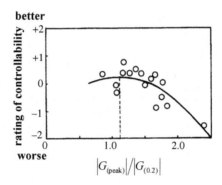

FIGURE 11.7

Controllability to frequency response gain ratio.

$\left|G_{(peak)}\right|/\left|G_{(0.2)}\right|$, to represent such changes of the shape of the gain, as shown in Figure 11.6. The relation between this ratio and the rating on the vehicle's controllability is examined by measuring this ratio for different normal passenger cars [4].

The result is shown in Figure 11.7. Looking at the figure, the controllability is rated best at a ratio around $\left|G_{(peak)}\right|/\left|G_{(0.2)}\right|$ value of 1.2. At the ratio above this, the controllability of the vehicle becomes rather poor. This is because when the $\left|G_{(peak)}\right|/\left|G_{(0.2)}\right|$ ratio is larger than one, the vehicle damping is insufficient; whereas, with a ratio smaller than one, delay is large, and the responsiveness becomes poor, which could be presumed from the general knowledge concerning the yaw rate frequency response. From the preceding, it is somehow confirmed that the $\left|G_{(peak)}\right|/\left|G_{(0.2)}\right|$ ratio is an important index for the vehicle motion characteristics that relates to the vehicle's controllability.

However, even if the $\left|G_{(peak)}\right|/\left|G_{(0.2)}\right|$ ratio value remains virtually unchanged with the change of the dynamic parameters, it is possible that there will be changes in the frequency of the peak gain, the reduction ratio of the gain at the high frequency region, and the phase lag.

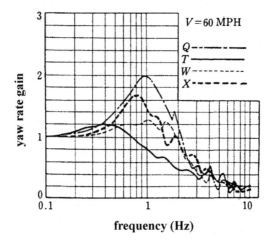

FIGURE 11.8

Four kinds of yaw rate gain.

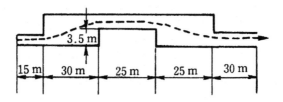

FIGURE 11.9

Evaluation test course.

Consequently, the vehicle's controllability, for example even under a fixed task, cannot be considered to depend only on the $|G_{(peak)}|/|G_{(0.2)}|$ ratio, and it cannot be judged by the value of the gain itself only.

The result obtained by W. Lincke is a good example that proves this [5]. Figure 11.8 is the gain of the yaw rate response to periodic steering inputs for four vehicles with different motion characteristics. W. Lincke then rates the controllability for the four vehicles based on the driver's rating with a four-point system by setting the course as shown in Figure 11.9 and the vehicle speed at 60 mph. The result is shown in Figure 11.10.

As could be seen from the figure, the vehicle's controllability coincides well with the frequency that gives the yaw rate gain peak, or in other words, the natural frequency of the vehicle motion. If the vehicle Q is compared with the vehicle W, W has a small $|G_{(peak)}|$ with large yaw damping, but the two vehicles have the same natural frequency and similar rating of controllability. If vehicles T and W are compared, the yaw damping of the two vehicles are about the same, but T has a smaller natural frequency and, thus, a poor rating of controllability. Consequently, in this kind of situation, the natural frequency is the vehicle motion characteristic that is more strongly related to the vehicle's controllability rather than the damping ratio.

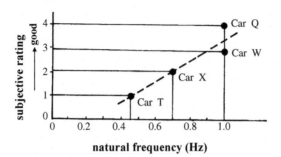

FIGURE 11.10

Controllability and natural frequency.

As described in Chapter 3, expressing the vehicle lateral motion by the two-degree-of-freedom vehicle model, i.e., by the side-slip angle, β, and yaw rate, r, the theoretical system natural frequency, ω_n, and damping ratio, ζ, could be written as the following approximated equations:

$$\omega_n = \frac{2(K_f + K_r)}{mV} \sqrt{\frac{l_f l_r}{k^2}} \sqrt{1 + AV^2} \qquad (3.67)'$$

$$\zeta = \frac{1 + k^2/l_f l_r}{2\sqrt{k^2/l_f l_r}} \frac{1}{\sqrt{1 + AV^2}} \qquad (3.68)'$$

In particular, this ω_n becomes the important characteristic expression that relates to the vehicle controllability.

Nevertheless, from Figures 11.4 and 11.7, it cannot be said that the damping ratio is not related to the vehicle's controllability at all. For the same reason as mentioned previously, the vehicle's controllability cannot be considered to depend only on the natural frequency, and the conclusion that a vehicle with a higher natural frequency always has a better controllability cannot be drawn.

The vehicle dynamic characteristics could be examined by the time lag of the motion response to a steering input. In other words, the responsiveness is expressed by the response time. Previous researches on general manual control systems show that the control objective with a large delay is more difficult to control. The vehicle yaw rate response to steering input is given by Eqn (3.78)'. If $l_f \approx l_r$ and $K_f \approx K_r$ from Eqn (3.78), this could be approximated by the following equation:

$$\frac{r(s)}{\delta(s)} = \frac{G_\delta^r(0)}{1 + t_r s}$$

where

$$t_r = \frac{mV}{2(K_f + K_r)} \left(\frac{k^2}{l_f l_r}\right) \qquad (11.1)$$

Hoffmann et al. measured the equivalent yaw rate response time, t_r, for different cars and investigated the relationship between this and the actual error rate (the vehicle touches the

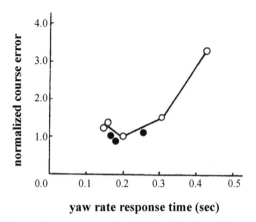

FIGURE 11.11

Course error to yaw rate response time.

cone and tumbles it) when the vehicle is required to pass through a course marked by cones without error (without touching the cones) [6]. The result is shown in Figure 11.11. From the figure, at the response time of around 0.2 s, the error rate is the lowest, and for response time larger than this, the error rate increases tremendously. It is assumed that with the increase of the response time, t_r, the vehicle becomes more difficult to control.

The side-slip response time is defined as the side-slip angle response delay to yaw rate. If $l_f K_f \approx l_r K_r$, the side-slip angle, β, response to the yaw rate, r, when the steering angle, δ, is fixed at zero could be expressed as follows from Eqn (3.12):

$$mV\frac{d\beta}{dt} + 2(K_f + K_r)\beta = -mVr$$

Hence, the side-slip angle response time t_β is expressed as follows:

$$t_\beta = \frac{mV}{2(K_f + K_r)} \tag{11.2}$$

Bergman measured the equivalent side-slip response time, t_β, for different cars and investigated the relation of this with the driver's rating on controllability based on a 10-point system [7]. The result is shown in Figure 11.12. The result shows that with the increase in the side-slip response time, the vehicle becomes less easily controlled.

From Eqns (11.1) and (11.2) and Eqns (3.73) and (3.74), if $k^2 \approx l_f l_r$, then t_β and t_r are about the same, and they are equal to the response time, t_R, described in Section 3.4.1.3.

$$t_r \approx t_\beta \approx t_R \approx \frac{mV}{2(K_f + K_r)} \tag{11.3}$$

In this manner, it is now understood that the vehicle motion response time is an important index for the expression of the vehicle motion characteristic that affects the vehicle's controllability. The response time is mainly dependent on the vehicle's front and rear cornering stiffness, the vehicle mass, and the traveling speed.

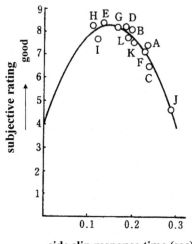

FIGURE 11.12

Controllability to side-slip response time.

The yaw rate responsiveness, as described in Section 3.4.2, could be expressed also by the following:

$$t_p = \frac{1}{\omega_n\sqrt{1-\zeta^2}}\left\{\pi - \tan^{-1}\left(\frac{\sqrt{1-\zeta^2}\,\omega_n T_r}{1-\zeta\omega_n T_r}\right)\right\} \tag{3.86}$$

Lincke et al. calls the product of the steady-state value of the side-slip angle relative to the unit lateral acceleration with a constant steer input given by the following:

$$\frac{G_\delta^\beta(0)}{G_\delta^{\dot{y}}(0)} = \frac{l_r}{V^2}\left(1 - \frac{m}{2l}\frac{l_f}{l_r K_r}V^2\right)$$

and the equivalent t_p, in other words:

$$\text{TB} = t_p G_\delta^\beta(0)\Big/G_\delta^{\dot{y}}(0) \tag{11.4}$$

as the TB factor, and shows the strong relation of this value with the vehicle's controllability [5]. The result is shown in Figure 11.13.

Based on the figure, when t_p is small and the $G_\delta^\beta(0)$ value is small, the yaw rate responsiveness is good, and the side slip at that instant is small: the vehicle is easier to control.

11.3.3 Response time and gain constant and controllability

As is seen from the previous, it is difficult to express the vehicle's controllability by just a single vehicle motion characteristic. Furthermore, there is also no guarantee that the driver evaluates

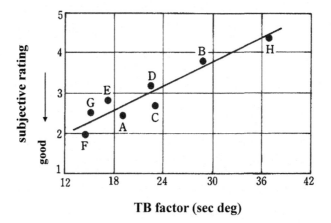

FIGURE 11.13

Controllability to TB factor.

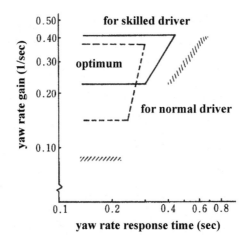

FIGURE 11.14

Controllability to yaw rate gain and response time.

the vehicle's controllability by just that particular motion characteristic. Consequently, it is not always suitable to judge the vehicle's controllability based on merely a single motion characteristic.

The vehicle yaw rate response to steering input given by Eqn (3.78)′ is dependent not only on the response time, t_r, but also on the gain constant $G_\delta^r(0)$. Here, Weir et al. measured the actual vehicle equivalent t_r and $G_\delta^r(0)$ and investigated the relation of these two values with the vehicle's controllability [8]. The result is shown in Figure 11.14.

From the figure, it is understood that the t_r and $G_\delta^r(0)$ of the vehicle that is easy to control occupies a certain region within the $t_r - G_\delta^r(0)$ plane. The region enclosed by the solid line is the region that is considered by experienced drivers as easy to control; while the region enclosed by the dotted line is the region considered by normal drivers as easy to control. If these two regions are compared for experienced drivers and normal drivers, the experienced drivers prefer a larger $G_\delta^r(0)$ even though t_r is a little large; while the normal drivers prefer a smaller t_r even though $G_\delta^r(0)$ is relatively small. The upper limit of $G_\delta^r(0)$ for both types of driver are virtually the same, which is precisely near the point where the vehicle exhibits NS characteristic.

This is an example that focuses on the vehicle yaw rate response to steering input. Bergman puts his focus on the vehicle lateral motion and tries to investigate the relationship between controllability and the vehicle lateral acceleration gain constant $G_\delta^{\ddot{y}}(0)$ and the side-slip response time, t_β [7]. From Eqn (11.3), t_β and t_r are two quantities that have about the same characteristics. $G_\delta^r(0)$ is expressed as follows:

$$G_\delta^r(0) = \frac{1}{1+AV^2}\frac{V}{l} \tag{3.81}$$

and $G_\delta^{\ddot{y}}(0)$ is the following:

$$G_\delta^{\ddot{y}}(0) = \frac{1}{1+AV^2}\frac{V^2}{l} \tag{3.89}$$

Therefore, it could be seen that $G_\delta^{\ddot{y}}(0)$ and $G_\delta^r(0)$ also have the same characteristic. Consequently, the combination of t_β and $G_\delta^{\ddot{y}}(0)$ could be seen as nearly the same as the combination of t_r and $G_\delta^r(0)$.

Bergman measured the equivalent t_β and $G_\delta^{\ddot{y}}(0)$ for a few vehicles and evaluated the vehicle controllability based on the driver's rating by a 10-point system. Then, he studied the relation of these two quantities with the vehicle's controllability. The result is shown in Figure 11.15. Looking at the figure, it is understood that the t_β and $G_\delta^{\ddot{y}}(0)$ of the vehicle that is easy to control occupies a certain region in the $t_\beta - G_\delta^{\ddot{y}}(0)$ plane, as in Figure 11.14. From the same figure, it is understood that the effect of the changes of t_β on the vehicle's controllability is more prominent than the effect of the changes of $G_\delta^{\ddot{y}}(0)$. In other words,

FIGURE 11.15

Controllability to lateral acceleration gain and side-slip response time.

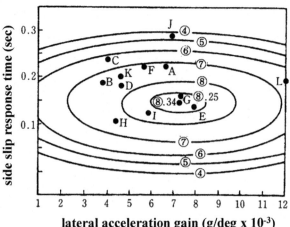

the side-slip response time is a vehicle motion characteristic that is more sensitive to the vehicle's controllability compared to the lateral acceleration gain constant.

Looking at the vehicle controllability related to the two vehicle motion characteristics leads to a relatively good understanding of the relationship of vehicle controllability with the vehicle motion characteristics. However, there is no theoretical background on which of the two vehicle motion characteristics should be considered to be able to express the vehicle controllability. Although, in the preceding examples, the response time and the steady-state gain have been selected. Consequently, when the parameters that govern the motion characteristics are changed, or when a new vehicle specification is given, it is not necessarily possible to judge that the vehicle is easy to control just because the vehicle $t_r - G_\delta^r(0)$ or $t_\beta - G_\delta^{\dot{y}}(0)$ falls inside the region shown in the example.

The vehicle has to also fulfill many other functional requirements other than controllability, and various considerations are needed. When other requirements are put at a higher priority than the controllability, then little importance may be laid on the vehicle controllability. Nevertheless, even if the problem is limited to the controllability of the vehicle motion presented in this book, a general theoretical method has not been provided for estimating the vehicle's controllability.

On the other hand, there is still no general method to measure qualitatively the actual driver's evaluation of the vehicle handling quality that is as accurate and comprehensive as the PR is for aircraft. The vehicle controllability evaluation obtained in practice lacks accuracy and comprehensiveness. Consequently, it is not sufficient to just try to theoretically understand the relation between vehicle motion characteristics and the vehicle controllability. While trying to establish the methodology to measure the controllability of the actual vehicle evaluated by human drivers, a theoretical progress and development for the investigation of the controllability is highly needed.

A discussion concerning these points will be extended in the next chapter.

REFERENCES

[1] Nakatsuka T. Vehicle safety in high speed driving and human characteristics. Jpn J Ergonomics 1968;4(4) [in Japanese].

[2] Saito Y, Mori S, Matsushita A. Effects of US and OS characteristics on driver handling behavior. In: JARI Report, No. 4; 1971 [in Japanese].

[3] Bergman W. Bergman gives new meaning to understeer and oversteer. SAE J 1965;73(12).

[4] Sugimoto, et al. Analysis on vehicle running stability subjected to side wind on highway. In: Proceedings of annual meeting of JSAE, No.761; 1976 [in Japanese].

[5] Linke W, Richter B, Schmidt R. Simulation and measurement of driver vehicle handling performance, SAE Paper730489.

[6] Hoffmann ER, Joubert PN. The effect of changes in some vehicle handling variables on driver steering performance. Hum Factors June, 1966:245–63.

[7] Bergman W. Relationships of certain vehicle handling parameters to subjective ratings of ease of vehicle control. In: Proceedings of the 16th FISITA congress, Tokyo; May, 1976.

[8] Weir DH, DiMarco RJ. Correlation and evaluation of driver/vehicle directional handling data, SAE Paper780010.

DRIVER MODEL-BASED HANDLING QUALITY EVALUATION

12.1 PREFACE

A road vehicle is called an automobile for the first time when the vehicle is controlled by a human driver on board. The vehicle dynamics controlled by the human driver are dealt with in Chapter 10, and the relation of handling quality evaluation by the driver to the vehicle handling dynamics is discussed in Chapter 11. However, no further discussion has been extended so far regarding the relation of the handling quality evaluation to the vehicle dynamics controlled by human drivers. This is because there is no fundamental established yet for reasoning the vehicle handling quality from the vehicle handling dynamics.

Though many investigations have been done, of course, handling quality evaluation by human drivers and its relation to the chassis design variables of the vehicles and the chassis control parameters have still been under ongoing debates, and a consistent, general way of understanding or explaining them has not been established yet. On the other hand, because it has recently been regarded that active vehicle motion controls are promising tools to improve vehicle handling and stability, it becomes more important than ever to have a general way of estimating or predicting the vehicle handling quality, which is possible to be controlled by the motion controls, evaluated by human drivers subjectively.

As shown in Chapter 11, there are many studies on the correlations between the subjective evaluations and objective measurements of the vehicle response or vehicle response parameters. However, the results are not necessarily generalized and directly applicable to the chassis design process. A primary reason for that seems to be the driver's subjective evaluation based on feeling that can't avoid psychological and mental effects, especially for ordinary drivers. The results are qualitatively reliable; yet, they are not necessarily quantitatively reliable and, thus, do not result in a generalized method.

In this chapter, some trials of applying a driver model and its parameter identification to dealing with the driver-vehicle system behavior will be introduced to generally and consistently understand what, how, and why the vehicle handling dynamics reflect the handling quality of the vehicle. This must be possible to be an effective and general method of realizing and estimating a relation of the handling quality evaluation with the characteristics of vehicle handling dynamics.

12.2 DRIVER MODEL AND PARAMETER IDENTIFICATION [1]

A driver is considered to determine the steering parameters not only depending on his/her inherent characteristics but also on adapting to the vehicle handling characteristics, as is discussed in Section 10.4, by changing the parameters during vehicle motion to keep the system response almost always the same, even with wide variations of the vehicle handling character-istics. Therefore, the vehicle performance controlled by the driver is unchanged even when the vehicle handling characteristics change. This is one of the primary reasons why the vehicle response itself does not always directly reflect the handling characteristics of the vehicle, and it is difficult to understand the handling quality from the objective measurement of the vehicle responses.

On the other hand, because a driver adapts his/her characteristics to the vehicle handling characteristics, the parameters in a driver model must directly reflect not only his/her inherent characteristics but also the vehicle handling characteristics. So, the idea arises that if the pa-rameters are reasonably identified, it must be possible to estimate the evaluation of the vehicle handling characteristics through the driver parameters identified. This section intends to show the driver model for this specific purpose and the method of how to identify the driver parameters by using experimental data.

12.2.1 DRIVER MODEL

The driver behavior during his/her motion control of the vehicle is dealt with in Chapter 10. As the transfer function of the driver steering behavior is described by Eqn (10.2) and the course error detected by the driver is expressed by (10.3), the steering angle of the driver is described as follows:

$$\delta(s) = -h(1 + \tau_D s)e^{-\tau_L s}\{y(s) + L\theta(s) - y_{OL}(s)\} \tag{12.1}$$

To simplify the model and reduce the number of parameters in the driver model, it is assumed that $dy/dt \approx V\theta$, as shown in Section 10.3.3, and the driver performs the equivalent derivative control action by looking ahead—a preview behavior. As a result, the driver derivative time, τ_D, is regarded to be almost zero, and Eqn (12.1) can be rewritten as follows:

$$\delta(s) = -he^{-\tau_L s}\left\{\left(1 + \frac{L}{V}s\right)y(s) - y_{OL}(s)\right\}$$

Moreover, assuming that τ_L is small and $e^{-\tau_L s} \approx 1/(1 + \tau_L s)$, in which τ_L is regarded to represent all the response delay of the driver, the following simplified driver model is used that describes the steering angle determined by the driver:

$$\delta(s) = -\frac{h}{1 + \tau_L s}\{(1 + \tau_h s)y(s) - y_{OL}(s)\} \tag{12.2}$$

Assuming that τ_h represents the effects of the look ahead (preview) behavior and includes the effect of the derivative control action of the driver, if any derivative control action is negligible, this is regarded to be the preview time, L/V. The block diagram of the simplified driver-vehicle model is shown in Figure 12.1, and the driver steering characteristics can be represented by the three parameters h, τ_h, and τ_L. Here, it is important to note that this model is applicable only to the driver making a sudden lane change on a straight road with constant lane width, y_{OL}.

12.2.2 PARAMETER IDENTIFICATION

Next is to study the experimental identification of the three parameters. The handling parameters in the driver model are identified here by using experimentally logged data of driver steering angle and vehicle trajectory. As a typical behavior of the driver-vehicle system can be observed

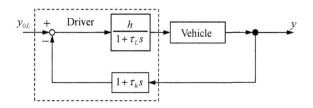

FIGURE 12.1

Block diagram of a simplified driver-vehicle model.

during a lane change on a straight road, the parameters can be identified from an experimentally measured time history of the driver's steering angle, δ^*, and lateral displacement, y^*, during the lane change on a straight road.

The error is defined between the measured steering angle, δ^*, and the driver's steering angle response to the lateral displacement, y^*, and the lane change width, y_{OL}, calculated by the driver model, Eqn (12.2):

$$e(s) = (1 + \tau_L s)\left[\delta^*(s) + \frac{h}{1 + \tau_L s}\left\{(1 + \tau_h s)y^*(s) - y_{OL}(s)\right\}\right]$$

$$= (1 + \tau_L s)\delta^*(s) + h\left\{(1 + \tau_h s)y^*(s) - y_{OL}(s)\right\}$$

(12.3)

The square integral of the weighted sum of the error and error rate is defined as the evaluation function:

$$J = \int_0^T e^2 dt = \int_0^T \left[\delta^* + \tau_L\frac{d\delta^*}{dt} + h\left\{y^* + \tau_h\frac{dy^*}{dt} - y_{OL}\right\}\right]^2 dt$$

(12.4)

where T is the time period long enough for the driver to finish the lane change. It is possible for us to find the parameters, h, τ_h, and τ_L that minimize J by solving the following equations:

$$\frac{\partial J}{\partial h} = 0, \quad \frac{\partial J}{\partial \tau_L} = 0, \quad \frac{\partial J}{\partial(h\tau_h)} = 0$$

(12.5)

These equations are first-order linear algebraic equations of h, τ_L, and $h\tau_h$, and they can easily be solved for h, τ_L and τ_h. The solved parameters are the identified driver parameters. With these, the driver steering angle can be described by the identified model and is as close as possible to the real driver steering angle measured.

12.3 DRIVER PARAMETERS REFLECTING DRIVER CHARACTERISTICS [1]

It seems that one of the typical examples of the dependence of driver handling behavior upon his/her inherent characteristics is the dependence on the age of the driver. For confirming this, there is a result of the parameter identification of young and aged drivers during a lane change with a small-size personal vehicle.

Figure 12.2 is a typical result of the driver steering behavior during the lane change, in which the time history of the measured steering angle is compared with that calculated by the driver

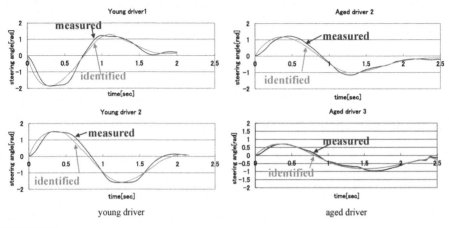

FIGURE 12.2

Driver steering behavior during lane change.

model with the identified parameters for the young and the aged drivers, respectively. They are in very good agreement with each other, showing that it is reasonable to describe the driver behavior by the simplified driver model introduced in Section 12.2.1 with the parameters identified by the method shown in Section 12.2.2.

Figure 12.3 shows how the driver parameters identified are distributed on a parameter plane depending on the age. The parameters of the aged drivers are distributed on relatively larger τ_h, larger τ_L, and smaller h on the parameter planes compared with those of the young drivers.

It is understood from Figure 12.3 that as smaller τ_L causes higher tension and more severe workload, especially on aged drivers, he/she adopts relatively larger τ_L and compensates for the delay of the driver response and stabilizes a less stable vehicle motion due to larger τ_L by adopting larger τ_h. This is equivalent to a preview behavior of the driver and is easy for drivers to change by changing the look ahead distance, L, with almost no workload for the driver to make a larger value of required τ_h, though its value would be limited by the lane change distance. Figure 12.3 also shows that the aged drivers adopt smaller gain, h, compared with the young drivers. This is supposed to be influenced by adopting τ_h and τ_L as described previously and is contributory to stabilizing the vehicle motion.

It is concluded from the preceding that it is possible to characterize the driver with the driver parameters identified in Eqn (12.2).

12.4 DRIVER PARAMETERS REFLECTING VEHICLE HANDLING CHARACTERISTICS

We have already understood in Chapter 3 that vehicle speed has significant effects on vehicle handling characteristics, and the tire cornering characteristics also have predominant effects as well. In this section, the effects of the vehicle speed and tire cornering stiffness on the identified

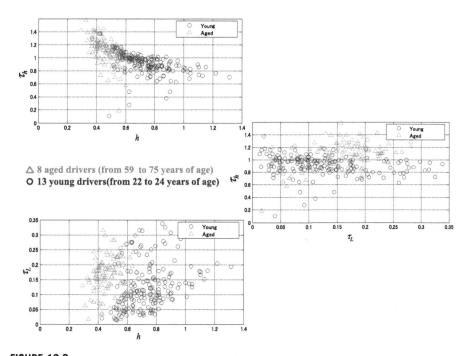

FIGURE 12.3

Identified handling parameters of young and aged drivers.

driver parameters will be examined as examples of the driver parameters reflecting vehicle handling characteristics.

12.4.1 DRIVER PARAMETERS TO VEHICLE SPEED [2]

As the vehicle speed has a significant effect on vehicle dynamics, the driver is assumed to adapt his (or her) characteristics to the change of vehicle speed, which is discussed in Section 10.4. To confirm this, the way that vehicle speed influences the identified driver parameters will be looked at in this section.

There is a result of applying the identification method in Section 12.2.2 to the experimental data obtained in the lane change test on a proving ground. The lane change test course is shown in Figure 12.4. The lane change width, d_C, is 3.0 m, and the lane change lengths, L_C, are 15, 22.5, and 30 m for vehicle speeds 40, 60, and 80 km/h, respectively.

In Figure 12.5, the calculated steering angle by the model, Eqn (12.2), is compared with the real driver steering angle measured. They agree well with each other, which means that the driver steering behavior is adequately described by Eqn (12.2).

The parameters identified are shown in Figure 12.6 and illustrate how the driver parameters change according to the increase in vehicle speed. It is understood that the proportional gain, h, significantly decreases with the increase of vehicle speed. In contrast, τ_h is always around

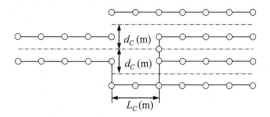

FIGURE 12.4

Lane change test course.

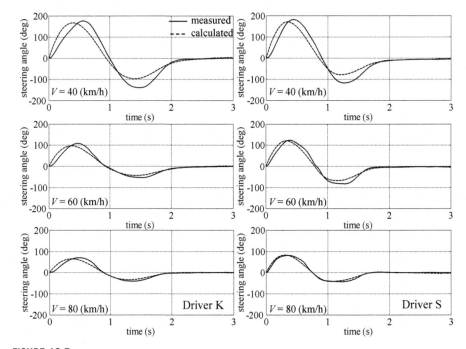

FIGURE 12.5

Comparison of measured and calculated driver steering angle.

$0.8 \sim 1.0$ s and does not change with increasing vehicle speed. So, if τ_h is approximated as the preview time, L/V, the preview distance (looking ahead distance), is almost proportional to the vehicle speed and equal to $0.8 \ V \sim 1.0 \ V$. This view corresponds with the driver adaptation to vehicle characteristics discussed in Section 10.4. The response delay, τ_L, decreases with the vehicle speed, which suggests that the increasing vehicle speed pushes the driver to achieve more stressful workloads on the steering control.

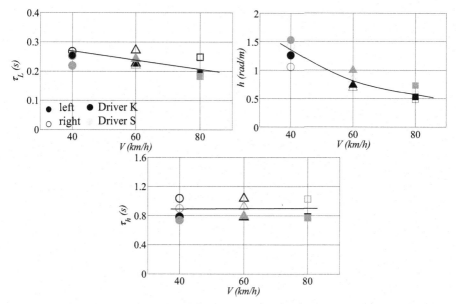

FIGURE 12.6

Change of driver parameters with increase of vehicle speed.

12.4.2 DRIVER PARAMETERS TO TIRE CHARACTERISTICS [3]

The vehicle handling characteristics strongly depend on tire cornering stiffness. We have already seen in Chapter 3 that the vehicle response parameters such as stability factor, yaw rate gain, natural frequency, damping ratio, yaw rate lead time, and response time are described as follows, and the considerable dependence of them to the tire is obvious:

$$A = -\frac{m}{2l^2} \frac{l_f K_f - l_r K_r}{K_f K_r}$$

$$G_\delta^r(0) = \frac{1}{1 + AV^2} \frac{V}{l}$$

$$\omega_n = \frac{2l}{V} \sqrt{\frac{K_f K_r}{mI}} \sqrt{1 + AV^2}$$

$$\zeta = \frac{m\left(l_f^2 K_f + l_r^2 K_r\right) + I(K_f + K_r)}{2l\sqrt{mIK_f K_r(1 + AV^2)}}$$

$$T_r = \frac{ml_f V}{2lK_r}$$

$$t_R = \frac{mV}{2(K_f + K_r)}$$

To experimentally confirm how the change of vehicle handling characteristics with changing the tire characteristics influences the driver parameters identified, the four different handling characteristic vehicles are provided by changing the front and rear tire cornering stiffness, as shown in Table 12.1. The driver steering behavior and vehicle motions are measured in the lane change tests on the same lane change course as shown in Figure 12.4 with the vehicle speed of 80 km/h. Some results are shown in Figure 12.7. As the drivers adapt to the change of vehicle response characteristics corresponding with the tire characteristics supposedly by changing their driver parameters, the driver-vehicle system behaviors are almost similar.

The driver parameters are identified using the measured date during the lane changes with four vehicles of different handling characteristics, respectively. Figure 12.8 shows a good agreement of the driver steering time history calculated by Eqn (12.2) using the parameters identified with the measured time history. This supports that it is reasonable to describe the driver behavior by Eqn (12.2).

Though the preview time, τ_h, of the drivers is almost unchanged from that observed in Section 12.4.1, we can see, in Figure 12.9, a distribution of the driver parameters, h and τ_L, depending on the respective vehicle handling characteristics controlled by the tire cornering stiffness. It is easy to see how the drivers vary their parameters according to the change of the vehicle handling characteristics, and the drivers are forced to change the time constant of the response delay, τ_L, widely with the variations of the handling characteristics.

As decreasing τ_L in the steering tasks makes the driver experience a more severe workload and if a larger τ_L is allowable for the driver to control the vehicle, the more relaxed the driver behaves during his (or her) maneuvering. In addition, changing h and τ_h, more or less, is easier for the driver, and the handling quality evaluation is assumed to correspond predominantly with τ_L. Figure 12.10 shows the correlation of the subjective handling quality evaluation by the drivers in the lane change with the identified response delay of the driver, τ_L. It seems reasonable that if a driver can behave with larger τ_L, then the driver feels it is easier to control the vehicle, which means a higher handling quality evaluation.

Table 12.1 Four cases of vehicle response characteristics at vehicle speed $V = 80$ km/h

Vehicles	K_f (kN/rad)	K_r (kN/rad)	A	$G_\delta^r (0)$ (1/s)	ω_n (rad/s)	ζ	t_R (s)
Baseline	66.1	81.9	0.00109	5.58	10.63	0.835	0.113
A	40.7	47.5	0.00162	4.78	6.86	0.767	0.190
B	40.7	81.9	0.00275	3.65	10.32	0.737	0.131
C	66.1	47.5	-0.000037	8.76	6.46	1.009	0.177

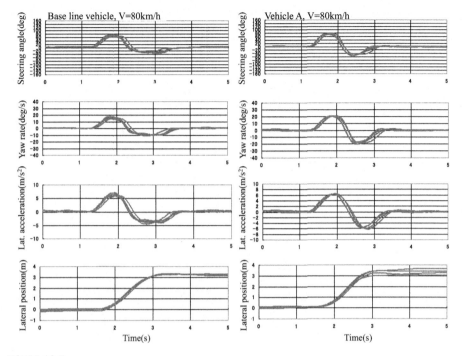

FIGURE 12.7

Driver steering behavior and vehicle motion during lane change.

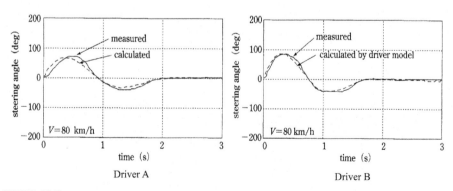

FIGURE 12.8

Comparison of measured and calculated driver steering behavior.

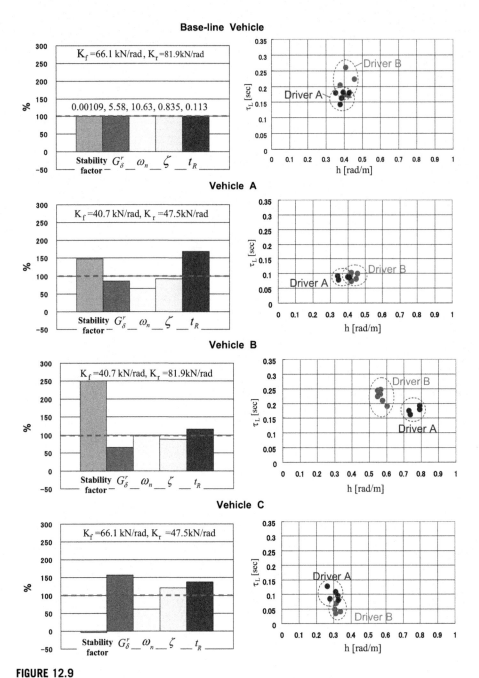

FIGURE 12.9

Relation of driver parameters to vehicle response parameters.

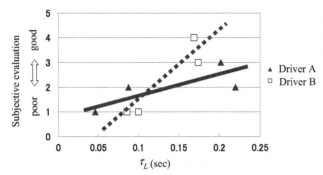

FIGURE 12.10

Relation of subjective evaluation to τ_L.

12.5 HANDLING QUALITY EVALUATION BASED ON DRIVER PARAMETER

As the handling quality evaluation seems strongly related to one of the identified driver parameters, τ_L, we will look at several examples of the handling quality evaluation by the parameter τ_L in this section.

12.5.1 HANDLING QUALITY TO NATURAL FREQUENCY AND DAMPING RATIO [4]

The natural frequency, ω_n, and the damping ratio, ζ, in the denominator of the transfer functions of side-slip and yaw rate responses to steering input described by Eqns (3.77)$'$ and (3.78)$'$ are important response parameters to dominate over the vehicle handling characteristics. They are described in Section 12.4.2, and it is difficult to independently change their values by changing specific vehicle parameters and chassis parameters such as tires and suspension. Therefore, a problem lies in experimentally examining the independent effects of ω_n and ζ on the handling quality evaluation by the driver.

A feed-forward type of front-wheel active steering control can equivalently solve this problem. From Eqns (3.77)$'$ and (3.78)$'$, the response of the vehicle with the active front-wheel steering control to steering wheel input is described as follows:

$$\beta(s) = G_\delta^\beta(0) \frac{1 + T_\beta s}{1 + \frac{2\zeta s}{\omega_n} + \frac{s^2}{\omega_n^2}} \frac{\delta}{\delta_h}(s)\, \delta_h(s) \tag{12.6}$$

$$r(s) = G_\delta^r(0) \frac{1 + T_r s}{1 + \frac{2\zeta s}{\omega_n} + \frac{s^2}{\omega_n^2}} \frac{\delta}{\delta_h}(s)\, \delta_h(s) \tag{12.7}$$

where $\delta/\delta_h(s)$ represents a transfer function of front-wheel steering angle to steering wheel angle input for the active steering control.

If the active control transfer function is set as follows:

$$\frac{\delta}{\delta_h}(s) = \frac{1 + \frac{2\zeta s}{\omega_n} + \frac{s^2}{\omega_n^2}}{1 + \frac{2\zeta^* s}{\omega_n^*} + \frac{s^2}{\omega_n^{*2}}} \tag{12.8}$$

then putting Eqn (12.8) into Eqns (12.6) and (12.7) brings us the response of the active steering vehicle to steering wheel angle input as follows:

$$\frac{\beta(s)}{\delta_h(s)} = G_\delta^\beta(0) \frac{1 + T_\beta s}{1 + \frac{2\zeta^* s}{\omega_n^*} + \frac{s^2}{\omega_n^{*2}}} \tag{12.9}$$

$$\frac{r(s)}{\delta(s)} = G_\delta^r(0) \frac{1 + T_r s}{1 + \frac{2\zeta^* s}{\omega_n^*} + \frac{s^2}{\omega_n^{*2}}} \tag{12.10}$$

where

$$\omega_n^* = \alpha_N \omega_n, \quad \zeta^* = \alpha_D \zeta \tag{12.11}$$

Here, α_N and α_D are the adjustment parameters, and the natural frequency, ω_n, and the damping ratio, ζ, are changed independently by changing the adjustment parameters appropriately around 1.0, respectively. If both parameters α_N and α_D are set equal to 1.0, the front-wheel active steering is eliminated, and the vehicle response becomes like that of the normal vehicle—the baseline vehicle.

Figure 12.11 shows the results of applying the previous idea to the driving simulator with a full dynamics model of the vehicle motion. The yaw rate frequency responses of the driving simulator vehicle with the front active steering control are compared in this figure to the vehicle yaw rate responses calculated by Eqn (12.10) for various increase and decrease of ω_n and ζ. It is confirmed that the natural frequency and the damping ratio are equivalently changed independently by the active front-wheel steering control.

Next is to experimentally examine the effects of the natural frequency and the damping ratio on the driver parameters, especially on the response delay time constant, τ_L, which correlates well with the handling quality evaluation. Using the driving simulator mentioned above, the driver model parameters during the same lane change as in Section 12.4.2 are identified by the method learned in Section 12.2.2. Figure 12.12 shows the measured time histories of lateral acceleration and yaw rate of the driving simulator motion-base compared with those calculated by the simulator software in the lane change. This guarantees fidelity of the driving simulator in terms of vehicle motion to be sensed by the drivers.

Figure 12.13 is the result of the identification. The identified response delay time constant of four drivers during the lane change with the vehicles of various combinations of the natural frequency, ω_n, and the damping ratio, ζ, is shown on the ω_n-ζ plane. The driver parameter, the time constant, τ_L, on the ω_n-ζ plane tells us in this figure that there is a combination of ω_n and ζ that gives us a peak value of τ_L for each driver. This implicates that there is an optimum value of ω_n and ζ for the vehicle handling characteristics that raises the vehicle handling quality evaluation to the peak. This agrees well with our normal experiences.

12.5.2 HANDLING QUALITY TO STEERING TORQUE [5]

Steering torque has a big effect on the handling quality evaluation. In this section, we will look at how the various steering torque characteristics to steering angle, shown in Figure 12.14, influence the parameter of the driver response delay, τ_L, identified, which eventually influences the handling quality evaluation of the driver during a lane change. Six cases of the steering reaction torque characteristics, as shown in Figure 12.14, are provided in the steering system of the same

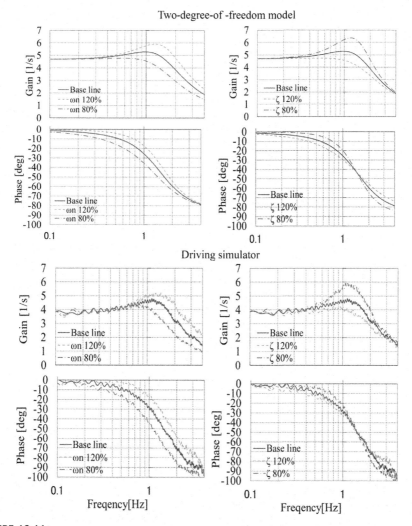

FIGURE 12.11

Yaw rate frequency responses of two-degree-of-freedom model and driving simulator.

driving simulator as was used in Section 12.5.1. The driver parameters are identified in the same way as the previous cases of the identifications during the lane change with the simulator vehicles equipped with the six various steering torque characteristics, respectively. The lane change course used is the same one as shown in Figure 12.4.

A result, shown in Figure 12.15, is the identified time constant, τ_L, of the 10 drivers for the six various steering torque characteristics. The time constants of the 10 drivers change with the variations of the steering reaction torque characteristics in a similar aspect, as

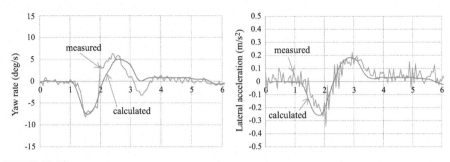

FIGURE 12.12

Calculated and measured motions of driving simulator in lane change.

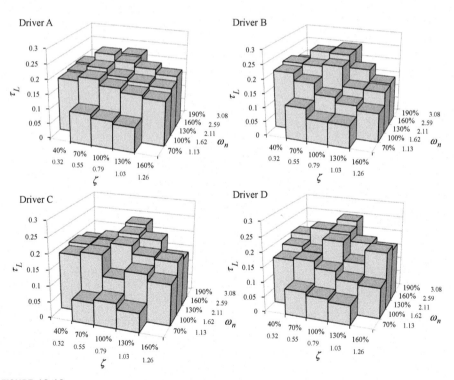

FIGURE 12.13

Identified driver parameter, τ_L, on ω_N-ζ plane ($V = 80$ km/h, $L_C = 45$ m).

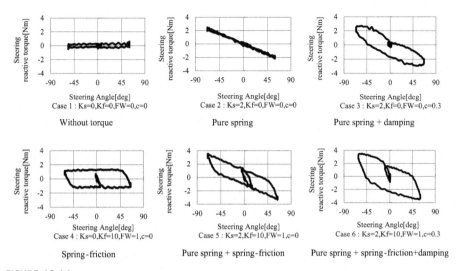

FIGURE 12.14

Reactive torque to steering angle characteristics.

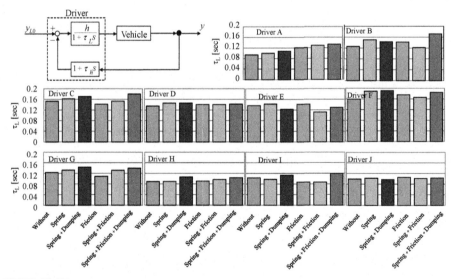

FIGURE 12.15

Identified time constant, τ_L, of 10 drivers for six various steering torque characteristics at $V = 100$ km/h and $L_C = 55$ m.

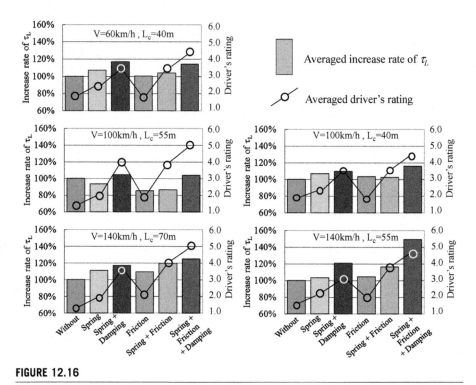

FIGURE 12.16

Averaged increase rate of τ_L and driver's rating.

shown. Figure 12.16 shows the increase rate of the averaged time constant of the 10 drivers according to the steering torque characteristics for the different vehicle speeds and lane change lengths. The average handling quality rated by the 10 drivers is also shown in the figure. The drivers adopted a higher time constant of the response delay, τ_L, during the lane change with the steering reaction torque of spring + damping and spring + friction + damping, and they gave a higher handling quality evaluation to them. This figure confirms that the identified time constant of the response delay, τ_L, strongly correlates with the handling quality evaluation.

The interesting point is that the steering torque has an effect on the driver parameters similar to vehicle response characteristics to steering angle input changes, even though there is no change in the vehicle response to steering angle input itself. Actually, it is clear that the steering torque has nothing to do with the open-loop transfer function in the control block diagram of steering angle control for the driver-vehicle system shown in Figure 12.1. Nonetheless, the drivers change their parameters according to the steering reactive torque as though the vehicle response characteristics to steering angle input was changed, and if they feel it is easy to handle, then they adopt a higher τ_L and give a good handling evaluation.

12.5.3 **HANDLING QUALITY TO DISTURBANCE SENSITIVITY [6]**

A road vehicle is normally operated under a wide variation in the number of passengers and/or various load conditions, which result in fairly large changes of the vehicle weight. The rise in the vehicle weight bring us the increase in vehicle specifications of mass and moment of inertia; meanwhile, the change of the vehicle weight causes variations in the vertical loads of the front and rear tires. Eventually, considerable variations in the cornering stiffness occur that depend on the vertical load. Thus, the vehicle handling characteristics will be changed considerably by the variations of the vehicle weight.

At the upper part of Figure 12.17, there is a calculated result of the change of one of the response parameters, the natural frequency, depending on five different tire sets in the dependence of cornering stiffness to vertical load, as shown in Figure 12.18. Also, the change of the natural frequency resulting from the rise in the weight of the vehicles with a respective tire set is shown.

As the vehicle weight as well as the tire characteristics have significant effects on the vehicle response characteristics, their corresponding effects on the driver's handling quality evaluation must emerge. To confirm this, there is a result of the identification of the driver time constant, τ_L, during a lane change using the previous driving simulator vehicles with various tires and vehicle weights. Figure 12.17, at the middle, shows the change of the identified τ_L averaged by four drivers for the vehicles with the five different tire sets and the different weights.

The increase of the time constant, τ_L, resulting from the change of the tire characteristics corresponds with the increase in the vehicle natural frequency shown in Figure 12.17. It seems reasonable from the study results in Sections 12.4.2 and 12.5.1, however, that there is a contradiction. Figure 12.17, at the top, also shows us that the time constant, τ_L, increases with the rise in the vehicle weight, even though the weight increment of the vehicle with any of the five tire characteristics results in the reduction of the natural frequency.

Because the natural frequency only is not necessarily enough to consistently explain these results, we will pay attention to a parameter other than the response parameters—a sensitivity parameter to the disturbance. As is discussed in the latter part of Section 4.2.3, the disturbance sensitivity parameter of the vehicle is defined here as the yaw angle gain to yaw moment disturbance input, and it is described as the following:

$$\theta_m = \frac{(K_f + K_r)\, V}{2l^2 K_f K_r \left\{ 1 - \frac{m(l_f K_f - l_r K_r)}{2l^2 K_f K_r} V^2 \right\}} \tag{12.12}$$

From the preceding equation, it is obvious that the greater the cornering stiffness, the less sensitive it is to the disturbance. In the bottom of Figure 12.17, we can see the calculated results of the change of disturbance sensitivity corresponding to the rise in the weight of the respective vehicles with five sets of the tire characteristics. A significant decrease in the sensitivity parameter with the increase in the vehicle weight corresponds well to the increase of τ_L with the weight increase shown in Figure 12.17. It seems that the driver feels a good handling quality strongly depending on the decrease in the vehicle's sensitivity parameter value, and the contradiction mentioned is now solved.

The natural frequencies of the vehicles equipped with tire-2 and tire-4 decrease less with the rise in the vehicle weight, as shown in Figure 12.17, because of more linear dependence of the

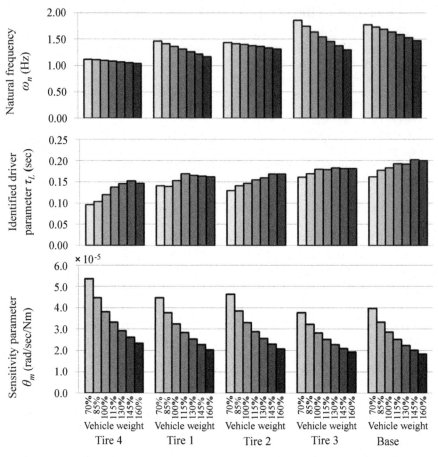

FIGURE 12.17

Vehicle parameters and driver parameter, τ_L, with respect to tire and vehicle weight.

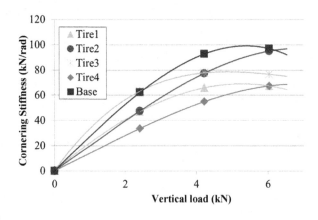

FIGURE 12.18

Five cases of dependence of tire cornering stiffness on vertical load.

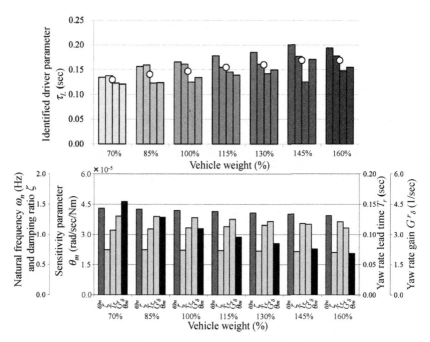

FIGURE 12.19

Four drivers' parameter τ_L and vehicle parameters with respect to vehicle weight.

cornering stiffness of the tires to the vertical load. Therefore, it is reasonable that the increase of the time constant, τ_L, with the rise in the vehicle weight is more considerable, especially for the vehicles with tire-2 and tire-4 in Figure 12.17 due to the decrease in the sensitivity parameter with the weight increase.

The bottom of Figure 12.19 shows the calculated result of the variation in the vehicle response parameters and the sensitivity parameter in more detail to the rise in the vehicle weight for the vehicle with tire-2. While all the response parameters change less, the sensitivity parameters decreases significantly with an increase in the vehicle weight, which coincides well with the increase in the identified τ_L of four drivers according to the vehicle weight increase, as shown at the top of the same figure. This supports the view that the vehicle handling quality depends not only on the natural frequency of the vehicle but also strongly on the sensitivity parameter, and the handling quality can be evaluated by the identified driver parameter, τ_L.

The sensitivity parameter itself is not the response parameter to steering angle input, and it has no explicit effect on the closed-loop characteristics of the driver-vehicle system of steering angle control input. Even so, it is interesting to see that the driver changes his/her control parameter, τ_L, of steering angle supposedly according to the change of sensitivity parameter during the lane change by feeling the sensitivity of the vehicle to the disturbance. The higher the sensitivity of the vehicle is, the lower τ_L the driver adopts to control the vehicle and the lower the handling quality evaluation becomes.

12.5.4 **EFFECT OF STABILIZING VEHICLE MOTION [1,2]**

As shown in Subsection 3.3.2.3, a body side-slip angle, β, of a normal understeer vehicle produces a positive yaw moment, $-2(l_f K_f - l_r K_r)\beta$, so far as the tire characteristics of the vehicle remain within a linear relation to side-slip angle. However, a saturation property of the tire lateral force to side-slip angle reduces the positive yaw moment with increase of the side-slip angle, which makes the vehicle less stable. In order to compensate for that, it is possible to exert the yaw moment produced by longitudinal forces of the tires on the vehicle body for stabilizing the vehicle motion. This is a kind of DYC that controls the tire longitudinal forces to give the vehicle the same yaw moment as linear tires can generate as much as possible, even when the vehicle motion gets into tire nonlinear ranges in order to augment the stability of the vehicle.

Figure 12.20 shows the measured time histories of the driver-vehicle system behavior during lane changes on a proving ground for different vehicles with and without DYC. Though the driver feels it is easy to control the vehicle with DYC and gives a better subjective handling quality evaluation, almost no specific difference that supports the driver's opinion is found between the measured time histories of the vehicles with and without DYC.

On the other hand, there are significant differences in the driver parameters identified by the same method as described in Section 12.2 using the time histories of driver-vehicle system during the lane changes. Figure 12.21 shows the identified driver parameters of the driver. It is found in the figures that the vehicle with DYC allows a larger time delay of the drivers, τ_L, and smaller

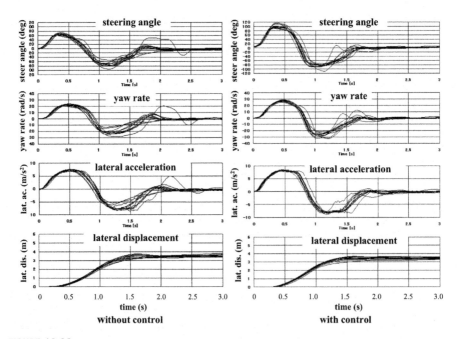

FIGURE 12.20

Vehicle responses during lane change.

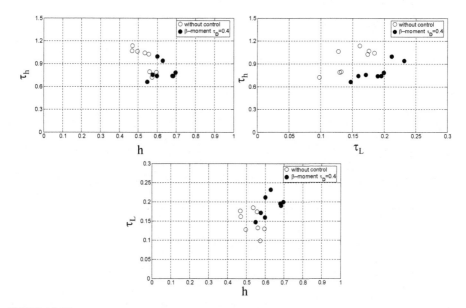

FIGURE 12.21

Identified driver parameters on parameter plane.

preview time, τ_h, compared with the identified parameters of the vehicle without control. Especially, a larger τ_L reasonably corresponds to the driver's evaluation on the vehicle handling quality [2].

REFERENCES

[1] Ishio J, Ichikawa H, Kano Y, Abe M. Vehicle handling quality evaluation through model based driver steering behavior. Veh Syst Dyn 2008;46(Suppl.).

[2] Abe M, Kano Y. A study on vehicle handling evaluation by model based driver steering behavior. In: Proceedings of FISITA2008, Munich; September 2008 [in CD].

[3] Abe M, Kano Y, Shibahata Y. Investigation of steering torque effects on handling quality evaluation based on steering angle control driver model. In: Proceedings of 21st IAVSD symposium; August 2009 [in CD].

[4] Aoki Y, Kano Y, Abe M. Variable stability vehicle with response parameters controlled by active chassis control devices. In: Proceedings of 22nd IAVSD symposium; August 2011 [in CD].

[5] Hibi M, Kobune T, Miki D, Kano Y, Abe M. A study on steering reactive torque for steering-by-WIRE vehicle using driving simulator. In: Proceedings of FISITA2012; November 2012 [Beijing China in CD].

[6] Miyamoto K, Suzuki T, Kano Y, Abe M. Effect of vehicle fundamental specification on steering characteristics relying on tire cornering characteristics. In: Proceedings of JSAE Annual Congress 47-20145286; 2014 [in Japanese].

Index

Note: Page numbers followed by "f" indicate figures.

Printed in the United States
By Bookmasters